Lecture Notes in Computer Science 4889

Commenced Publication in 1973
Founding and Former Series Editors:
Gerhard Goos, Juris Hartmanis, and Jan

Alexander Pasko   Valery Adzhiev
Peter Comninos (Eds.)

Heterogeneous Objects Modelling and Applications

Collection of Papers
on Foundations and Practice

Volume Editors

Alexander Pasko
Valery Adzhiev
Peter Comninos
Bournemouth University, The National Centre for Computer Animation, UK
E-mail: {apasko,vadzhiev,peterc}@bournemouth.ac.uk

Library of Congress Control Number: 2008927522

CR Subject Classification (1998): H.2.1, H.2.5, H.2.8

LNCS Sublibrary: SL 6 – Image Processing, Computer Vision, Pattern Recognition, and Graphics

ISSN 0302-9743
ISBN-10 3-540-68441-7 Springer Berlin Heidelberg New York
ISBN-13 978-3-540-68441-1 Springer Berlin Heidelberg New York

Springer is a part of Springer Science+Business Media

springer.com

Printed in Germany

Typesetting: Camera-ready by author, data conversion by Scientific Publishing Services, Chennai, India
Printed on acid-free paper SPIN: 12274291 06/3180 5 4 3 2 1 0

Preface

Heterogeneous object modelling is a new and quickly developing research area. It shows great promise in a number of important application areas such as in volume modelling and rendering, in the modelling of objects with multiple materials in CAD as well as in rapid prototyping and fabrication, in the simulation of different physical properties and in various other areas.

One of the main goals of computer simulation is to digitally represent reality as precisely as possible in order to provide us with a better understanding of the simulated material or phenomenon and to allow us to reliably experiment with it, in a virtual fashion, applying to it various modifications. A computer model of a real object should be able to represent many different aspects of the object, such as its outer shape, its general functionality, its internal structure, an ability to move and to interact with its environment, and perhaps other properties of the object. Man-made objects are often nearly uniform in their internal structure; for example, mechanical parts are usually made of a single type of metal. In contrast, natural objects are rarely homogeneous, having a complicated internal distribution of various materials.

During their relatively short history, such application areas of computing as computer graphics, geometric and solid modelling, and computer-aided design have primarily been concerned with the digital representation and visualization of homogeneous objects. The main reasons for this were a preoccupation of the researchers in the field with man-made objects and the limited computing power available at the time. The next generation of mathematical models and supporting software systems should provide means for modelling artifacts that may be heterogeneous in material, dimension, and other geometrical or physical properties.

Two-dimensional texture mapping in computer graphics can be considered a first attempt to cope with an object's optical properties, such as its color and reflectance. Volume graphics took the next step in introducing a discrete volumetric distribution of density and other object properties that were simulated or sampled using special volume scanning devices. Emerging composite material technology and fabrication processes allowing the deposition of different materials at a given spatial location have stimulated research in modelling functionally graded materials. However, we can observe that currently the modelling of two opposite clusters of objects is supported, namely, objects that have a fairly complex geometry and a rather simple internal distribution of properties, or alternatively objects that have a fairly simple geometry and an elaborate internal distribution of properties. There is a need for a universal model capable of representing the full spectrum of heterogeneous objects with both complex geometric shapes and complex variations in their internal properties.

Although real objects are essentially three-dimensional, it is useful to consider them as lower-dimensional entities in some modelling applications. Including one- and two-dimensional objects in a model allows both for a higher abstraction level and for model simplification. An important issue is the combination of entities of different dimensions into a single mixed-dimensional heterogeneous object model. Various types of topological complexes have been proposed for this purpose. An open research question is the modelling of objects where a heterogeneous topology is combined with heterogeneous materials or other property attributes.

It is well-known that there is no single mathematical model which is universally the best for the representation of different shapes and their transformations. The same is true for heterogeneous object modelling. One of the ways to deal with this problem is by introducing hybrid models, where the most appropriate object representation is applied to the entire object or each of its elements at every stage of the modelling process. Here, the key idea is that there are several representations supported by the modelling system and a conversion method between these representations is applied either on user request or when some predefined conditions are satisfied. A challenging problem for a hybrid modelling system is the support of topological complexes with entities that have different mathematical representations.

Heterogeneous object modelling is still considered as an emerging research topic. Recently, there have been quite a few journal publications and conference papers devoted to different aspects of this broad area. This book is one of the first attempts to systematically cover the most relevant themes and problems of this new and challenging subject area. Our objective is to provide a first-hand description of the modern state of the art and to outline the most interesting directions for future research.

This book is a collection of invited papers and papers co-authored by the editors. Each chapter presents either new research results or a survey on the following topics:

Formal models and abstractions of heterogeneous objects including geometric, topological, discrete and continuous models, operations forming special algebras and conversions between different model types.

Data structures and algorithms for representing, modifying and computing with heterogeneous objects. Computational techniques for the design, reconstruction, optimization, analysis and simulation of heterogeneous objects that incorporate information on shape, material and physical behavior using a common framework.

Applications of heterogeneous object modelling in engineering and scientific areas, including geophysical, biomedical, artistic and multi-material fabrication applications.

The editors are grateful to all the authors who have presented their most recent work and hope that this book will prove useful in advancing this novel, challenging and important research and development area.

January 2008

Alexander Pasko
Valery Adzhiev
Peter Comninos
The National Centre for Computer Animation
Bournemouth University, UK

Table of Contents

An Implicit Complexes Framework for Heterogeneous Objects Modelling

Elena Kartasheva[1], Valery Adzhiev[2], Peter Comninos[2], Oleg Fryazinov[2], and Alexander Pasko[2]

[1] Moscow Institute for Mathematical Modelling, Russian Academy of Science, Russia
ekart@imamod.ru

[2] The National Centre for Computer Animation, Bournemouth University, Poole UK
{vadzhiev, peterc, ofryazinov, apasko}@bournemouth.ac.uk

Abstract. In this paper we further develop a novel approach for modelling heterogeneous objects containing entities of various dimensions and representations within a cellular-functional framework based on the implicit complex notion. We provide a brief description for implicit complexes and describe their structure including both the geometry and topology of cells of different types. Then the paper focuses on the development of algorithms for set-theoretic operations on heterogeneous objects represented by implicit complexes. We also describe a step-by-step procedure for the construction of a hybrid model using these operations. Finally, we present a case-study showing how to construct a hybrid model integrating both boundary and function representations. Our examples also illustrate modelling with attributes and dynamic modelling.

1 Introduction

Heterogeneous object modelling is becoming an important research topic in different application areas such as volume modelling and rendering, modelling of objects with multiple and varying materials in CAD and in rapid prototyping [12]. Such objects may represent mechanical parts or assemblies, geological and medical models, the results of physical simulations as well as time-dependant models of an artistic nature, and are heterogeneous in terms of their internal structure and their dimensionality. Various approaches based on boundary representations, topological decompositions and constructive procedural methods have been developed for the description of such heterogeneous objects [16]. The subject of our particular interest is topological subdivisions which describe an object or its boundary by a collection of disjoint open point sets or by a collection of quasi-disjoint closed point sets (which are glued together at their common boundaries). Such representations provide all information necessary for numerical simulation, for contact and friction analysis and for other CAD/CAM applications, but it is difficult to parameterize them and to ensure their validity (especially so in the case of the boundary representations). Moreover the implementation of many operations on objects described by topological decompositions is accompanied by a considerable loss of precision and performance.

A. Pasko, V. Adzhiev, and P. Comninos (Eds.): HOMA, LNCS 4889, pp. 1–41, 2008.

In contrast, constructive representations are always valid and can easily be parameterized. They are very efficient for model editing and lead to considerable storage savings in comparison with topological decompositions. On the other hand, constructive models provide very little explicit information about the associated objects. Consequently, they have to be converted into equivalent topological subdivisions to make it possible to apply numerical methods for the purpose of analysis of the described objects. All of this was the motivation for introducing in [1] a hybrid cellular functional model based on the notion of implicit complexes (IC), which allows for the flexible combination of a cellular representation and a constructive function representation altogether with attribute models.

ICs describe composite heterogeneous objects consisting of several components that can differ in their dimensionality, geometric representation, and non-geometric attributes. Within the IC framework we proposed a way of constructing a hybrid model which is not supposed to serve just for the combined usage of separate representations but it is genuinely unified. The IC-based framework provides a unified description of the geometry, topology, and attributes of a heterogeneous object. An object is described as the union of cells of various representation types and dimensionalities along with the relations between them. The main relations characterizing mutual locations of cells are the boundary and the containment relations. Non-geometric attributes are independently described by functional or cellular models and are associated with IC's cells by means of attribute relations.

In our previous papers [1, 15], we introduced an IC's general structure and described some basic procedures of the IC-based model construction along with suitable rendering methods. In this paper, we give a systematic description of the IC-based hybrid model, consider algorithms for the construction and the discretization of ICs, and examine in detail set-theoretic operations on heterogeneous objects within the IC framework.

The paper structure is as follows. Section 2 reviews some related works concerning the modelling of heterogeneous objects. We present the main features of IC based models in section 3 and then in Section 4 we give a rather detailed outline of a formal IC framework along with a description of the IC topology, IC geometry, IC attribute model and some other basic material. In Section 5, the basic operations on ICs and their elements are described. The set-theoretic operations on objects described by ICs are presented in Section 6. Section 7 includes some case studies illustrating practical applications of the IC framework. Finally, some conclusions are made and future work is outlined.

2 Related Works

In this section we briefly discuss some previous works on modelling dimensionally heterogeneous objects, objects with varying distribution of material and other attributes, and approaches to combining various geometrical representations. Both topological subdivisions and constructive procedural methods are used for the description of heterogeneous objects.

Various topological stratifications and complexes are used to describe spatial subdivisions [21, 30]. Types of stratifications and complexes differ in the constraints imposed on the topology of their elements and on the connections between these elements. Multidimensional simplicial complexes are used in [23] for dimension-independent geometric modelling for various applications. A Selective Geometric Complex (SGC) [32] is a non-regularised non-homogeneous point set represented through enumeration as the union of mutually disjoint connected open subsets of the real algebraic variety. A SGC provides a framework for representing objects of mixed dimensionality possibly having internal structures and incomplete boundaries.

The Djinn API for solid modelling [4] is based on objects partitioned in a cellular way and containing mutually disjoint cells which are manifold point-sets of differing dimensionality in 3D. In [10] an extension of B-spline surfaces to surfaces of arbitrary topology is proposed. Polyhedral complexes are used to describe the surface topology. A procedure for designing cellular models based on CW-complexes with an emphasis on the topological validity of the resulting shapes is considered in [19, 22]. Selective Nef complexes were proposed in [11].

The work on constructive topological representations [28] introduces the stratified structure that is quite different from the topological complexes. This stratification is defined on n-dimensional solids using 'natural' topology based on the neighbourhood concept. It considers only the n-dimensional atoms and ignores the lower dimensional ones as well as the connectivity characteristics of the atoms and of the corresponding solid. Such a model can not be used for describing heterogeneous objects containing components of different dimensionalities.

To specify non-geometric properties of objects, spatial subdivisions are also used in computer graphics and in finite element analysis (FEA) as the underlying structures for piecewise analytical descriptions of attribute functions. Usually a basic topological subdivision is selected, which can be described by a topological stratification [4, 17, 32], a cell complex [18, 7], or a voxel model [6]. Different types of functions can be used to describe attributes [13, 24, 20, 34]. A detailed survey on modelling heterogeneous objects consisting of multiple materials can be found in [16].

Constructive approaches are applied to the description of heterogeneous object models combining different geometrical representations. In the STC framework [31], a composite object is defined using a combination of layers each of which is described by a geometric complex, which is homogeneous with respect to the representations of its components. A model for objects with fixed dimensionality and heterogeneous internal structure (i.e., multidimensional point sets with multiple attributes or constructive hypervolumes) was proposed in [26]. This model supports uniform constructive modelling of point set geometry and attributes using real functions of point coordinates.

A feature based design methodology is represented in [27]. Under this methodology, heterogeneous objects consisting of multiple materials are constructed by engineering significant high-level components called form features and material features. Form features describe the shape of the objects and material features are used for modelling material variation. The relationships between form features and material features in heterogeneous objects are examined in [27] along with constructive feature operations. Set-theoretic operations for modelling functionally graded materials associated with a BRep geometry model are discussed in [33].

Constructive Volume Geometry (CVG) [6] combines geometry and attributes in a systematic manner. The model is presented as an algebra of 3D spatial objects utilizing voxel arrays and continuous scalar fields for representing both geometry and photometric attributes (such as opacity, color, etc.). A distance field based approach for heterogeneous object modelling, in which the space is parametrized by distance to the geometry boundaries is proposed in [5].

The HybridTree [2] is a constructive tree with leaves defined by a number of representations. Both the function evaluation and the surface mesh generation are provided for modelled objects. Depending on a user query, corresponding conversions between representations are applied. A hybrid constructive tree in [8] has leaves with both implicit and parametric representations. To polygonize the surface of a complex object, surface meshes of primitives are classified against the subtree defining function, trimmed, and then merged into the resulting mesh. However, both of these approaches do not support heterogeneous objects with components of different dimensionalities and do not provide a description of the topological structure of the object being modelled.

Some discussion and motivation for this work follow in the next section.

3 Main Features of the Hybrid IC-Based Model

From the careful examination of the literature mentioned above one is forced to conclude that the topological decomposition approach is not suitable for describing heterogeneous objects consisting of overlapping components which differ in their geometrical representations and attributes, because the application of such an approach leads to the subdivision of the initial components with loss of the their initial representations. In contrast, the constructive approach allows for the description of overlapping components but provides insufficient topological information about represented objects. In [1] we introduced a hybrid cellular-functional model based on the notion of an Implicit Complex (IC) which combines the advantages of both the topological and constructive representations. Thus it provides a valid topological description of heterogeneous objects and allows for the flexible combination of cellular and functional representations of both the geometry of objects and their attributes.

Let us outline the main features of this framework. It allows for representing a heterogeneous object by a union of high-level components that are significant for a given application. For example, such components can describe mechanical parts or animated characters. We allow the components to overlap each other but we introduce special constraints on the description of the mutual dispositions of these components. Thus we provide the representation with a distinctive structure that can easily be reduced to a cellular topological subdivision. The intersections of the components are described by constructive methods which preserve the precision of the representation. The representation can be quite compact if it involves only those entities which are necessary for the descriptions of the initial components and their mutual dispositions.

In [15], we have shown how this framework can be exploited to represent some heterogeneous models without using set-theoretic operations. In this paper, we first concentrate on advancing the structure of an IC and then on developing the methodology for constructing the ICs especially using set-theoretic operations. Set-theoretic operations on polyhedral topological complexes have been discussed previously in [11]. Here we stress the specifics of these operations caused by the heterogeneous structure of ICs containing BRep and FRep components in particular.

Data structures used for the descriptions of topological subdivisions of objects or their boundaries are typically represented by the adjacency graphs whose nodes correspond to vertices, edges, faces, connected volume regions or combinations of these, and whose links capture information related to adjacency, orientation, and ordering [30]. Instead of the frozen combination of the adjacency graphs optimized for a narrow range of applications we have elaborated more flexible data structures based on the concept of relations developed in discrete mathematics and widely used in computer science. Relations provide a unified description of various connections between individual cells, collection of cells, and entire complexes. They are also used for assigning attributes. The use of well established operations on relations makes it possible to change the data structures dynamically thus adapting them to specific applications. Relations based tools allow us to realize various operations on ICs in a compact form. A flexible mechanism of dynamically creating and removing relations provides an IC programming implementation that is effective for the design of complicated assemblies as well as for working with cellular complexes and meshes.

4 Implicit Complexes

In this section we provide a brief description of the theoretical framework that is based on the Implicit Complex (IC) notion.

4.1 The IC Basic Definition

We consider a hybrid model defined in the Euclidian modelling space E^3 as follows. Let $g_i^{q_i} \subset E^3$ be a closed point set called a *cell*, where i is its index number and q_i is its dimension. Then, a geometric object D is defined as the union of cells $g_i^{q_i}$ under the following conditions:

1) The boundary of each cell $g_i^{q_i}$ is the union of a finite number of cells of lower dimensions;

2) Cells can overlap each other but the intersection of any two cells is either the union of a finite number of cells or is empty. (Note that we call the cells satisfying conditions 1 and 2 as properly joined cells.)

3) Each $g_i^{q_i}$ is unambiguously described by some known geometric representation which provides a set of tools for geometrically and topologically correct discretization of the cell. So, a variety of representations can be used for the description of

the cell shapes. However, all of these representations should guarantee a conversion into a mesh described by a polyhedral complex.

A collection K of cells satisfying the above conditions is called an implicit complex (i.e.: $K = \{g_i^{qi}\}_{i=1}^N$). The dimension of the IC is the maximal dimension of its cells. In accordance with the above IC definition, polyhedral, cellular, and CW complexes can also be represented in the IC framework.

The above conditions actually ensure the ability to convert an arbitrary IC K into a polyhedral complex, which approximates both geometrically and topologically the object D being modeled thus ensuring the validity of the IC model. In fact, the reducibility of each representation used into a polyhedral one guarantees a correct execution of any operation on objects described by the various representations. However, we strive to exploit advantages of the different types of representation. That is why we keep the initial representations for components of the model and use meshes only for the implementation of the various numerical procedures applied in topology analysis, computational geometry, and finite element analysis. Fig. 1 shows an example of an IC describing a hybrid model combining functionally represented (FRep) and boundary represented (BRep) components.

This IC consists of two 3D cells g^3_{bunny}, g^3_{turtle}, three 2D cells $g^2_{sufr\_bunny}$, $g^2_{sufr\_turtle}$, g^2_{zone} and a 1D cell g^1_{line}. Geometric types of cells are explained below in Section 4.4. In this example, the 3D cell g^3_{bunny} describes the Bunny's body (represented initially by a BRep) and the 2D cell $g^2_{sufr\_bunny}$ which represents its boundary defined by a triangular mesh. The turtle's body (defined initially as an FRep) and its boundary are represented by a 3D cell g^3_{turtle} and a 2D cell $g^2_{sufr\_turtle}$, correspondingly. Finally, the intersection of the Bunny and the Turtle models, which forms a contact zone is described by a 2D cell g^2_{zone}. The contact zone is illustrated by Fig 1b. The boundary of the contact zone is described by the 1D cell g^1_{line}. The cells $g^2_{sufr\_bunny}$, $g^2_{sufr\_turtle}$ contain the cells g^2_{zone} and g^1_{line}.

The support of overlapping cells allows the insertion of components of a composite object into its IC model without the need for subdivision. The IC definition conditions are satisfied by including additional cells describing the mutual intersections of the components. This allows for the preservation of the initial representations of the components, which is useful for heterogeneous object modelling.

We denote the point set represented by a cell g_i^{qi} as $|g_i^{qi}|$. Correspondingly the point set union of all cells of the IC K is denoted by $|K|$ and called a carrier of K. This term allows us to distinguish the discrete set of cells $K = \{g_i^{qi}\}_{i=1}^N$ from the point set $|K| = \bigcup_{i=1}^N |g_i^{qi}|$. Thus formally, the hybrid representation for a geometric object $D \subset E^3$ is defined as follows: $D = |K|$, or $D = \{X \mid X \in |K| \subset \Omega \subseteq E^3\}$, where K is an implicit complex and Ω is a modelling space.

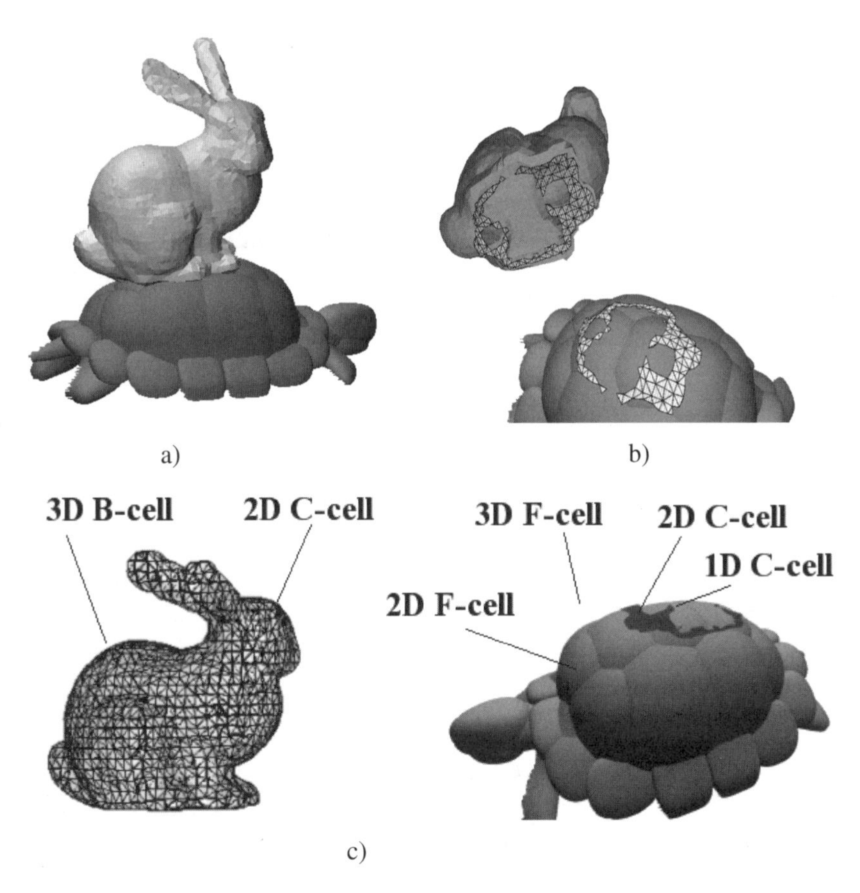

Fig. 1. The unified hybrid model combining an FRep Turtle (courtesy of G. Pasko) and a BRep Bunny (Stanford 3D Scanning Repository); a) The general view of the model; b) The contact zone on the surfaces of both objects; c) the IC structure and the types of the cells; the IC consists of the following cells g^3_{bunny}, g^3_{turtle}, $g^2_{sufr\_bunny}$, $g^2_{sufr\_turtle}$, g^2_{zone}, g^1_{line} (cell types are described in 4.4)

An implicit complex provides a consistent description of both the geometry and the topology of the modelled object. The geometry is represented by the geometry of the individual cells and the topology is described by means of the relations between cells.

4.2 The IC Topology

The main relations defining the topology of an IC are the boundary relation and the containment relation. According to the first two conditions of the IC definition the mutual disposition of any of the IC cells can be evaluated through queries to its main relations.

We denote a boundary relation between p-dimensional cells and s-dimensional cells of an IC as Rb^{ps}, $s < p$. By definition the boundary relation Rb^{ps} consists of a set of pairs (g_i^p, g_j^s) where the point set $| g_j^s |$ belongs to the boundary of the point set $| g_i^p |$ and does not lie in the interior of any other boundary cell of g_i^p. The containment relation between p-dimensional cells and s-dimensional cells of an IC is denoted by Rc^{ps}, $s \le p$. The pair (g_i^p, g_j^s) belongs to Rc^{ps} if $| g_j^s | \subset | g_i^p |$ and $| g_j^s |$ does not lie on the boundary of $| g_i^p |$.

For the example shown in Fig. 1 the IC topology is described by the following relations:

$$Rb^{32} = \{(g^3_{bunny}, g^2_{surf\_bunny}), (g^3_{turtle}, g^2_{surf\_turtle})\} ; Rb^{21} = \{(g^2_{zone}, g^1_{line})\}$$

$$Rc^{22} = \{(g^2_{surf\_bunny}, g^2_{zone}), (g^2_{surf\_turtle}, g^2_{zone})\};$$

$$Rc^{21} = \{(g^2_{surf\_bunny}, g^1_{line}), (g^2_{surf\_turtle}, g^1_{line})\}$$

For an unambiguous definition of a 3D IC, it is necessary to describe three boundary relations $Rb = \{Rb^{10}, Rb^{21}, Rb^{32}\}$ and nine containment relations $Rc = \{Rc^{ps}\}$, ($s \le p$, p=1,2,3, s=0,1,2,3) for all cells of dimensions from 0 to 3. Other boundary relations can be calculated using the composition operation (denoted by a symbol '$\circ$'). For example, $Rb^{20} = Rb^{21} \circ Rb^{10}$. Thus the description of an implicit complex K consists of the collection $G = \{g_i^{qi}\}_{i=1}^N$ of cells and the sets of the corresponding boundary *Rb* relations and containment *Rc* relations, i.e. $K = < G, Rb, Rc >$. For notational convenience we use the IC name *K* as a prefix for indicating to what complex some set of cells or relation belongs. For example, $K.G$, $K.Rb^{ps}$, $K.Rc^{ps}$.

Some additional relations can be useful for implementing those operations on the ICs that require a faster access to the information about the mutual disposition of its cells. The most often used additional relations are the co-boundary, the "to be contained", the incidence and the adjacency relations. These relations can be derived from the boundary and the containment relations using various operations on relations. In particular, considering the relations between p-dimensional and s-dimensional cells of an IC *K* we can conclude that the co-boundary relation denoted as Rcb^{ps} is the inversion of the corresponding boundary relation Rb^{sp}, so $Rcb^{ps} = (Rb^{sp})^{-1}$ ($p < s$, $p = 0,1,2$, $s = 1,2,3$), and the "to be contained" relation denoted as Rcc^{ps} is the inversion of the corresponding containment relation Rc^{sp}, $Rcc^{ps} = (Rc^{sp})^{-1}$. The incidence relation denoted as Rin^{ps}, $p \ne s$, $p = 0,1,2,3$, $s = 0,1,2,3$ is defined as follows:

$$Rin^{ps} = \begin{cases} Rb^{ps}, & if \ p > s \\ (Rb^{sp})^{-1}, & if \ p < s \end{cases}.$$

The adjacency relation denoted by Rd^{psp}, $p \neq s$, $p = 0,1,2,3$, $s = 0,1,2,3$ is defined as follows. A pair (g_i^p, g_j^p) of p-dimensional cells of the complex K belongs to the adjacency relation Rd^{psp} if there exists a cell $g_l^s \in K$, $s \neq p$ which is incident to both the cells g_i^p and g_j^p. The adjacency relation Rd^{psp} is calculated as the composition of the corresponding incidence relations Rin^{ps} and Rin^{sp}, $Rd^{psp} = Rin^{ps} \circ Rin^{sp}$. Note, that if $s < p$ then $Rd^{psp} = Rb^{ps} \circ (Rb^{ps})^{-1}$, otherwise $Rd^{psp} = (Rb^{sp})^{-1} \circ Rb^{sp}$.

Note that in principle it is not necessary to have all the possible relations explicitly described and stored. As to queries to the relations, they can be evaluated using the described dependences between relations. If needed, the additional relations can be dynamically computed on the basis of the main relations and then discarded (if not needed again).

4.3 Relationships between Implicit Complexes

Here we define relations between ICs and introduce several types of implicit complexes which are to be used further for describing different algorithms.

One can define not only relations within the IC cells but also between different complexes. They describe the mutual disposition of cells belonging to different ICs and are introduced by analogy with the corresponding relations between the cells of a single complex. So we introduce the *equivalence relation* and the containment relations between implicit complexes.

Let us consider two ICs A and B. We denote the cells of these ICs as $A.G = \{a_i^{q_i}\}_{i=1}^{I}$ and $B.G = \{b_j^{r_j}\}_{j=1}^{J}$. Then we denote the equivalence relation between the s-dimensional cells of A and B as Rq_{AB}^{s}, and we denote the containment relation between the p-dimensional cells of A and s-dimensional cells of B by Rc_{AB}^{ps}, $s \leq p$. By our definition the relation Rq_{AB}^{s} consists of pairs (a_i^p, b_j^p) of equivalent cells. We assume that any two cells are equivalent if they have the same carrier. The relation Rc_{AB}^{ps} consists of pairs (a_i^p, b_j^s) where cells $a_i^p \in A.G$ and $b_j^s \in B.G$ which comply with the following conditions: (i) a_i^p contains b_j^s; (ii) the carriers of cells a_i^p and b_j^s are not equal; (iii) b_j^s does not belong to the boundary of a_i^p. These relations are used by various operations on the implicit complexes.

Next, we introduce the concept of a subcomplex, which is a means of aggregation inside the IC. Formally, an IC L, is called a *subcomplex* of IC K, if the collection of cells of IC L is a subset of the collection of cells of IC K. Correspondingly the complex K is called a *supercomplex* of L. The equivalence relations unambiguously describe the complex L on the basis of its supercomplex K. Each subcomplex is a subset of the entire set of the given complex. Note that not every collection of the cells of a complex K is a subcomplex of K but only the collection which satisfies the conditions of the IC definition. For example, the following cells collections define the subcomplexes of the IC shown in Fig. 1:

$$L_1 = \{g^3_{bunny}, g^2_{surf\_bunny}\},\ L_2 = \{g^3_{turtle}, g^2_{surf\_turtle}\},\ L_3 = \{g^2_{zone}, g^1_{line}\},$$

$$L_4 = \{g^3_{bunny}, g^2_{surf\_bunny}, g^2_{zone}, g^1_{line}\}$$

We also introduce the concept of properly joined complexes. We call two implicit complexes *properly joined* if their cells altogether satisfy conditions 1 and 2 of the IC definition. Thus for the IC based model shown in Fig. 1, the subcomplexes L_1 and L_3 introduced above are properly joined but L_1 and L_2 are not because their cells overlap each other but their intersection is not represented by any cells of the complexes.

Another very important type of IC is a *nested* IC. Implicit complex B is nested into IC A if for any cells a_i^p and b_j^s which have common internal points it follows that a_i^p contains b_j^s. According to our definitions, if the IC B is nested into the IC A then the ICs A and B are properly joined. For the IC shown in Fig. 1 the subcomplex L_3 is nested into the other subcomplexes L_1, L_2, and L_4.

Note, that if the intersection of the carriers of any two ICs A and B is equal to the carrier of an IC C which is nested into both ICs A and B, then the ICs A, B, and C altogether from a group of properly joined complexes. Let us again consider the example shown in Fig. 1. The IC L_1, describing the model of the Bunny, and the IC L_2, describing the model of the Turtle, are not properly joined. The intersection of the Bunny with the Turtle is represented by the IC L_3 which is nested into ICs L_1 and L_3. The collection of the cells belonging to all the ICs L_1, L_2 and L_3 satisfy the conditions of the IC definition and form the entire hybrid model combining the Bunny with the Turtle. Further we exploit this important property of nested complexes for the realization of the set theoretic operations on objects represented by ICs.

4.4 The IC Geometry

In [1] we have proposed a basic IC structure. Here we introduce more elaborated structural classification in the form of five types of IC cells that differ in their geometric representations but are topologically uniformly related to each other:

- The P-cell, which is an explicit cell representing a simple polyhedron.
- The B-cell, which is a cell representing a manifold defined by its boundary. B-cells describe segments of parametric curves, patches of parametric surfaces and boundaries defining 3D solids. A 1D (2D) B-cell is defined by its supporting curve (surface) and by its oriented boundary. A 3D B-cell is defined by its oriented boundary only. In the general case the boundary of a B-cell can consist of cells of all the other types supported in the IC framework.
- The F-cell, which is an implicit cell described by the FRep that is a constructive representation by real-valued functions in the form of an inequality $F(X) \geq 0$ [25]. We restrict a valid variety of 2D and 1D FRep objects by s-dimensional

F-cells (s<3) to be those which are represented as subsets of the boundaries of 3D manifolds. Thus each 2D F-cell is defined as a patch of an implicit surface which is the boundary of an FRep 3D manifold M. It is described by a pair of continuous real functions of point coordinates (F, F_M), such that the point set of the cell is described by the inequality $F(X) \geq 0$ and the underlying 3D manifold is defined in the form $F_M(X) \geq 0$. Accordingly, each 1D F-cell is defined as a segment of an implicit curve belonging to the boundaries of two FRep 3D manifolds M_1 and M_2. So such a cell is described by a triple of continuous real functions of point coordinates (F, F_{M1}, F_{M2}) representing the cell and the underlying 3D manifolds.
- The C-cell, which is a composite cell aggregating cells of various types. Each C-cell is defined as a carrier of an implicit complex T differing from the complex K containing this C-cell. The complex T can consist of the cells of all types supported in the IC framework. In a particular case T can be a simplicial or a polyhedral mesh. The complex T is not a subcomplex of K. Its cells are not properly joined with respect of the cells of K.
- The T-cell which is a cell described by a constructive tree. Its leaves represent objects described by cells of all the other types. The tree nodes represent operations admissible for the IC – in particular, some bijective geometric transformations, non-regularized set-theoretic operations and trimming by 3D manifolds. The T-cells allow for the description of the result of applying set-theoretic operations to cells of different types without the need for converting between representations.

Various types of cells are illustrated by the example shown in Fig. 1. In the IC based model combining the BRep Bunny and the FRep Turtle, a 3D B-cell describes the Bunny's body and a 2D C-cell represents its boundary. The Turtle's body and its boundary are represented by a 3D F-cell and a 2D F-cell, correspondingly. The contact *zone* is described by a 2D C-cell. The boundary of the contact zone is described by the 1D C-cell.

4.5 An Illustrative Example

Let us consider an example illustrating both the geometric and topological features of an IC based model with cells of different types and with basic relations between them. Figure 2 shows a 2D object consisting of two 2D components (the rectangles *DEQS* and the disk *LHKF*) and a 1D component (the segment *OM*). We assume that the components are represented by different methods and have different non-geometric properties (attributes). One of the possible IC representations of this heterogeneous model is described by the complex K consisting of the following cells:

1. *2D cells*: the rectangle g^2_{DEQS}, the disk g^2_{LHKF}, and the half-disk g^2_{LKF};
2. *1D cells*: the closed polyline g^1_{DEQS}, the circle g^1_{LHKF}, the arc g^1_{KFL}, and the segments $g^1_{KL}, g^1_{OM}, g^1_{ON}$;
3. *0D cell*: the points $g^0_K, g^0_L, g^0_O, g^0_N, g^0_M$.

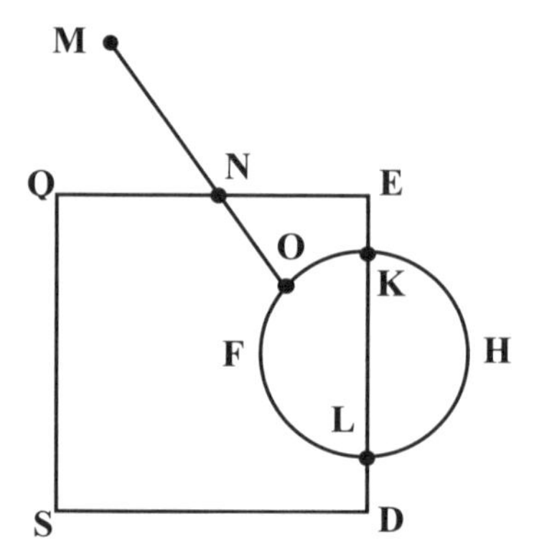

Fig. 2. A 2D object consisting of two 2D components (the rectangle *DEQS* and the disk *LHKF*) and a 1D component (the segment *OM*)

According to the IC definition, the complex *K* includes the cells representing the initial components g^2_{DEQS}, g^2_{LHKF}, and g^1_{OM}, their boundaries g^1_{DEQS}, g^1_{LHKF}, g^0_O, and g^0_M, and the cells describing the mutual intersections between the listed initial components. For example, the cell g^2_{KFL} represents the intersection of the initial components g^2_{DEQS}, g^2_{LHKF}. We assume that initially the rectangle *DEQS* is defined by a boundary representation, the disk *LHKF* is described functionally and the segment *OM* is specified explicitly by its end points. Consequently we represent these components by the B-cell g^2_{DEQS}, the F-cell g^2_{LHKF} and the P-cell g^1_{OM}, correspondingly. Other cells are added to satisfy the constraints of the IC definition. Table 1 shows these cell types and descriptions.

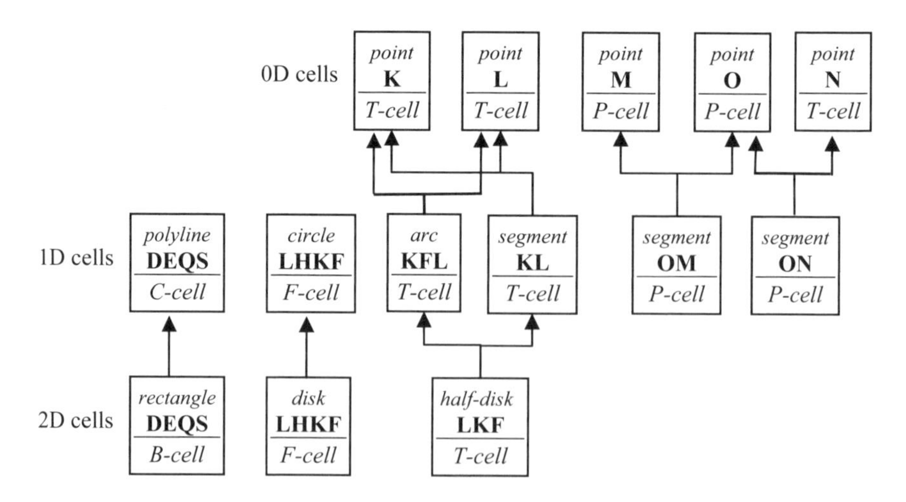

Fig. 3. The graph representing the boundary relations for an IC of the object shown in Fig. 2

Table 1. Types and descriptions of the cells of the IC representing the object shown in Fig. 2

object	*cell*	*description*
rectangle *DEQS*	2D B-cell g^2_{DEQS}	boundary representation
disk *LHKF*	2D F-cell g^2_{LHKF}	function representation $F_{disk} = R^2 - \sum(x_i - xcenter_i)^2$
half-disk *LKF*	2D T-cell g^2_{LKF}	constructive representation $\mid g^2_{LKF} \mid = \mid g^2_{DEQS} \mid \cap \mid g^2_{LHKF} \mid$
polyline *DEQS*	1D C-cell g^1_{DEQS}	mesh representation
circle *LHKF*	1D F-cell g^1_{LHKF}	function representation (F_{circle}, F_{disk}), where $F_{circle} = -(F_{disk})^2$
arc *KFL*	1D T-cell g^1_{KFL}	constructive representation $\mid g^1_{KFL} \mid = \mid g^1_{LHKF} \mid \cap \mid g^2_{DEQS} \mid$
segment *KL*	1D T-cell g^1_{KL}	constructive representation $\mid g^1_{KL} \mid = \mid g^1_{DEQS} \mid \cap \mid g^2_{LHKF} \mid$
segment *OM*	1D P-cell g^1_{OM}	a simple line segment
segment *ON*	1D T-cell g^1_{ON}	constructive representation $\mid g^1_{ON} \mid = \mid g^1_{OM} \mid \cap \mid g^2_{DEQS} \mid$
point *M*	0D P-cell g^0_M	3D coordinates
Point *O*	0D P-cell g^0_O	3D coordinates
point *N*	0D T-cell g^0_N	constructive representation $\mid g^0_N \mid = \mid g^1_{OM} \mid \cap \mid g^1_{DEQS} \mid$
point *K*	0D T-cell g^0_K	constructive representation $\mid g^0_K \mid = \mid g^1_{DEQS} \mid \cap \mid g^1_{LHKF} \mid \cap upper\_half\_plane(FH)$
point *L*	0D T-cell g^0_L	constructive representation $\mid g^0_L \mid = \mid g^1_{DEQS} \mid \cap \mid g^1_{LHKF} \mid \cap lower\_half\_plane(FH)$

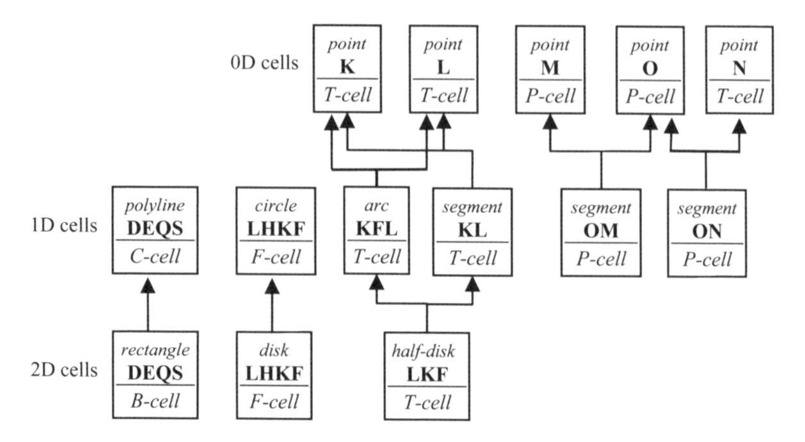

Fig. 4. The graph representing the containment relations for the IC of the object shown in Fig.2

Note, that if some geometric entities (for example, points *S, D, E,* and *Q*) describing shapes of the initial components are not essential for the representation of mutual dispositions of the components, then the resulting IC does not include them.

Figure 3 shows a graph representing the basic boundary relations in the IC *K*. Here the graph nodes represent the cells of the IC *K*, and the graph edges represent connections between cells in the boundary relations.

The containment relations for the IC *K* are represented by a graph shown in the Figure 4. Here once again the graph nodes represent the cells of the IC *K*, and the graph edges represent the connections between the cells in the containment relations.

4.6 The Attribute Model

In this section we consider a cellular-functional representation of the attributes associated with an IC. The attributes of an object and its geometry are described independently. Each attribute Λ is described by a set N_Λ of its values embedded into a multi-dimensional real number space $\Re^{m_\Lambda}$ of a proper dimension m_Λ. The interpretation of an attribute's value depends on its nature and specifics. For the sake of uniformity, we assume that the value set N_Λ of each attribute is supplemented with a special "empty" value θ and so all the attributes are defined at each point of the modelling space Ω. The attribute values are assigned to geometric object points described by the IC *K* using a collection of *attribute functions* $S_\Lambda = \{S_{\Lambda j}\}$, *j=1,…, J*, and a set of *attribute relations* $R_\Lambda = \{Rs^3_\Lambda, Rs^2_\Lambda, Rs^1_\Lambda, Rs^0_\Lambda\}$. Thus, an attribute Λ of an IC *K* is represented as $\Lambda = < S_\Lambda, R_\Lambda >$. Each function $S_{\Lambda j}$ of an attribute Λ maps the modelling space Ω into the attribute value set N_Λ, $S_{\Lambda j} : \Omega \to N_\Lambda$. Attribute functions can be analytic, piecewise analytic or be defined by an interpolation method [1]. Note that for defining interpolated attribute functions, we can use a space partition different from the one associated with the object. This means in the general case introducing another complex different from *K*. Note that it is possible to introduce various operations on attributes in our model. As attribute value sets belong to mathematical spaces $\Re^m$, all these operations are reduced to operations on real numbers and vectors. The attribute functions are defined on the entire modelling space. So we can construct new attribute functions on the basis of known attribute functions using various mixing functions.

The relations Rs_Λ^p (*p=0, 1, 2, 3*) associate functions of an attribute Λ with cells of the IC *K*, that is $Rs_\Lambda^p \subset G^p \times S_\Lambda$, where G^p is a set of all p-dimensional cells of the complex K. If $(g_i^p, S_{\Lambda j}) \in Rs_\Lambda^p$ then the value of the attribute Λ at any point $X \in | g_i^p |$ is defined as $S_{\Lambda j}(X)$.

Only one function of each attribute can be associated with a cell. Taking into account that IC cells can overlap each other, we propose *priority* and *additive schemes* for calculating the value of an attribute Λ at an arbitrary point X of the object represented by the IC *K*. According to the *priority scheme*, we look through all the cells associated with the attribute Λ and containing the point X and select one cell of the lowest dimension which does not contain other cells associated with Λ within it. The value of the attribute function defined on that cell is used for calculating the attribute value at the given point. According to the *additive scheme* the value of the attribute Λ at the point X is calculated as a blend of the attribute functions associated with all cells containing the point *X*.

4.7 Comparison with Other Approaches

Let us compare the IC based approach with the topological decomposition and the constructive representation using a simple example shown in Fig. 5. This example illustrates the construction of a heterogeneous object combining a white rectangle and a dark grey disk. We propose that the area where the rectangle intersects the disk becomes grey in color. If we use the topological decomposition then we should represent the combined object as a collection of non-intersecting cells. Thus we lose the initial representations of the components. However the topological decomposition provides full information about the topology of the object assembly which is very important for the numerical analysis. The application of the constructive approach results in the tree-like representation shown in the bottom part of Fig. 5. The tree preserves the descriptions of the initial objects but does not contain any information about mutual disposition of the components. Finally, the IC represents the combined object by a collection of properly joined overlapping cells. The IC preserves the initial descriptions and provides full topological and geometrical information about the assembly. The intersection of the initial components is described by the T-cell represented by a constructive tree. Thus the ICs allow us to combine different approaches to modelling heterogeneous objects and to describe complicated models consisting of a number of components defined by various geometrical representations.

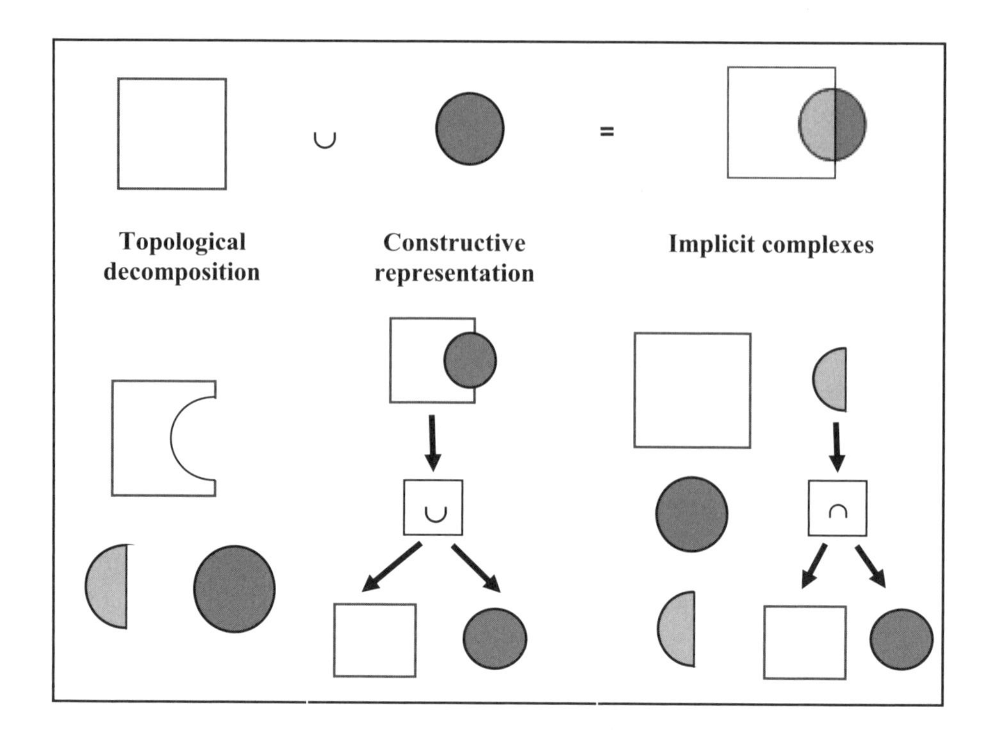

Fig. 5. A comparison of the different approaches to heterogeneous object modelling

4.8 Model Implementation

In this section we describe a software implementation for the cellular-functional modelling of heterogeneous objects within an object-orientated framework. Let us outline here only the principal classes which are directly derived from the presented theoretical description.

The basic *IComplex* class represents an implicit complex data structure. Its attributes represent six boundary relations and nine containment relations as well as references to the cell's description. Only the base relations are permanently supported, while other relations are calculated automatically by means of private methods of the *ICcomplex* class. The methods of the *IComplex* class realize the operations on the cells of the complex (see Section 5) as well as operations on ICs (see Section 5 and 6). The *IComplex* class includes operations for modifying the relations as well as inquiry operations on the relations.

Each relation is described by an object of the *Relation* class which contains all the pairs of numbers of related cells. The operations of the *Relation* class allow us to get the indices of all the related cells as well as to add and delete pairs of cells.

The IC geometry within the *IComplex* class is specified using objects of classes inherited from the abstract *Shape* class that contains virtual operations for defining the point membership as well as for rendering and discretization.

The *ICattribute* class represents an attribute data structure. It contains attribute relations and references to the descriptions of the attribute functions represented by classes inherited from the abstract template class *ICFunction.* The objects of the *IC-complex* and *ICattribute* classes are also linked by mutual references.

5 Operations in the IC Framework

In this section we introduce operations on both cells and entire implicit complexes. These operations are especially important in the context of constructing and manipulating implicit complexes.

5.1 Some Basic Operations in the IC Framework

Here we consider some basic operations such as adding and removing cells, adding attributes, etc. These operations will allow us to design various computational algorithms and to introduce some high-level functionality. For each operation, there are constraints on the input data that should be checked before their actual evaluation. These basic operations are implemented as procedures or procedure-quires.

The procedure *add_cell* adds a new cell to an IC. A cell g^r can be added to IC *K* if g^r is properly joined to all the cells of *K* and the boundary of g^r is represented as the union of cells of *K*. The input data of this procedure includes the cell geometry description and the list of cells of *K* related to the cell being added including: boundary cells, cells containing the added cell, and cells lying inside the added cell.

The procedures *add_attribute, add_function*, and *set_attribute* are used to establish attributes of the ICs. The first of these procedures adds a new attribute to an IC, the second one adds a function to an IC's attribute, and the last binds cells of the IC with attribute functions.

The procedure *remove_cell* deletes a cell from an IC. It follows from the IC definition that a cell g_c^r can not be deleted if at least one of the following conditions are satisfied: (i) the cell g_c^r has co-boundary cells in *K*; (ii) the cell g_c^r represents the intersection of some other cells of *K*. These conditions are checked automatically using IC relations. The procedure *cut_cell* removes a cell with all its boundary cells from an IC. It is implemented under the same restrictions as the *remove_cell* operation.

The following procedure-queries are used to acquire cells relationship information: *get_boundary_cells, get_coboundary_cells, get_adjacent_cells, get_cells_inside,* and *get_containing_cells.* These procedures take the dimension and the identifiers of an input cell as well as the dimension of the queried cells and return the list of the IC cells linked with the input cell by means of the corresponding relation.

5.2 The Basic Geometric Operations

In this section we introduce the basic geometric operations necessary for IC construction, rendering and numerical calculations of different kinds, such as IC transformation, the evaluation of a point membership and the intersection of an IC with a ray.

The *IC_transform* function implements *bijective geometric transformations* on ICs including affine transformations and nonlinear bijective transformations. This function is implemented by applying the corresponding transformation to a given cell.

The *Cell_ray_intersection* function computes the point of intersection of a given ray with a given cell. The *Cell_point_classify* function being applied to a given individual cell realises the point membership classification of the cell. These functions are optimally implemented for each particular cell type. Similar functions dealing with an entire IC have also been implemented: these are the *IC_ray_intersection* function dealing with intersection of the given ray with the object represented by the given IC, and the *get_cell_bypoint* function evaluating the point membership classification for that object.

The implementation of the *IC_ray_intersection* function is based on applying the *Cell_ray_intersection* function to the individual cells of the IC. The cells are examined in the order of increasing dimensionality. Among the cells of the same dimensionality we first analyze those which do not contain other as yet unprocessed cells. The processing of cells stops as soon as the first intersection with the ray is found. The *get_cell_bypoint* function is implemented in a similar manner. It returns the number of the first found cell for which the *Cell_point_classify* test gives a positive result.

5.3 The Subdivision of ICs

The *IC_meshing* procedure implements the conversion of an IC representation into a simplicial one. The discretization of IC models is guaranteed by the IC definition according to which the appropriate mesh generation methods have to be available for all types of IC cells. The discretization of an implicit complex K is implemented as an iterative process. The mesh generation of each kind of cell is implemented using specific meshing algorithms. An extensive survey of discretization methods is presented in [9]. We subdivide IC cells in the order of increasing dimensionality. Among the cells of the same dimensionality we first subdivide those which do not contain other as yet unprocessed cells. Thus, at the moment of the meshing of a cell we already know the discretization of its boundary and the subdivision of all the cells lying inside the one being considered. Then we subdivide the cell into mesh elements such that they are compatible with other meshes which already belong to it. Corresponding incremental mesh generation approaches that allow for preserving existing mesh elements can be found in [2,14] and in the references contained in these works.

For the purpose of illustration let us consider the application of an *IC-meshing* operation to the object introduced in Section 4.7. The corresponding meshes are shown in Fig. 6. According to the described meshing algorithm initially all the nodes $g^0_K, g^0_L, g^0_O, g^0_N, g^0_M$ are included into the mesh and then other cells are subdivided in the following order:

1) 1D cells $g^1_{KL}, g^1_{ON}, g^1_{KFL}$ (where g^1_{KFL} is subdivided taking into account the contained point g^0_O)
2) 1D cell g^1_{OM} (is compatible to the subdivision of the cell g^1_{ON});
3) 1D cell g^1_{DEQS} (is compatible to the subdivision of the cell g^1_{KL} and the point g^0_O);

4) 1D cell g^1_{LHKF} (is compatible to the subdivision of the cell g^1_{KFL});
5) 2D cell g^2_{LKF} ;
6) 2D cell g^2_{LHKF} (is compatible to the subdivided contained cell g^2_{LKF} it contains);
7) 2D cell g^2_{DEQS} (is taking into account the subdivision of the cells g^2_{LKF} and g^1_{ON}).

The last three steps of the meshing procedure are shown in Fig. 6

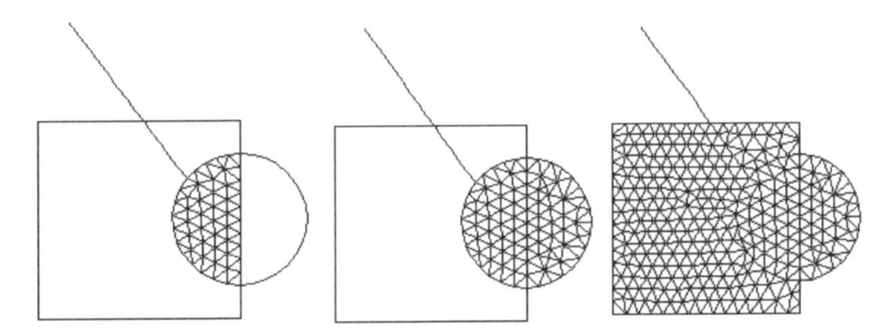

Fig. 6. The sequence of the 2D cell discretization steps for the IC containing overlapping cells (see Section 4.7 for a detailed description of the IC structure)

5.4 The IC Simplification Operations Dealing with Cells

Here we introduce operations *Removing_embedded_cells* and *Merging_cells* that reduce a number of cells of an IC preserving the carrier of the complex as well as the information representing its attributes and relations with other complexes. These simplification operations can be applied to an individual cell of an IC, to all the cells of the same dimensionality and to all the cells of a particular IC.

The operation *Removing_embedded_cells* removes all the cells that are contained in other cells with the same properties. We call an *r*-dimensional cell g^r_i of an IC *K* an *embedded* cell if the IC includes an s-dimensional cell g^s_j ($s \geq r$) containing g^r_i. The embedded cell g^r_i can be removed from *K* if the following conditions are satisfied: (i) both the cell g^r_i and the cell g^s_j that contains it have the same attribute values; (ii) all the cells containing g^r_i also contain g^s_j; (iii) all the co-boundary cells of g^r_i belong to the set of co-boundary cells of g^s_j.

The operation *Merging_cells* combines two cells with identical properties. Two *r*-dimensional cells g^r_i , g^r_j of the IC *K* can be merged into one if the following conditions are complied with: (i) the shape of the point set $|g^r_i| \cup |g^r_j|$ can be described by one of the representations supported by the IC. This means that it is possible to construct the cell g^r_m such that $|g^r_m| = |g^r_i| \cup |g^r_j|$; (ii) the boundary of g^r_m is represented by a subset of the union of boundary cells of g_i^r with boundary cells of g_j^r; (iii) the cells g_i^r, g_j^r have identical values for each attribute defined on K.

The operation of merging cells g_i^r, g_j^r includes the following steps:

1) Adding g_m^r to *K*. Note that the cells g_i^r, g_j^r lie inside g_m^r and the boundary cells of g_i^r, g_j^r which do not belong to the boundary of g_m^r are contained in g_m^r;

2) Deleting the cells g_i^r, g_j^r from *K* by means of the *remove_cell* operation;

3) Removing redundant cells contained in g_m^r by applying the *Removing_embedded_cells* operation.

We illustrate the *Removing_embedded_cells* and *Merging_cells* operations with an example of simplification of the complex shown in Fig.7. The initial IC *K* consists of two 2D cells (rectangles *ABEF* and *BCDE*), seven 1D cells (segments *AB*, *BC*, *CD*, *DE*, *EF*, *FA*, *BE*) and six 0D cells (points *A*, *B*, *C*, *D*, *E*, *F*). First we merge the 2D cells. We assume that both 2D cells have the same attributes. So we get a new combined 2D cell *ACDF* bounded by the segments *AB*, *BC*, *CD*, *DE*, *EF*, *FA*. The segment *BE* does not belong to the boundary of the new 2D cell, so it is registered as being contained in *ACDF*. Then we merge the 1D cells assuming that all the segments have the same attributes. According to the rules imposed by the *Merging_cells* operation we can merge all the boundary segments of the cell *ACDF* but we can not merge the segment *BE* with other segments because they have the co-boundary 2D cell but *BE* does not. After merging the 1D cells we get an IC consisting of one 2D cell *ACDF*, two 1D cells (the contour *ACDF* and the segment *BE*), and six 0D cells.

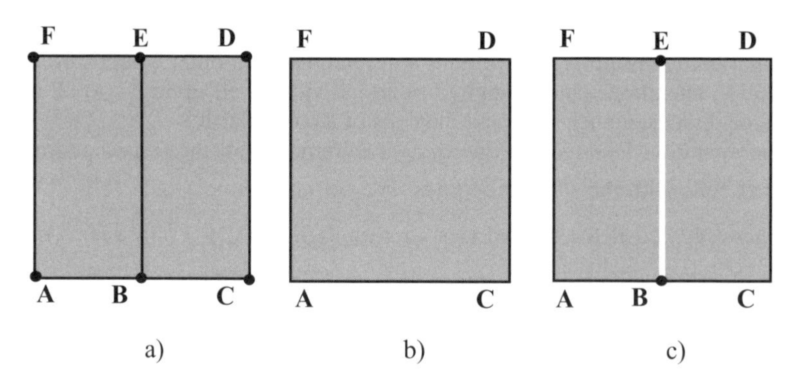

Fig. 7. Example of IC simplification. a) the initial IC; b) the simplified IC in the case where *BE* has the same attribute value as the cells *ABEF* and *BCDE*; c) the simplified IC in the case where the attribute values of the cells *BE*, *ABEF* and *BCDE* are different.

After that we apply the *Removing_embedded_cells* function to the resulting IC assuming that the 0D cells have the same attribute values as the segments. Let us consider two cases depending on the attribute value assigned to the segment *BE*. In the first case we assume that the segment *BE* has the same attribute value as the 2D cells of the IC *K*. In this case we apply the *Removing_embedded_cells* operation to *BE* and then to all the 0D cells. As a result we get an IC consisting of one 2D cell and one 1D cell, as shown in Fig.7b. In the second case we assume different attribute values for the segment *BE* and the 2D cells of the IC *K* we cannot remove *BE*. In this case by applying the *Removing_embedded_cells* operation to the IC nodes we delete the points *A*, *C*, *D*, *F* as embedded in the 1D cell *ACDF*. However, we cannot remove the points *B* and *E* because they do not satisfy the third conditions of the *Removing_embedded_cells* operation. Thus in this case we get the simplified IC consisting of one 2D cell, two 1D cells, and two 0D cells, as shown in Fig.7c. Here the 1D cell *BE* is registered as being contained in the 2D cell *ACDF* and the 0D cells *B* and *E* are registered as being contained in the 1D cell *ACDF*.

5.5 Summation Operations on Properly Joined Complexes

Let us remind the reader that we call two implicit complexes *properly joined* if their cells altogether satisfy conditions 1 and 2 of the IC definition. The summation operation on such ICs is defined as follows. Given any two properly joined implicit complexes *C* and *T*, the sum of C and T is a complex *M* consisting of all the cells of the complexes C and T, this sum is denoted as $M = C \oplus T$. The resulting IC *M* is properly joined to the complexes *C* and *T*. The summation operation is used for constructing ICs and for the realization of the set-theoretic operations on objects represented by ICs.

This algorithm described further is implemented as a procedure *IC_adding* which takes two initial ICs altogether with the equivalences and the containment relations between them and returns an IC consisting of all the cells belonging to the initial ICs. This procedure also creates the relations between the resulting IC and the initial ICs. These relations are used for establishing the attributes and can also be required for the implementation of some other operations on these complexes.

The steps of this algorithm are as follows:

1) The cells of *C* are copied into the initially empty IC *M*, and the equivalence relations between the ICs *M* and *C* are created.
2) Those cells of *T* which do not have equivalent cells in *C* are added to *M*. At the same time the equivalence relations between the ICs *M* and *T* are created.
3) The boundary relations and the containment relations of *M* are automatically formed on the basis of the same relations defined on the complexes *T* and *C* using the equivalence relations and the containment relations between these initial ICs. In doing so, we obey the following rules. The cells g_i^p and g_j^s of IC *M* are connected by the boundary relations of *M* if their equivalent cells in the initial ICs are themselves associated by the boundary relation defined on either IC *C* or IC *T*. The pair (g_i^p, g_j^s) belongs to the containment relation on M, if either (i) both cells g_i^p and g_j^s have equivalent cells in the same complex *T* or *C*, and are associated within that complex through the containment relation or (ii) one of the cells g_i^p and g_j^s has one equivalent cell in *T* and another in *C*, and they are related through the containment relation between the complexes *T* and *C*.
4) Finally, the containment relations between the resulting complex *M* and the initial complexes *T* and *C* are formed only if they are required for the implementation of

some other operations on these complexes. The containment relations between the ICs M and C are constructed according to the following rule. If the pair (g_i^p, g_j^s) belongs to the containment relation in the IC M and the cell g_j^s has an equivalent cell c_k^s in C, then we add the pair (g_i^p, c_k^s) to the containment relation between the ICs M and C. The relations between the ICs M and T are constructed in a similar manner.

5) Attributes are established on the IC M by combining the attributes defined on the initial complexes C and T using an appropriate mixing function. Suppose an attribute Λ is defined on both the ICs C and T. To set this attribute on M we look through all the cells of M, and for each cell g_m^r of M we find its equivalent cells in the complexes C and T. If g_m^r has only one equivalent cell, then it inherits an attribute function from this cell. Otherwise, the attribute function is calculated by blending the attribute functions defined on the cells of the ICs T and C equivalent to the cell g_m^r.

6 The Set-Theoretic Operations on Objects Described by ICs

Set-theoretic operations are the main mechanism for constructing composite geometric objects starting from more primitive ones. Set-theoretic operations applied to solids, point sets, cellular and geometrical complexes have been discussed previously in [3, 11, 29, 31, 32]. Here we address the specifics of these operations related to the heterogeneous structure of ICs containing overlapping components with different representations.

In the IC framework, we introduce the union and intersection operations on ICs with attributes. Given two implicit complexes A and B, the *intersection* of A and B is an implicit complex C whose carrier is equal to the set-theoretic intersection of the carriers of A and B, $C = A \cap B$. The *union* of two implicit complexes is a complex whose carrier is equal to the set-theoretic union of the carriers of the initial complexes. We denote the union operation as $C = A \cup B$. Note that we use the non-regularized set-theoretic operations.

The *difference* operation is more problematic as a mere set-theoretic difference of the carriers results in non-closed objects. So we consider a restricted version of the difference operation, namely, *trimming* with a 3D manifold. Let the IC B represent a 3D manifold; then the result of trimming the IC A by the complex B is a complex C whose carrier is equal to the set-theoretic intersection between the carrier A and the point set described as the closure of the inversed carrier of B (that is the cavity in the universal solid space). We denote this operation as $C = A - B$.

We outline four procedures for the implementation of the set-theoretic operations. Each of these takes two input ICs A and B, and returns an IC C. The intersection and the trimming operations are implemented by the procedures *IC_intersection* and *IC_trimming*, correspondingly. There are two procedures implementing the union operation, namely, *IC_union* and *IC_subtractive_union*.

Next we consider how to construct an IC C describing the result of the set-theoretic operations on the carriers of A and B. First we will consider the intersection operation as other operations rely on it. Then we describe algorithms for the trimming and the union operations. Finally, we consider how to establish the attributes of the resulting IC C combining those defined on the initial complexes A and B.

6.1 The Intersection Operation

Here we outline the implementation of the *IC_intersection* operation. We denote the cells of the IC's as $A.G=\{a_i^{qi}\}_{i=1}^{I}$, $B.G=\{b_j^{rj}\}_{j=1}^{J}$ and $C.G=\{c_l^{sl}\}_{l=1}^{L}$ respectively. By our definition, the carriers of the implicit complexes A, B, and C, that represent the intersection of A and B, have to satisfy the condition $|C|=|A|\cap|B|$. Thus the carrier of C is formed by the point sets $|a_i^{qi}|\cap|b_j^{rj}|$ which we denote as $|a_ib_j|$. Let V be the collection of all such pairs belonging to $|C|$. Analyzing the intersection of the two arbitrary point sets $|a_ib_j|$, $|a_lb_m|$ of V, one can note that according to the IC definition $|a_i|\cap|a_l|=\bigcup_{k=1}^{K\le I}|a_k|$, $|b_j|\cap|b_m|=\bigcup_{n=1}^{N\le J}|b_n|$. Therefore $|a_ib_j|\cap|a_lb_m|=\bigcup_{k=1}^{K}|a_k|\cap\bigcup_{n=1}^{N}|b_n|=\bigcup_{n=1}^{N}\bigcup_{k=1}^{K}(|a_k|\cap|b_n|)$. Thus the intersection of any two point sets of V is equal to the union of the point sets of the same collection. This means that the collection V satisfies the second condition of the IC definition. However some of the point sets $a_i b_j$ can be dimensionally heterogeneous. Such point sets have to be subdivided into several dimensionally homogeneous connected components. Thus, the general procedure for implementing the intersection operation on ICs can be described by the following pseudocode:

ICcomplex C = empty
for q from 0 to the dimension of A {
for s from 0to the dimension of B {
while (there exist non-updated q-dimensional cells of A){
find cell a_i^q *that does not contain non-updated cells of A*
while (there exist non-updated s- dimensional cells of B) {
find cell b_j^r *that does not contain non-updated cells of B*
ICcomplex L = empty
Mesh M = cells_intersection(a_i^q , b_j^r *)*
analyse connected_components of M
aggregate cells of M
convert aggregated cells of M into C- cells of L
establish relations between L and A, L and B
C = IC_adding (C, L)
mark cells a_i^q *and* b_j^r *as updated*
}}}}

As a result of this procedure we get the required IC $C=A\cap B$ along with the relations between C and the initial complexes A and B. Based on our construction, one can claim that the final IC C is properly joined to both complexes A and B. The relations between C and, A and B allow us to establish attributes on the IC C using attributes defined on the initial ICs A and B, and an appropriate attributes mixing function.

As can be seen from the above procedure, we build the IC C iteratively. The pairwise intersections of the cells are considered in the order of increasing dimensionality. Among the cells of the same dimensionality we first analyze those which do not contain other as yet unprocessed cells. Thus, at the moment of the actual evaluation of the intersection between a_i^q and b_j^r, the complex C already contains the cells describing the intersections of the boundaries of a_i^q and b_j^r, as well as the intersection a_i^q with all the cells lying inside the cell b_j^r along with the intersections of b_j^r with all the cells lying inside the cell a_i^q. This allows for constructing the intersection of a_i^q with b_j^r in the form of a set of properly joined cells of the IC C.

As to the actual algorithm of the intersection of a_i^q and b_j^r, its choice depends of the cell types. The most general method is based on the cells discretization. Depending on the IC definition appropriate algorithms are available for IC cells of any type.

In the general case we first perform the discretization of the initial cells a_i^q and b_j^r compatible to all the cells of IC C lying either inside a_i^q or b_j^r or on their boundaries (the list of such cells is provided by the relations between the ICs A, B and C). Thus we convert the cells a_i^q and b_j^r into the simplicial complexes properly joined with the IC C. Then we evaluate the non-regularized set-theoretic intersection between these simplicial complexes. To do this, one can apply methods similar to those developed for geometric complexes [32].

However, sometimes one can apply a more effective approach to evaluating the intersection of cells depending on the particular cell types. For instance, if the cell b_j^r is a 3D F-cell then it is enough to discretize only the cell a_i^q and to apply a procedure for trimming with an FRep solid. Moreover, 3D mesh generation is not required if we evaluate the intersection of two 3D cells represented by their boundaries.

In any case, after the evaluation of the cell intersection we get a simplicial complex representing the point set $|a_i^{qi}| \cap |b_j^{rj}|$. Then we analyze the topology of this complex and separate connected dimensionally-homogeneous components of the complex. After that, we aggregate the simplexes of each component of the resulting complex into subcomplexes using the following rules: any two n-dimensional cells can be merged only if they have a common boundary, the same co-boundary cells and if they are shared by the same cells in the initial ICs. As a result we convert the simplicial complex M into the IC L consisting of aggregated C-cells corresponding to the subcomplexes of M.

In all the previously described operations we maintain information about the relationships between complexes A, B and C. So, for each aggregated C-cell it is known what were its initial cells in the implicit complexes A and B. Preserving such information allows us to build a constructive or functional description of the aggregated cells depending on the types of the initial cells whose intersection these aggregated cells represent. If both of the initial cells are of the F-type, then the resulting cell is of the F-type too, otherwise it is of the T-type. Note that the polygonal representations of the cells a_i^q and b_j^r as well as the cells of the IC C can be preserved for a subsequent usage. Finally, we connect new cells of the IC C with the initial cells a_i^q and b_j^r by adding the corresponding pairs to the containment relations between the ICs A and C, and to the containment relations between the ICs B and C. Then all the cells of IC A containing the cell a_i^q are also linked with the new cells of the IC C by the containment relation between the ICs A and C. The same operations are performed on cells of the ICs B and C.

To illustrate the described algorithm, let us consider the intersection of two 2D objects shown in Fig. 8. The fist object is represented by the IC A consisting of one 2D cell $a^2_{polygon}$ and one 1D cell $a^1_{polyline}$, which is the boundary of $a^2_{polygon}$. The second object is a rectangle represented by the IC B consisting of one 2D cell b^2_{rect} and one 1D cell b^1_{frame}, which is the boundary of b^2_{rect}. Initially we evaluate the intersection of the cells $a^1_{polyline}$ and b^1_{frame}, $a^1_{polyline}$ and b^2_{rect}, and b^1_{frame} and $a^2_{polygon}$. As a result we get the IC C containing nine 0D cells representing points M, N, C, D, E, F, G, H, K, and the 1D cells describing the segments EF, MH, NK, HK, MN, CD, and the polyline CLD. An analysis of the intersection of $a^2_{polygon}$ and

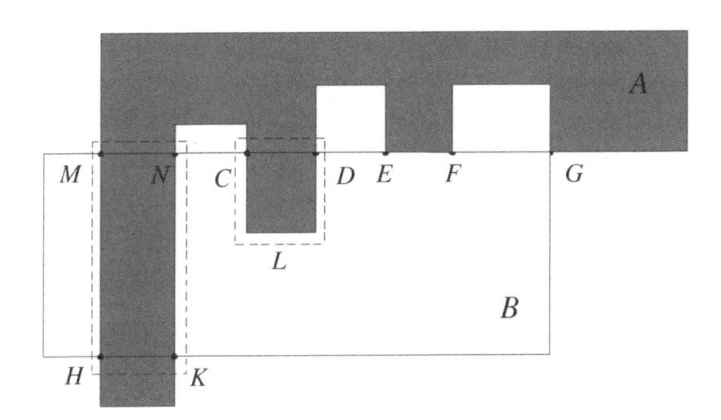

Fig. 8. The intersection of two 2D ICs (the IC A describes a grey polygon and the IC B represents a white rectangle). The bounding boxes of the components are shown in dashed lines.

b^2_{rect} allows us to find four connected components: the rectangles $MNKH$ and CLD, the segment EF, and the point G. The segment EF and the point G as well as the boundary elements of the 2D components had already been found at the previous stage; so we only have to add two 2D cells to the IC C. Let us denote these 2D cells as c^2_{MNKH} and c^2_{CLD}.

To build an unambiguous constructive description of the new cells, one needs to separate the corresponding components from each other. This can be done using, for instance, the bounding boxes as shown by the dashed lines in Fig. 9. Then we get the following description of the cells:

$|c^2_{MNKH}| = |a^2_{polygon}| \cap |b^2_{rect}| \cap |bounding\_box(MNKH)|$

$|c^2_{CLD}| = |a^2_{polygon}| \cap |b^2_{rect}| \cap |bounding\_box(CLD)|$. If the bounding boxes of different components intersect, then we build the bounding polyhedrons using hierarchic multiresolutional grids, as shown in Fig. 9.

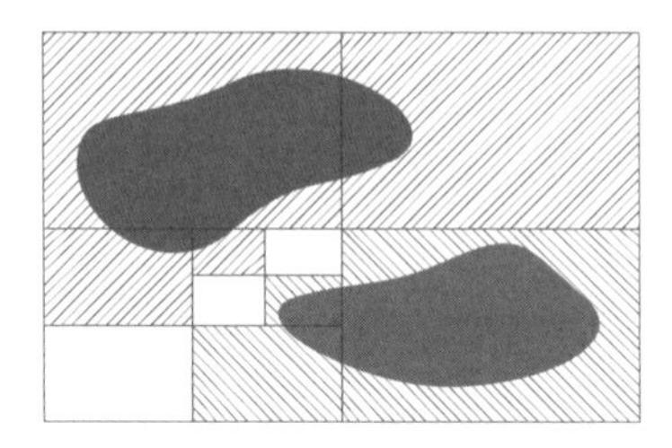

Fig. 9. Creating bounding polyhedrons for two objects

Let us consider the example in Fig. 10 showing an intersection of objects represented by ICs containing overlapping cells. Object *A* is represented by an IC consisting of one 2D cell a^2_{rect} and one 1D cell a^1_{frame}. Here a^1_{frame} is the boundary cell of a^2_{rect}. Object *B* includes two 2D cells b^2_{triang} and b^2_{disk} and two 1D cells $b^1_{polyline}$, b^1_{circle}. Pairs of cells (b^2_{triang}, $b^1_{polyline}$) and (b^2_{disk}, b^1_{circle}) belong to the boundary relation of the IC *B*. Pairs $(b^2_{triang}, b^2_{disk})$, $(b^2_{triang}, b^1_{circle})$ belong to the containment relations defined on the IC *B*.

The procedure for evaluating the intersection between the objects *A* and *B* includes the following steps.

1. The intersection of cells a^1_{frame} and b^1_{circle}. The resulting IC *C* consists of 0D cells c^0_M and c^0_N. The containment relations between the ICs *A* and *C* links the cell a^1_{frame} with the cells c^0_M and c^0_N. The containment relations between the ICs *B* and *C* links the cells b^1_{circle}, $b^1_{polyline}$ and b^2_{triang} with the cells c^0_M and c^0_N.

2. The intersection of cells a^1_{frame} and $b^1_{polyline}$. New 0D cells c^0_S and c^0_T are added to the resulting IC *C*. The pairs (a^1_{frame}, c^0_S) and (a^1_{frame}, c^0_T) belong to the containment relation between the ICs *A* and *C*. The pairs $(b^1_{polyline}, c^0_S)$ and $(b^1_{polyline}, c^0_T)$ belong to the containment relation between the ICs *B* and *C*.

3. The intersection of cells a^1_{frame} and b^2_{disk}. Using the containment relations connecting the IC *C* with the initial ICs we get the cells c^0_M and c^0_N belonging to the

point set $|a^1_{frame}| \cap |b^2_{disk}|$. Then we subdivide the cells a^1_{frame} and b^2_{disk} compatible with points *M* and *N*. This means that the points *M*, *N* coincide with the nodes of meshes representing the cells a^1_{frame} and b^2_{disk}. After evaluating the intersection we get the cell $c^1_{angle\_MN}$ whose boundary is formed by the cells c^0_M and c^0_N. The cell $c^1_{angle\_MN}$ is connected with the initial cells a^1_{frame} and b^2_{disk}, and the cell b^2_{triang} by means of the containment relations.

4. The intersection of cells a^1_{frame} and b^2_{triang}. The containment relations between the ICs *A*, *B* and *C* allow us to get the cells of the IC *C* belonging to the point set $|a^1_{frame}| \cap |b^2_{triang}|$. So we subdivide the cells a^1_{frame} and b^2_{triang} taking into account the already created cells c^0_M, c^0_N, c^0_S, c^0_T and $c^1_{angle\_MN}$. As a result we form a new cell c^1_{SMNT}. According to the construction procedure the cells c^0_S and c^0_T are the boundary cells of c^1_{SMNT}. These cells are linked by the boundary relation defined on the IC *C*. The remaining cells c^0_M, c^0_N and $c^1_{angle\_MN}$ are initially extracted as belonging to the point set $|a^1_{frame}| \cap |b^2_{triang}|$ and are marked as contained in the cell c^1_{SMNT}. So we add the corresponding pairs to the containment relations defined on the IC *C*. The cell c^1_{SMNT} is also linked with the cells a^1_{frame} and b^2_{triang} by means of the containment relations between the ICs *A*, *B* and *C*.

5. The ntersection of cells a^2_{rect} and b^1_{circle}. We add a new cell $c^1_{arc\_MN}$ to the IC *C* and connect it with the cells a^2_{rect}, b^1_{circle}, and b^2_{triang} by the containment relations.

6. The intersection of cells a^2_{rect} and $b^1_{polyline}$. We add a new cell c^1_{ST} to the IC *C* and connect it with the initial cells by the containment relations.

7. The intersection of cells a^2_{rect} and b^2_{disk}. Using the containment relations we get the cells of the IC *C* lying on the boundary of the initial cells. Evaluation of the cells intersection produces the cell $c^2_{sect\_MN}$ contained in the initial cells and in the cell b^2_{triang}.

8. The intersection of cells a^2_{rect} and b^2_{triang}. We know that the cells of IC *C* form the boundary of the point set $|a^2_{rect}| \cap |b^2_{triang}|$. As a result of the boundary evaluation, we add the cell c^2_{SMNT} to the IC *C*. Using the containment relations between the ICs *A*, *B* and *C* we find that the point set $|a^2_{rect}| \cap |b^2_{triang}|$ also contains the cell $c^2_{sect\_MN}$. So we add the pair (c^2_{SMNT}, $c^2_{sect\_MN}$) to the containment relation

defined on the IC C. Then we link the cell c^2_{SMNT} with the cells a^2_{rect} and b^2_{triang} by means of the containment relations between the ICs A, B and C.

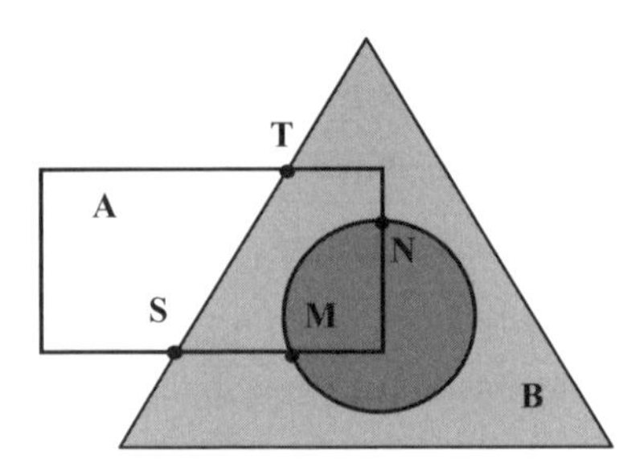

Fig. 10. The intersection of two 2D ICs. The IC A describes a transparent rectangle and the IC B represents an object consisting of two components which are a grey triangle and a dark grey disk.

6.2 The Trimming Operation

Let us construct the implicit complex C that represents the result of the trimming operation applied to the ICs A and B, $C = A - B$. This operation is valid only if the IC B represents an nD manifold, where n is the dimensionality of the modelling space denoted here by Ω . The corresponding algorithm is implemented as the procedure *IC_trimming*.

We construct the complex C through intersection operations as follows. First, we evaluate the boundary of the IC B, invert the orientation of its cell and get the IC which describes the boundary representation of a cavity in the universal set. We denote this IC by D. According to our construction, D is properly joined to B. We also form the relations between D and B. They are created during the boundary evaluation process. After that, we calculate the required IC C as the intersection between A and D. According to our intersection algorithm, C is properly joined to A and D; therefore it is properly joined to B. The relations between C and B are established as the composition of the corresponding relations between C and D, and D and B. The resulting IC C inherits attributes from the IC A. The trimming operation is illustrated by Fig. 11.

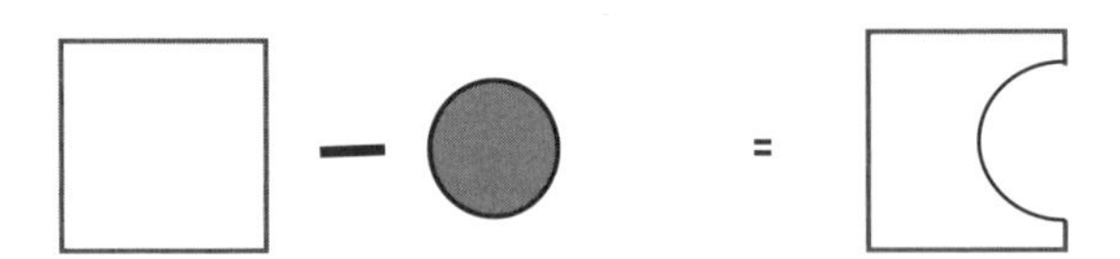

Fig. 11. The trimming object represented by the IC A (white rectangle) by the object represented by the IC B (gray disk)

6.3 The Union Operation

Let us construct an implicit complex C that represents the union of A and B, $C = A \cup B$. We propose two procedures for the implementation of the union operation: *IC_union* and *IC_subtractive_union.* Each of these takes two input ICs and returns an IC representing the union of the carriers of the initial IC. Let us consider these two procedures in more detail.

1. The procedure *IC_union*. The required IC *C* is calculated as the sum of two properly joined ICs, $C = (A+T)+B$, where *T* is an IC describing the intersection between the initial ICs, $T = A \cap B$. Thus, the resulting IC *C* involves all the cells of the input ICs *A* and *B* and additional cells representing their intersection. For the example shown in Fig. 12 the resulting IC *C* returned by the procedure *IC_union* contains 2D cells: the triangle *CED*, the disk *IGHF* and the half-disk *FHG*; 1D cells: the polyline *CED*, the circle *IGHF*, the segment *FG* and the arc *FHG*, and 0D cells: the points *F* and *G*. Let us show that the summation operation can be applied to the ICs $M = A+T$ and *B*. It follows from our IC intersection algorithm that the IC *T* is properly joined to *A* and *B*. Then, the intersection of any cell from the IC $M = A+T$ with any cell from *B* belongs to the complex $T = A \cap B$ and consequently is contained in *M*. So the IC *M* is properly joined to *B*.

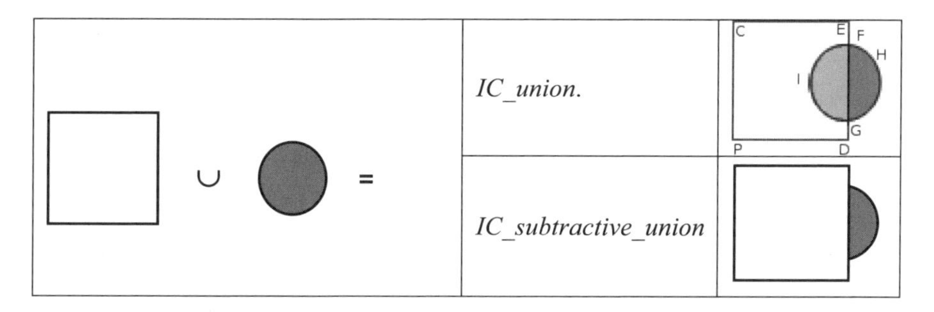

Fig. 12. The union of the objects represented by the ICs *A* (white rectangle) and *B* (dark gray disk)

2. The procedure *IC_subtractive_union*. The required IC *C* is calculated as $C = A+(B-A)$. Thus, the resulting IC *C* does not include those points of the IC *B* which lie inside the carrier of *A*. According to our trimming algorithm, IC $L = B-A$ is properly joined to the IC *A*; so we can apply the sum operation to the ICs *L* and *A*. This implementation of the union operation is available only if the IC *A* represents a 3D manifold. The bottom right part of Fig. 12 illustrates this case.

6.4 Establishing Attributes in Set-Theoretic Operations

The attributes are associated with the resulting IC *C* through the relations between *C* and the initial complexes *A* and *B*. These relations are created, as already described, by the procedures implementing the set-theoretic operations.

The attribute functions associated with the cells of *C* belonging to both ICs *A* and *B* are calculated as a blend of the initial attributes functions associated with the same cells on the ICs *A* and *B*. Other cells of *C* inherit attributes from their preimages in the initial complexes.

Let us consider the example in Fig 12. Here the attribute "color" is defined on the initial ICs A and B. Then in the resulting IC C produced by the *IC_union* operation (shown in the top right part of Fig. 12), the cells representing the rectangle and the disk preserve those attribute values that were defined on the initial ICs. The attribute

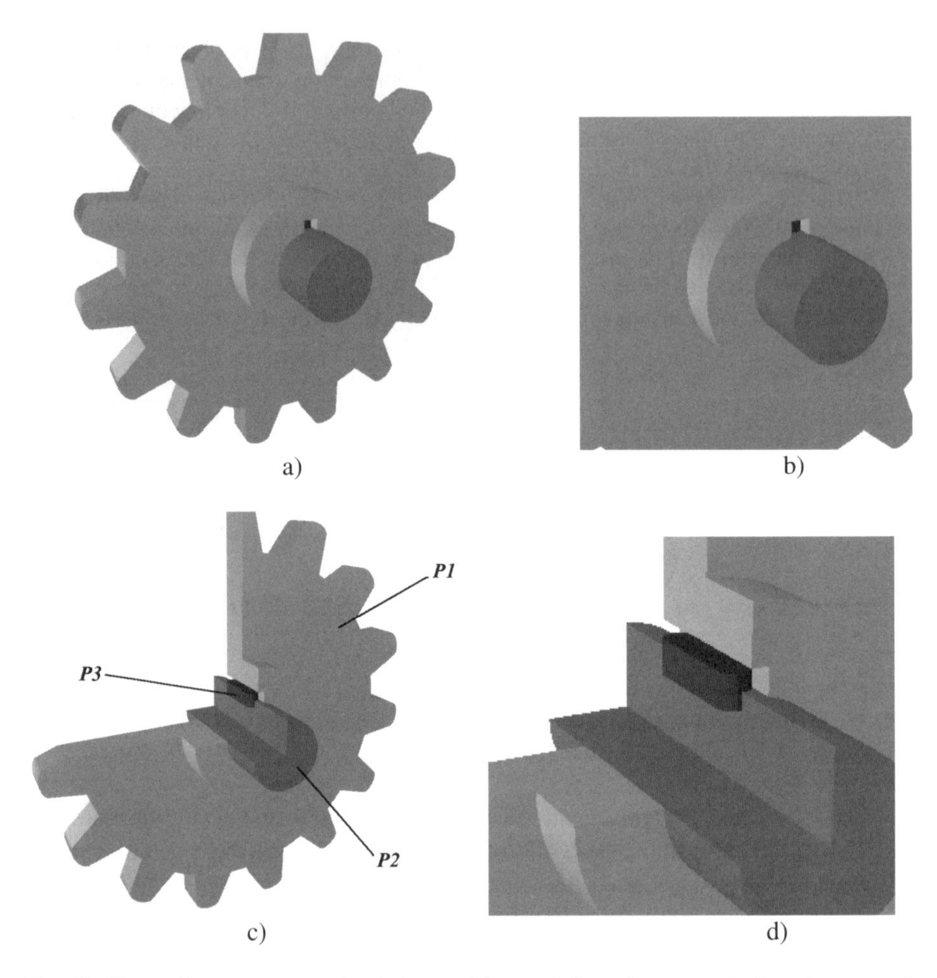

Fig. 13. The multicomponent mechanical assembly consisting of a gear, a shaft, and a two end-round key: a) general view; b) zoom to the shaft; c) and d) sections

value of the cell describing the half-disk, generated as a result of the intersection of the initial cells, is calculated as a blend of the attribute values defined on the original cells.

7 Case Studies

In this section, we present case studies that allow us to show how to model some heterogeneous objects within the IC framework using the set-theoretic operations on ICs. We consider two examples. The first example concerns modelling mechanical assemblies and the second illustrates the construction of heterogeneous objects for animation.

7.1 The Modelling of Mechanical Assemblies

This example illustrates how multicomponent mechanical assemblies are described using the IC framework. Fig. 13 shows an assembly consisting of three main parts: a gear, a shaft, and a two end-round key.

The IC based model of the assembly is illustrated by Fig. 14. Let us consider the cells forming the IC representing this assembly. The 3D cells g^3_{P1}, g^3_{P2} and g^3_{P3} describe the initial components: the gear, the shaft, and the two end-round key. Their surfaces are represented by the cells g^2_{dP1}, g^2_{dP2} and g^2_{dP3} respectively. The intersection area of the gear and the shaft is described by the 2D cell g^2_{Z1}, and the boundary contour of that area is described by the 1D cell g^1_{C1}. Then, the cells g^2_{Z2} and g^1_{C2} describe the intersection area of the two end-round key and the shaft as well as its boundary curve. The contact zone of the gear and the two end-round key consists of two rectangles *Z3* and *Z4*. These rectangles are represented by the cells g^2_{Z3} and g^2_{Z4}, while their boundaries are described by the cells g^1_{C3} and g^1_{C4}. The boundaries of the intersection areas of the initial components intersect along the segments *L1* and *L2*. The nodes *T1*, *T2*, *T3* and *T4* represent the ends of these segments. The segments and their end points are represented by the cells g^1_{L1}, g^1_{L2}, g^0_{T1}, g^0_{T2}, g^0_{T3} and g^0_{T4}.

The following boundary and containment relations describe the IC structure:

$$Rb^{32} = \{(g^3_{P1}, g^2_{dP1}), (g^3_{P2}, g^2_{dP2}), (g^3_{P3}, g^2_{dP3})\}$$

$$Rb^{21} = \{(g^2_{Z1}, g^1_{C1}), (g^2_{Z2}, g^1_{C2}), (g^2_{Z3}, g^1_{C3}), (g^2_{Z4}, g^1_{C4})\}$$

$$Rb^{10} = \{(g^1_{L1}, g^0_{C1}), (g^1_{L1}, g^0_{T2}), (g^1_{L2}, g^0_{T3}), (g^1_{L2}, g^0_{T4})\}$$

$$\begin{aligned} Rc^{22} = \{&(g^2_{dP1}, g^2_{Z1}), (g^2_{dP2}, g^2_{Z1}), (g^2_{dP2}, g^2_{Z2}), (g^2_{dP3}, g^2_{Z2}), \\ &(g^2_{dP1}, g^2_{Z3}), (g^2_{dP1}, g^2_{Z4}), (g^2_{dP3}, g^2_{Z3}), (g^2_{dP3}, g^2_{Z4})\} \end{aligned}$$

$$\begin{aligned} Rc^{21} = \{&(g^2_{dP1}, g^1_{C1}), (g^2_{dP2}, g^1_{C1}), (g^2_{dP2}, g^1_{C2}), (g^2_{dP3}, g^1_{C2}), \\ &(g^2_{dP1}, g^1_{C3}), (g^2_{dP1}, g^1_{C4}),, (g^2_{dP3}, g^1_{C3}), (g^2_{dP3}, g^1_{C4}), \\ &(g^2_{dP1}, g^1_{L1}), (g^2_{dP1}, g^1_{L2}), (g^2_{dP2}, g^1_{L1}), (g^2_{dP2}, g^1_{L2}), (g^2_{dP3}, g^1_{L1}), (g^2_{dP3}, g^1_{L2})\} \end{aligned}$$

$$Rc^{11} = \{(g^1_{C1}, g^1_{L1}), (g^1_{C2}, g^1_{L1}), (g^1_{C1}, g^1_{L2}), (g^1_{C2}, g^1_{L2}), (g^1_{C3}, g^1_{L1}), (g^1_{C4}, g^1_{L2})\}$$

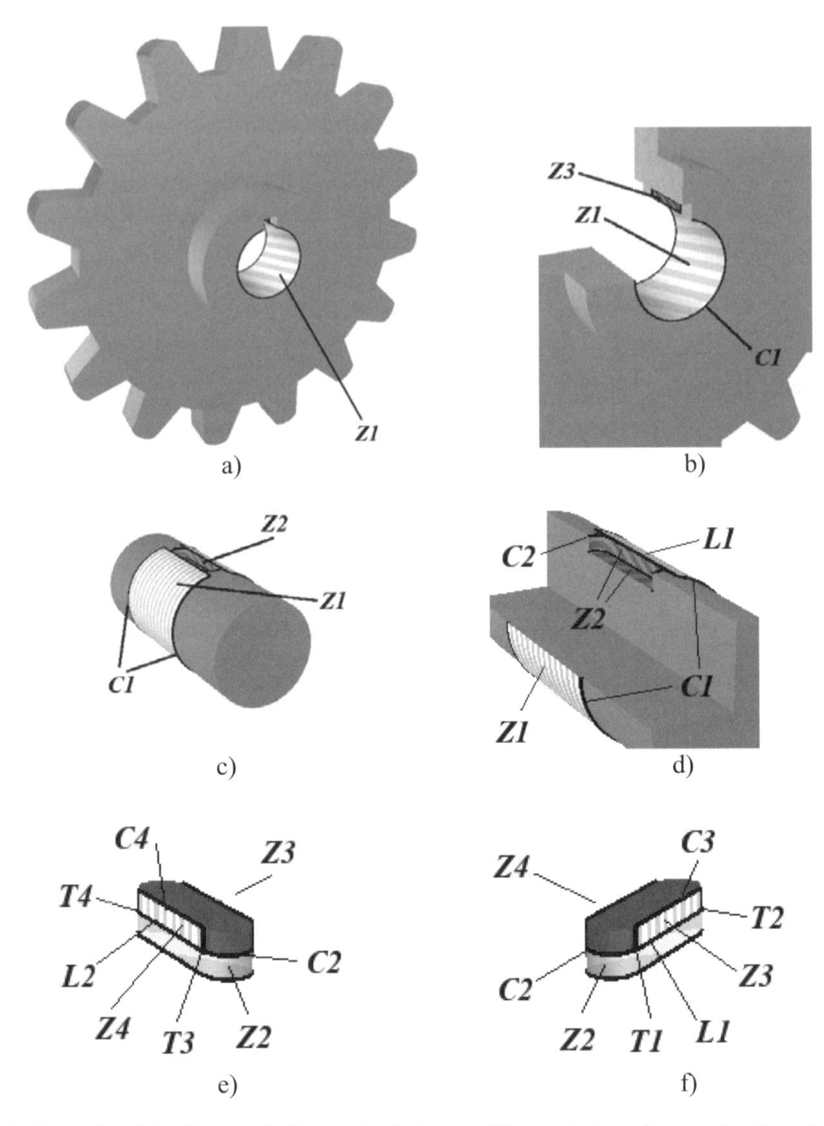

Fig. 14. The cells of the IC model of a mechanical assembly consisting of a gear (a, b), a shaft (c, d), and a two end-round key (e, f)

$$
\begin{aligned}
Rc^{20} = \{ & (g^2_{dP1}, g^0_{T1}), (g^2_{dP1}, g^0_{T2}), (g^2_{dP1}, g^0_{T3}), (g^2_{dP1}, g^0_{T4}), \\
& (g^2_{dP2}, g^0_{T1}), (g^2_{dP2}, g^0_{T2}), (g^2_{dP2}, g^0_{T3}), (g^2_{dP2}, g^0_{T4}), \\
& (g^2_{dP3}, g^0_{T1}), (g^2_{dP3}, g^0_{T2}), (g^2_{dP3}, g^0_{T3}), (g^2_{dP3}, g^0_{T4}) \} \\
Rc^{10} = \{ & (g^1_{C1}, g^0_{T1}), (g^1_{C2}, g^0_{T1}), (g^1_{C1}, g^0_{T2}), (g^1_{C2}, g^0_{T2}), \\
& (g^1_{C1}, g^0_{T3}), (g^1_{C2}, g^0_{T3}), (g^1_{C1}, g^0_{T4}), (g^1_{C2}, g^0_{T4}), \\
& (g^1_{C3}, g^0_{T1}), (g^1_{C3}, g^0_{T2}), (g^1_{C4}, g^0_{T3}), (g^1_{C4}, g^0_{T4}) \}
\end{aligned}
$$

From the above example one can clearly see the advantages of the IC based approach. In particular, the model contains a description of all the components in their initial state (without sections) along with the contact zones described by the additional cells that are constructively defined with respect to the initial components. Such a description is very effective for the purpose of subsequent optimisation of the assembly process. At the same time, the IC framework guarantees such a description of both geometry and topology that provides all the necessary data for the subsequent discretisation, finite element analysis, and other numerical calculations on the assembly.

7.2 Constructing a Hybrid Model Combining Brep and Frep Components

Next we consider an example with very important practical applications. Namely a hybrid model unifying objects represented in a mixture of boundary and functional representations. Here we deal with two relatively simple initial objects (see Fig. 15); a Triceratops BRep model and a Mouse FRep model. We will show in a step-by-step manner how to construct a hybrid model representing the Triceratops with the head of the mouse. First we construct the ICs representing each of the initial objects. Then we build a unified model using set theoretic operations on ICs.

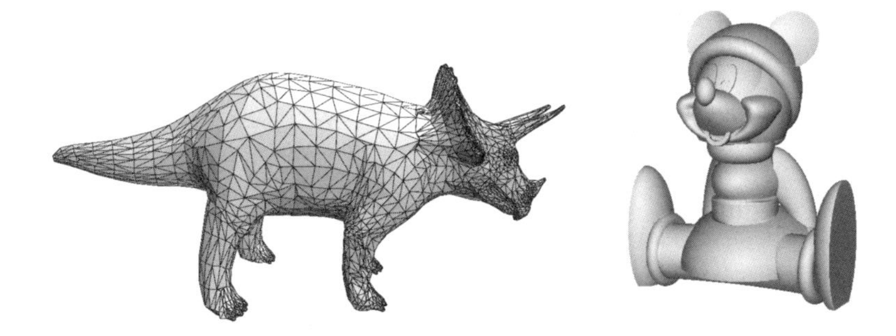

Fig. 15. A BRep object Triceratops and an FRep object Mouse

7.2.1 The Construction of the ICs Representing the Initial Objects

We denote by T and M the ICs that describe the models of the Triceratops and the Mouse respectively. Initially the Triceratops is described by a surface mesh loaded from a file and the Mouse is described functionally in the form of the inequality $F_M(x, y, z) \geq 0$. We also associate material attributes with these models. We assume that the material of the Triceratops is described by the functions, S_{tric_in} (for internal points) and S_{tric_on} (for boundary points), and the material of the Mouse is specified by the function S_{mouse}.

We use the *add_cell* procedure to construct the initial ICs. The input data of this procedure includes cell type and dimension, its shape description, and the following: the lists of its boundary cells, the list of the IC cells containing the cells being added, and the list of the IC cells lying inside the cells being added. The *add_cell* procedure returns the ID of each added cell. Attributes are established using the procedures *add_attribute*, *add_function* and *set_attribute*. The procedure *add_attribute* takes the name of the attribute being added and the dimension of its values space and returns the ID of the added attribute. The *add_function* adds the function to the IC attributes and returns its ID, and the *set_attribute* binds cells with attributes functions taking their IDs as input data.

The following pseudocode describes the initial ICs construction:

```
IC T=empty, M=empty;
// Set up of the initial data
MeshSurface Triceratops(filename);
FrepSolid Mouse(FunM);
FrepSurface MouseBnd(-(FunM*FunM), FunM);
// Construction of the ICs
Cell_ID tric, tric_surf, mouse_surf, mouse;
tric_surf = T.add_cell(type = 'Ccell', dim = 2, shape = Triceratops, bnd_cells = empty, containing_cells = empty, cells_inside = empty);

tric = T.add_cell(type = 'Bcell', dim = 3, shape = empty, bnd_cells = {tric_surf}, containing_cells = empty, cells_inside = empty);

mouse_surf = M.add_cell(type = 'Fcell', dim = 2, shape = MouseBnd, bnd_cells = empty, containing_cells = empty, cells_inside = empty);

mouse = M.add_cell(type = 'Fcell', dim = 3, shape = Mouse, bnd_cells = {mouse_surf}, containing_cells = empty, cells_inside = empty);

// Setting of the attributes
IC_Attribute T_mat, M_mat;
Attribute_function_ID Stric_in, Stric_on, Smouse;

T_mat = T.add_attribute(attribute_name = 'material', values_space_dim = 1);
Stric_in = T_mat.add_function(MaterialTricInside);
Stric_on = T_mat.add_function (MaterialTricSurf);
T.set_attribute(cell = tric_surf, function = Stric_on);
T.set_attribute(cell = tric, function = Stric_in);

M_mat = M.add_attribute(attribute_name = 'material', values_space_dim = 1);
Smouse = M_mat.adding_function( MouseMaterial);
M.set_attribute(cell = mouse_surf, function = Smouse);
M.set_attribute(cell = mouse, function = Smouse);
```

As a result of performing this procedure, we get the ICs *T* and *M*. The IC *T* consists of the 2D composite cell (C-cell) $t^2_{tric\_surf}$ and the 3D boundary cell (B-cell) t^3_{tric}. These cells are related by the boundary relation Rb^{32}. The cell t^3_{tric} is associated with the attribute function S_{tric_in} and the material attribute is described on $t^2_{tric\_surf}$ by the function S_{tric_on}.

The IC M includes the 2D functional cell (F-cell) $m^2_{mouse\_surf}$ and the 3D F-cell m^3_{tmouse}. The shape of $m^2_{mouse\_surf}$ is described by the pair of functions $(F = -(F_M)^2, F_M)$ and the shape of m^3_{tmouse} is described by the function F_M. These cells are related by the boundary relation Rb^{32}. Both cells of the IC M are associated with the attribute function S_{mouse}.

The containment relations of the ICs T and M are empty.

7.2.2 The Construction of the Headless Triceratops

We create an IC A representing a Triceratops without a head as the result of the intersection between the IC T and the additional IC D describing a spherical cavity in the modelling space, $A = T \cap D$ (see Fig. 16). The cavity is represented functionally. The construction process of the IC A is described by the following pseudocode:

```
IC D = empty, A = empty;
FrepSolid Cavity(-FunSphere);
FrepSurface Sphere(-(FunSphere * FunSphere), FunSphere);
Cell_ID cavity, cavity_surf;
cavity_surf = D.add_cell(type = 'Fcell', dim = 2, shape = Sphere, bnd_cells = empty, containing_cells = empty, cells_inside = empty);
cavity = M.add_cell(type = 'Fcell', dim = 3, shape = Cavity, bnd_cells = {cavity_surf}, containing_cells = empty, cells_inside = empty);
A = IC_intersection(T,D);
```

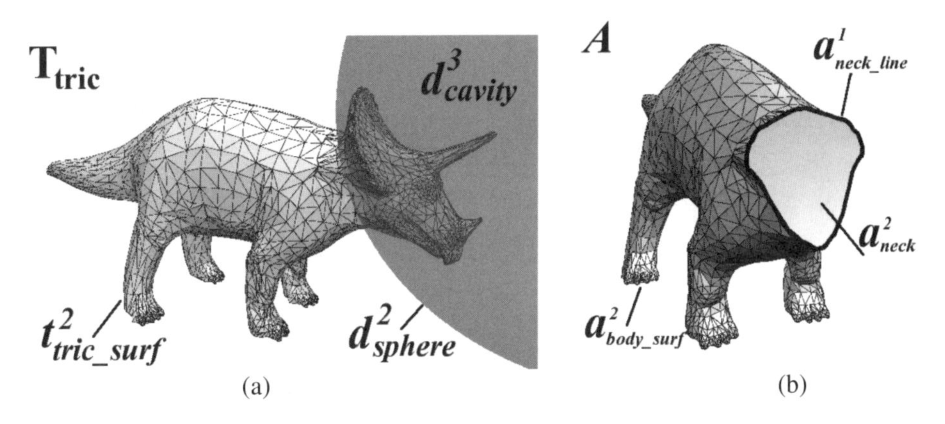

Fig. 16. The intersection operation: (a) Input objects: ICs T and D; (b) The resulting IC A, cyan $=S_{tric\_on}$, yellow $= S_{tric\_in}$

The IC A is generated automatically using the procedure *IC_intersection*. The complex A consists of the following constructive cells (T-cells): $|a^1_{neck\_line}| = |t^2_{tric\_surf}| \cap |d^2_{sphere}|$;

$|a^2_{neck}| = |t^3_{tric}| \cap |d^2_{sphere}|$;

$|a^2_{body\_surf}| = |t^2_{tric\_surf}| \cap |d^3_{cavity}|$;

$| a^3_{body} |=| t^3_{tric} | \cap | d^3_{cavity} |$

The cells of the IC A are related by the following boundary relations:

$A.Rb^{32} = \{ (a^3_{body}, a^2_{body\_surf}), (a^3_{body}, a^2_{neck}) \}$

$A.Rb^{21} = \{ (a^2_{body\_surf}, a^1_{neck\_line}), (a^2_{neck}, a^1_{neck\_line}) \}$

The boundary relations $A.Rb^{10}$ and all the containment relations of IC A are empty.

The cells of A automatically inherit the material attribute functions from its pre-images in IC T.

7.2.3 Combining the Headless Triceratops with the Mouse

We apply the subtractive union operation to the ICs M and A. As a result we calculate the IC $H = A + (M - A)$. This operation is possible because A describes a 3D manifold. The IC H is generated automatically by the operation

IC H = IC_subtractive_union(A,M)

The complex H involves cells equivalent to all the cells of IC A: $h^1_{neck\_line} = a^1_{neck\_line}$, $h^2_{body\_surf} = a^2_{body\_surf}$, $h^2_{neck} = a^2_{neck}$ and $h^3_{body} = a^3_{body}$.

Additionally, the IC H includes the following constructive T-cells (see Fig. 17): $h^1_{left\_line}$, $h^1_{right\_line}$, $h^1_{m\_neck\_line}$, $h^2_{left\_track}$, $h^2_{left\_surf}$, $h^2_{right\_track}$, $h^2_{right\_surf}$, $h^2_{m\_neck}$, $h^2_{head\_surf}$, h^3_{left}, h^3_{right} and h^3_{head}.

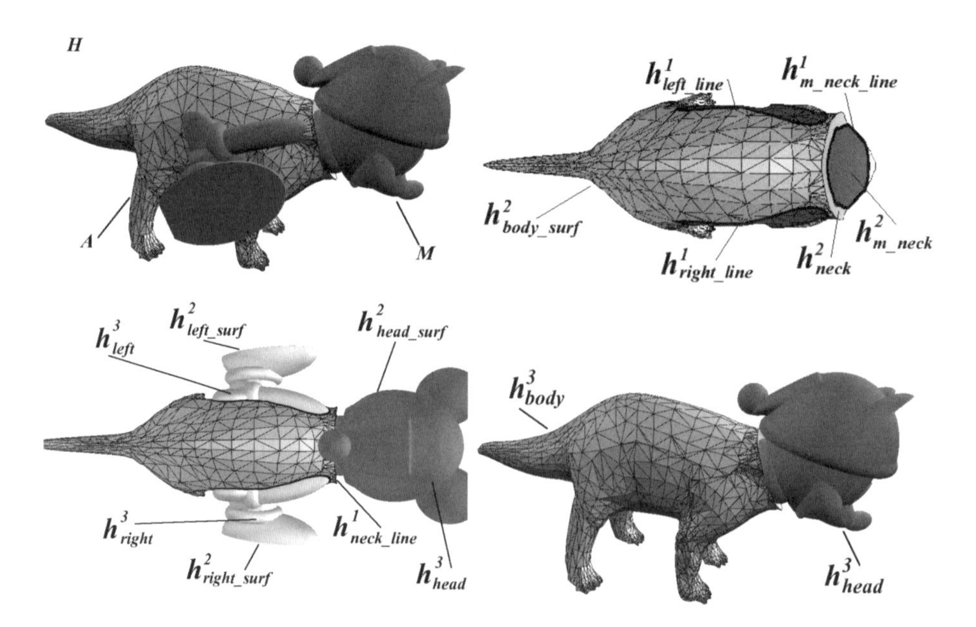

Fig. 17. Modelling of a Triceratops with a mouse head. (a) input IC models A and M; (b) 2D and 1D cells of the IC $H = A \cup M$; (c) 3D cells of the IC H shown in different colors; (d) the resulting IC model after cutting cells h^3_{left} and h^3_{right}.

7.2.4 Cutting Unnecessary Cells from the Model

We cut the unnecessary cells h^3_{left} and h^3_{right} from *H*. One can identify these cells using, for example, the points selected on the boundary mesh of the Mouse's legs. This is illustrated by the following pseudocode:

```
Points3D P1, P2; // points on the Mouse's legs
Cell_ID left_cell, right_cell;
left_cell = H.select_cell_by_point(dim = 3, point = P1);
right_cell = H.select_cell_by_point(dim = 3, point = P2);
H.cut_cell(left_cell);
H.cut_cell(right_cell);
```

Note that the *cut_cell* operation removes h^3_{left} and h^3_{right} together with their boundary cells. After that the IC *H* represents the required model of a Triceratops with the mouse head. The final IC *H* contains the cells $h^1_{neck\_line}$, $h^2_{body\_surf}$, h^2_{neck}, h^3_{body}, $h^1_{m\_neck\_line}$, $h^2_{m\_neck}$, $h^2_{head\_surf}$ and h^3_{head}. The first four cells are inherited from the IC *A*. The last four cells are described as follows:

$$|h^3_{head}| = (|m^3_{mouse}| - (|t^3_{tric}| \cap |d^3_{cavity}|)) \cap BPh_{head}$$

$$|h^2_{head\_surf}| = (|m^2_{mouse\_surf}| - (|t^3_{tric}| \cap |d^3_{cavity}|)) \cap BPh_{head}$$

$$|h^2_{m\_neck}| = (|m^3_{mouse}| \cap (|t^3_{tric}| \cap |d^2_{sphere}|)) \cap BPh_{head}$$

$$|h^1_{neck\_line}| = (|m^2_{mouse\_surf}| \cap (|t^3_{tric}| \cap |d^2_{sphere}|)) \cap BPh_{head}$$

where BPh_{head} is the bounding polyhedron which separates the point set h^3_{head} from the point sets h^3_{left} and h^3_{right}.

The cells of the IC *H* are related by the following boundary relations:

$$H.Rb^{32} = \{ (h^3_{body}, h^2_{body\_surf}), (h^3_{body}, h^2_{neck}), (h^3_{head}, h^2_{head\_surf}), (h^3_{head}, h^2_{m\_neck}) \}$$

$$H.Rb^{21} = \{ (h^2_{body\_surf}, h^1_{neck\_line}), (h^2_{neck}, h^1_{neck\_line}),$$
$$(h^2_{head\_surf}, h^1_{m\_neck\_line}), (h^2_{m\_neck}, h^1_{m\_neck\_line}) \}.$$

and the following containment relations:

$$H.RC^{22} = \{ (h^2_{neck}, h^2_{m\_neck}) \} \quad H.RC^{21} = \{ (h^2_{neck}, h^1_{m\_neck\_line}) \}.$$

Following the procedure introduced above the attributes are defined on the basis of relations between the complexes. Thus, the material attribute of complex *H* is described by the functions $S = \{S_{mouse}, S_{tric\_on}, S_{trick\_in}\}$. The function S_{mouse} is associated with the cells $h^2_{m\_neck}$, $h^2_{head\_surf}$ and h^3_{head}. The function S_{tric_in} is associated with the cells h^2_{neck} and h^3_{body}, and the cells $h^1_{neck\_line}$ and $h^2_{body\_surf}$ are related to the function S_{tric_on}. In Figs.17 and 18 we use shades of grey to represent different material attributes.

7.2.5 Animation

Finally, we show an example of IC based dynamic modelling. We use the resulting IC model H from above to create animation sequences describing a rotation of the model's head. Fig. 18 shows four frames for four angles of the rotation, namely 30°, 60°, 120°, and 330°. The head represented by the cell h^3_{head} altogether with its boundary cells $h^2_{m\_neck}$, $h^2_{head\_surf}$ and $h^1_{m\_neck\_line}$ is rotated around the axis passing through the centroid of the cell h^2_{neck} in the direction of the normal vector to the surface of h^2_{neck}. Other cells of the IC H remain stationary. The IC's topology described by the relations does not change because the contact zone between the body and the head is represented by the cells h^2_{neck}, $h^2_{m\_neck}$, $h^1_{neck\_line}$ and $h^1_{m\_neck\_line}$ whose mutual orientation is preserved during rotation.

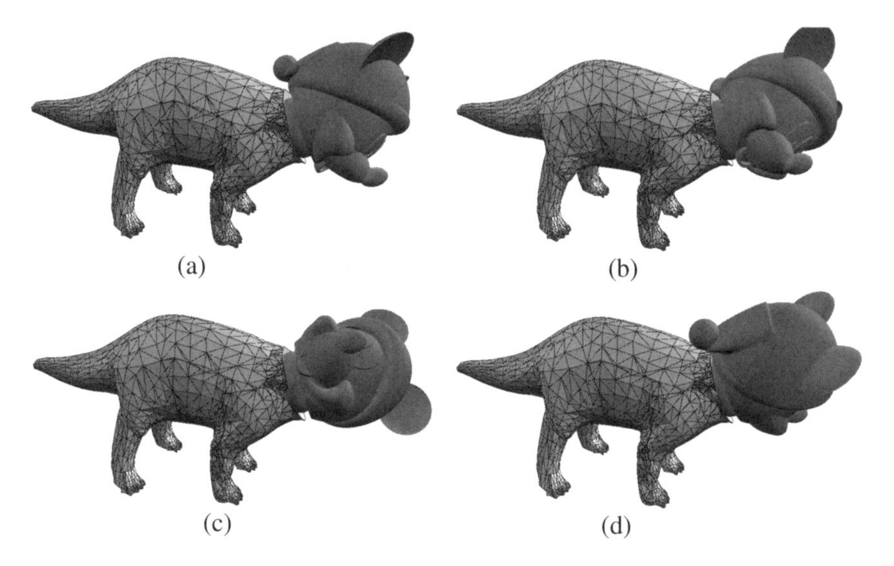

Fig. 18. Rotation of the head (clockwise): (a) angle = 30°; (b) angle = 60°; (c) angle = 120°; (d) angle = 330°

8 Conclusion

Implicit complexes provide a novel framework that makes it possible to model heterogeneous objects exploiting hybrid representation schemes. In this paper we extended the theoretical framework of ICs and introduced a number of new types of cells. We have also described the algorithms for the operations of intersection, union and trimming by 3D manifolds. This allowed us to develop a general step-by-step procedure for the construction of hybrid models. As a case-study, we presented an IC-based model for a multicomponent mechanical assembly. We have also implemented an important particular case involving the integration of BRep and FRep components into a unified hybrid model. We have shown how non-geometric attributes can be modeled within the same framework. Finally, we have generated an animation sequence using of time-dependent IC-based hybrid model.

The main advantages of this way of modelling complex assemblies include the preservation of the original representations of all the components (however different they may be) and the ability to guarantee topologically correct definitions for all parts and relations of the hybrid model (in particular for problematic regions lying on the boundaries). This approach also allows us to handle conformity between the object's geometry and its attributes representing non-geometric properties that are crucial for heterogeneous modelling.

At first sight, our approach, relying on non-trivial topological concepts may appear as too abstract and overcomplex to the end user of modelling software. As we have shown in the case studies, this inherent complexity of the method is eventually hidden from the user by the provision of a set of library routines. Thus, the end user is blissfully unaware of the underlying complexity and only deals with conceptually simple high-level features of the model whose internal structure is generated automatically and is made transparent through a number of high-level queries.

Future work directions include the development of specific operations applicable to entire implicit complexes, an extension of the model to time-dependent implicit complexes; further development of the multidimensional version of the model and its applications, and the implementation of a specialized modelling and animation language which uses this novel modelling technique.

References

1. Adzhiev, V., Kartasheva, E., Kunii, T., Pasko, A., Schmitt, B.: Hybrid cellular-functional modelling of heterogeneous objects. Journal of Computing and Information Science in Engineering, Transactions of the ASME 4(2), 312–322 (2002)
2. Allegre, R., Galin, E., Chaine, R., Akkouche, S.: The HybridTree: mixing skeletal implicit surfaces, triangle meshes and point sets in a free-form modelling system. Graphical Models 68(1), 42–64 (2006)
3. Arbab, F.: Set models and Boolean operations on solids and assemblies. IEEE Computer Graphics and Applications 10(6), 76–86 (1990)
4. Armstrong, C., Bowyer, A., Cameron, S., et al.: Djinn. A Geometric interface for solid modelling. In: Information Geometers, Winchester, UK (2000)
5. Biswas, A., Shapiro, V., Tsukanov, I.: Heterogeneous material modelling with distance fields. Technical Report, Spatial Automation Lab, University of Wisconsin-Madison (2002)
6. Chen, M., Tucker, J.: Constructive volume geometry. Computer Graphics Forum 19(4), 281–293 (2000)
7. Cutler, B., Dorsey, J., McMillan, L., Mueller, M., Jagnow, R.: A procedural approach to authoring solid models. In: SIGGRAPH 2002: Proceedings of the 29th annual conference on Computer Graphics and interactive techniques, vol. 21(3), pp. 302–311 (2002)
8. Fougerolle, Y., Gribok, A., Foufou, S., Truchetet, F.: Boolean operations with implicit and parametric representation of primitives using R-functions. IEEE Transactions on Visualization and Computer Graphics 11(5), 529–539 (2005)
9. Frey, P.J., George, P.L.: Mesh Generation Application to Finite Elements, p. 816. Hermes Science Publishing, Oxford, Paris (2000)

10. Grimm, C.M., Hughes, J.F.: Modelling surfaces of arbitrary topology using manifolds. In: SIGGRAPH 1995: Proceedings of the 22th annual conference on Computer Graphics and interactive techniques, vol. 29, pp. 359–368 (1995)
11. Hachenberger, P., Kettner, L.: Boolean Operations on 3D selective Nef complexes: optimized implementation and experiments. In: Proc. of 2005 ACM Symposium on Solid and Physical Modelling (SPM), pp. 163–174 (2005)
12. Pasko, A., Shapiro, V. (eds.): Heterogeneous object models and their applications. Computer-Aided Design (Special issue), 37(3) (2005)
13. Jackson, T., Liu, H., Patrikalakis, N., Sachs, E., Cima, M.: Modelling and designing functionally graded material components for fabrication with local composition. Control, Materials and Design 20(2/3), 63–75 (1999)
14. Kartasheva, E., Adzhiev, V., Pasko, A., Fryazinov, O., Gasilov, V.: Surface and volume discretization of functionally-based heterogeneous objects. Journal of Computing and Information Science in Engineering, Transactions of the ASME 3(4), 285–294 (2003)
15. Kartasheva, E., Adzhiev, V., Comninos, P., Pasko, A., Schmitt, B.: Construction of implicit complexes: a case-study. In: Skala, V. (ed.) Proc. 13th International Conference in Central Europe on Computer Graphics, Visualization and Computer Vision WSCG 2005, University of West Bohemia, Plzen, Czech Republic, pp. 219–226 (2005) ISBN 80-903100-7-9
16. Kou, X.Y., Tan, S.T.: Heterogeneous object modeling: A review. Computer-Aided Design 39(4), 284–301 (2007)
17. Kumar, V., Dutta, D.: An approach to modelling multi-material objects. In: Fourth Symposium on Solid Modelling and Applications, ACM SIGGRAPH, pp. 336–345 (1997)
18. Kumar, V., Burns, D., Dutta, D., Hoffmann, C.: A framework for object modelling. Computer-Aided Design 31(9), 541–556 (1999)
19. Kunii, T.: Valid computational shape modelling: design and implementation. International Journal of Shape Modelling 5(2), 123–133 (1999)
20. Martin, W., Cohen, E.: Representation and extraction of volumetric attributes using trivariate splines: a mathematical framework. In: Anderson, D., Lee, K. (eds.) Sixth ACM Symposium on Solid Modelling and Applications, pp. 234–240. ACM Press, New York (2001)
21. Middleditch, A., Reade, C., Gomes, A.: Point-sets and cell structures relevant to computer aided design. International Journal of Shape Modelling 6(2), 175–205 (2000)
22. Ohmori, K., Kunii, T.: Shape modelling using homotopy. In: Proc. International Conference on Shape Modelling and Applications, pp. 126–133. IEEE Computer Society, Los Alamitos (2001)
23. Paoluzzi, A., Bernardini, F., Cattani, C., Ferrucci, V.: Dimension-independent modelling with simplicial complexes. ACM TOG 12(1), 56–102 (1993)
24. Park, S.M., Crawford, R., Beaman, J.: Volumetric multi-texturing for functionally gradient material representation. In: Anderson, D., Lee, K. (eds.) Proc. Sixth ACM Symposium on Solid Modelling and Applications, pp. 216–224. ACM Press, New York (2001)
25. Pasko, A., Adzhiev, V., Sourin, A., Savchenko, V.: Function representation in geometric modelling: Concepts, implementation and applications. The Visual Computer 11(8), 429–446 (1995)
26. Pasko, A., Adzhiev, V., Schmitt, B., Schlick, C.: Constructive hypervolume modelling. Graphical Models 63(6), 413–442 (2001)
27. Qian, X., Dutta, D.: Feature-based design for heterogeneous objects. Computer-Aided Design 36, 1263–1278 (2004)

28. Raghothama, S.: Constructive topological representations. In: Proc. the ACM Symposium on Solid and Physical Modelling, pp. 39–51 (2006)
29. Requicha, A.A.G., Voelcker, H.B.: Boolean operations in solid modelling: boundary evaluation and merging algorithms. Proc. IEEE 73(1), 30–44 (1985)
30. Rossignac, J.: Through the cracks of the solid modelling milestone. In: Coquillart, S., Strasser, W., Stucki, P. (eds.) From Object Modelling to Advanced Visualization, pp. 1–75. Springer, Heidelberg (1994)
31. Rossignac, J.: Structured Topological Complexes: A feature-based API for non-manifold topologies. In: Proc. the ACM Symposium on Solid Modelling, pp. 1–9 (1997)
32. Rossignac, J., O'Connor, M.: SGC: A dimension independent model for pointsets with internal structures and incomplete boundaries. In: Wozny, M., Turner, J., Preiss, K. (eds.) Geometric modelling for product engineering (1990)
33. Shin, K., Dutta, D.: Constructive representation of heterogeneous objects. Journal of Computing and Information Science in Engineering 1(3), 205–217 (2001)
34. Siu, Y.K., Tan, S.T.: "Source-based" heterogeneous solid modelling. Computer-Aided Design 34, 41–55 (2002)

Heterogeneous Object Design: An Integrated CAX Perspective

X.Y. Kou and S.T. Tan

Department of Mechanical Engineering
The University of Hong Kong, Pokfulam Road, Hong Kong
kouxy@hku.hk, sttan@hku.hk

Abstract. CAD *modeling*, *analysis of properties* and *fabrication* of heterogeneous objects have been extensively studied in the past few decades. Conventionally these topics are separately investigated in CAD, CAE and CAM communities. Such explicit separations, however, suffer from some apparent limitations. This article presents an alternative scheme to consider the heterogeneous object design problem in an integrated CAX (CAD/CAE/CAM) framework. The motivation is to design heterogeneous objects which not only *look right*, but also *functionally work right*. In addition to the data representation, model constructions and visualizations, other considerations such as data communications to and from CAE/CAM modules, fabrication efficiency in layered manufacturing etc., are also considered. The presented CAX based design method facilitates design tools integration and enhances the interoperability in the entire design process.

1 Introduction

Tremendous research efforts have been made in modeling objects with designed material heterogeneities. Throughout the whole design process, the following questions are frequently raised by the end users: (1) How to represent the heterogeneous objects' geometries and material distributions? (2) Does the designed object fulfill the users' functional requirements? How to validate its functional properties? and (3) Is it physically realizable? How to make it? Conventionally these topics are separately investigated in CAD, CAE and CAM communities; however, such explicit separations suffer from some apparent limitations. CAD modelers are usually unable to ascertain whether the modeled objects can really meet the end users' functional requirements, as they only concentrate on the data representations, model constructions and object visualizations. CAE engineers focus on using analytical and numerical approaches to simulate the behaviors of the objects, however, due to the lack of powerful CAD models, only objects with simple (e.g. unidirectionally graded [1]) material distributions were vigorously studied [2-6]. Separate studies in CAD and CAM of heterogeneous object also impede the interoperability required at the process planning and fabrication stages: the data structures of the CAD models were seldom fully utilized to improve the manufacturing efficiency and product qualities, only direct one to one data conversions are conducted, resulting in degraded fabrication performances or productivities.

Traditional design approaches emphasized the modularity and maintainability of heterogeneous object design; however they failed to answer *all* the questions the end users are concerned with. From a practical point of view, the answers to *all* of these questions are indispensable to assure the design qualities and failures in either one may

A. Pasko, V. Adzhiev, and P. Comninos (Eds.): HOMA, LNCS 4889, pp. 42 – 59, 2008.

undermine the design feasibility and authenticity. This article is motivated to address the heterogeneous object design from an integrated collaborative perspective. In addition to the modular design methodologies, we emphasize the data communications and effective use of CAD models in the downstream CAE and CAM modules.

The subsequent sections of this article are organized as follows. CAD modeling of heterogeneous objects is first described in Section 2, where the key concept and usage of our extended Heterogeneous Feature Tree (eHFT) structure are presented. Based on the eHFT model, the Finite Element Analysis (FEA) and Rapid Prototyping (RP) of heterogeneous objects are discussed in Section 3 and Section 4. Section 5 describes the implementation details of the integrated CAX approach and finally concluding remarks and discussion are offered in Section 6.

2 CAD Modeling of Heterogeneous Object

CAD modeling of heterogeneous objects received considerable research focus in the past and there have been a variety of models in the literature. Among them, the voxel model [7, 8], volume mesh model [9], implicit function model [10, 11], explicit function model [3, 4], control point based model [12, 13], control feature based model[14-18], assembly model[19, 20], cellular model[21, 22] and composite hybrid model [23, 24] are most widely used [25]. In this article, we utilize the Heterogeneous Feature Tree (HFT) structure and a Heterogeneous Cellular Representation (HC-Rep), which fall in the scope of the *control feature based model* and the *cellular model*, to present our integrated CAX perspective on heterogeneous object design.

2.1 The Heterogeneous Feature Tree (HFT) Representation

Heterogeneous objects are generally characterized as having location *dependent* material compositions [17]. The idea of the HFT representation is to represent the material

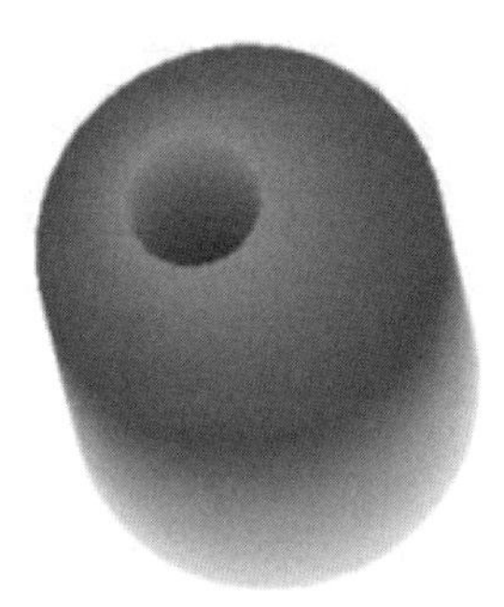

Fig. 1. A heterogeneous object with bi-directional material gradations[1]

[1] The figure was reprinted from Computer-Aided Design, 37(3), Kou, X.Y. and S.T. Tan, *A hierarchical representation for heterogeneous object modeling*, pp. 307-319, Copyright (2005), with permission from Elsevier.

heterogeneities by encoding the material variation *dependencies* with a proper data structure. A tree structure is selected because of its hierarchical nature [18] and the capability of representing complex (e.g. 2D or 3D dependent) material heterogeneities.

For instance, the heterogeneous object in Fig. 1, constructed sequentially as shown in Fig. 2, can be conceptually represented by a simplified heterogeneous feature tree structure as shown in Fig. 3.

The Heterogeneous Feature Tree (HFT) structure maintains the material variation dependencies with different hierarchies [18, 26]. By definition, the material composition of a feature in a higher level is dependent on (or determined by) its child features'

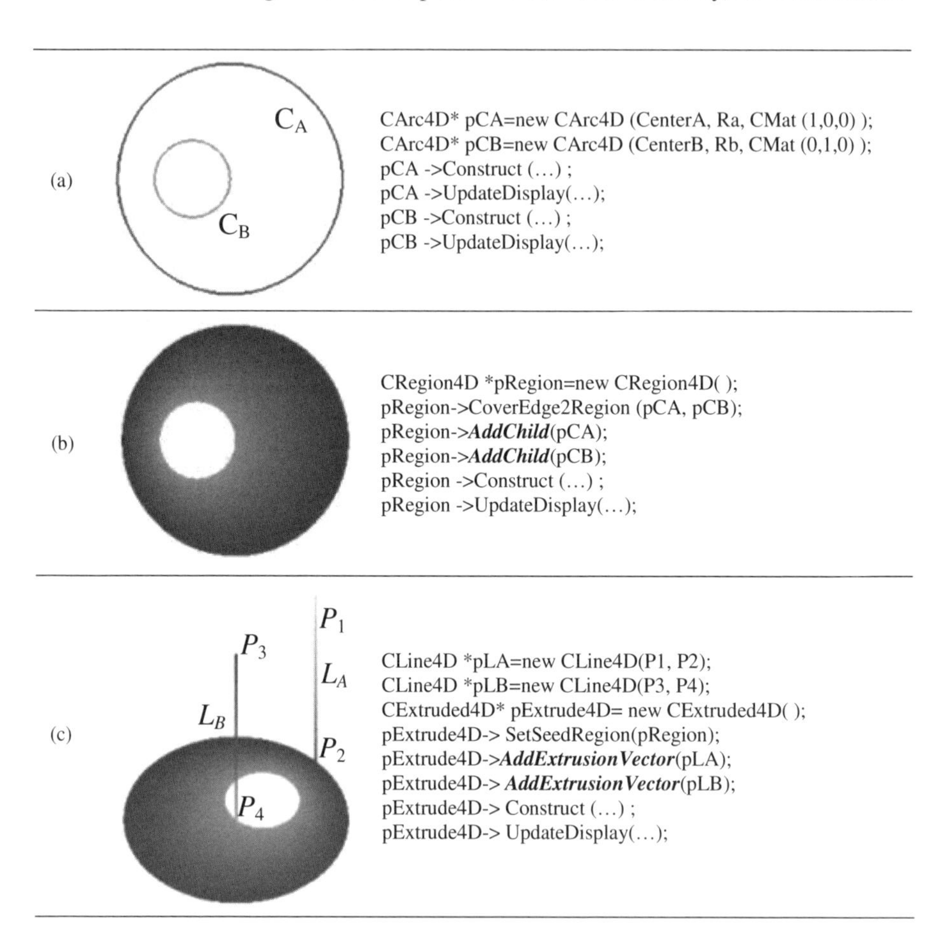

Fig. 2. The modeling process and the pseudo code for the heterogeneous object construction. The bold italic function calls are subroutines used for the construction of the heterogeneous feature tree structures, reprinted from [18] with permission from Elsevier[2].

[2] The figure was reprinted from Computer-Aided Design, 37(3), Kou, X.Y. and S.T. Tan, *A hierarchical representation for heterogeneous object modeling*, pp. 307-319, Copyright (2005), with permission from Elsevier.

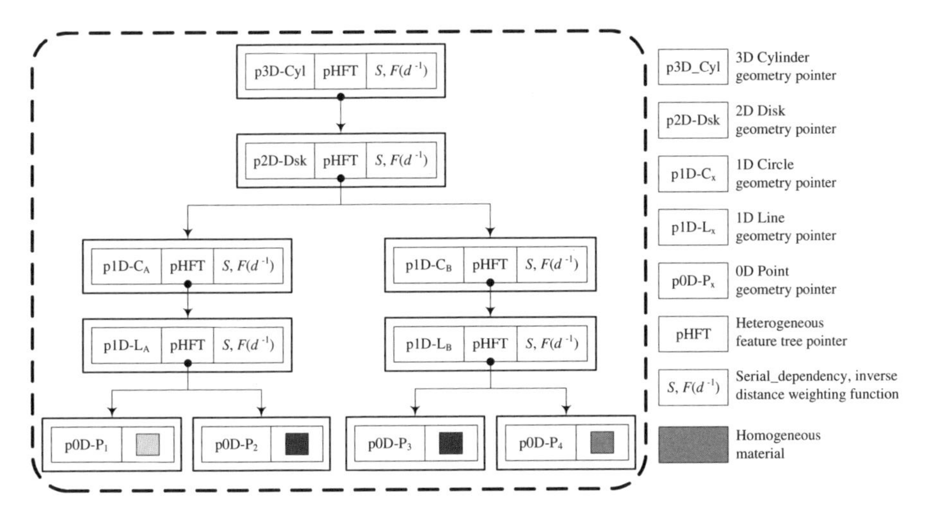

Fig. 3. A simplified HFT representation for the object in Fig. 1. Colors are used to represent different materials

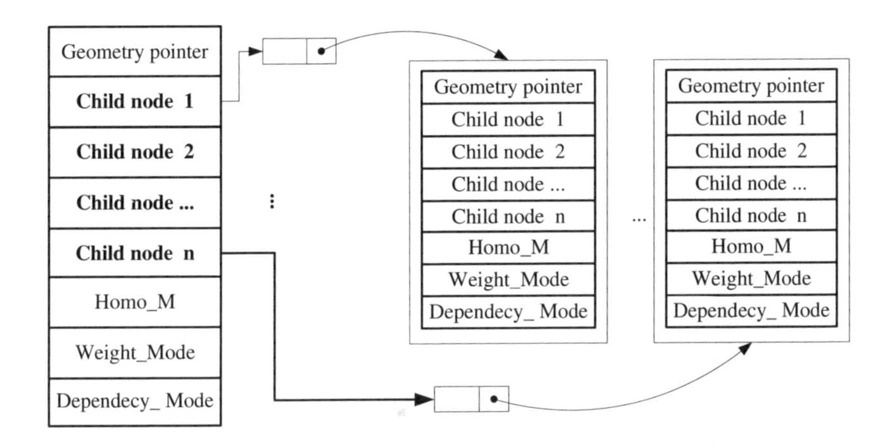

Fig. 4. The Heterogeneous Feature Tree (HFT) structure, reprinted from [18] with permission from Elsevier[3]

geometries and material definitions. The material compositions evaluated from each child tree are interpolated/blended at their parent level, and are used to represent the parent feature's material distributions [26].

For an arbitrary point P inside this heterogeneous cylinder, its material is dependent on the base 2D region's material composition (since the object is directly extruded from the 2D region, see Fig. 2 (c)). Each section perpendicular to the extrusion vector is a

[3] The figure was reprinted from Computer-Aided Design, 37(3), Kou, X.Y. and S.T. Tan, *A hierarchical representation for heterogeneous object modeling*, pp. 307-319, Copyright (2005), with permission from Elsevier.

heterogeneous 2D region, which is similar to the base 2D region in both geometry and material distributions. For each section, the material composition is determined by two bounding contours, (Fig. 2 (a)). Along the extrusion directions, these contours' material variations are constrained (regulated) by the other two heterogeneous vectors (L_A and L_B, Fig. 2 (c)). The material distribution of these extrusion vectors, in turn, are dependent on the composition of their bounding vertices (i.e. P_i, i=1, 2, 3, 4).

It can be seen that the constructed HFT structure in Fig. 3 faithfully conveys such material variation dependencies as described above. Note that the HFT structure in Fig. 3 is only a graphic representation which qualitatively depicts the material variation dependencies across different hierarchies. The complete HFT structure also embraces other information such as blending functions between child-parent features and other enumeration data, as shown in Fig. 4.

2.2 Extended Heterogeneous Feature Tree (eHFT) Structure and the Heterogeneous Cellular Representation (HC-Rep)

The HFT representation is initially proposed to model 2D or 3D dependent material gradations with hierarchical tree structures [18]. However, in real application, only few objects have such regular material gradations. It is commonplace that hybrid homogeneous, Functionally Graded Material (FGM) and other distributions coexist in different portions of a complex heterogeneous object. Fig. 5 demonstrates an example of such object. For such objects, the HFT representation is still inadequate.

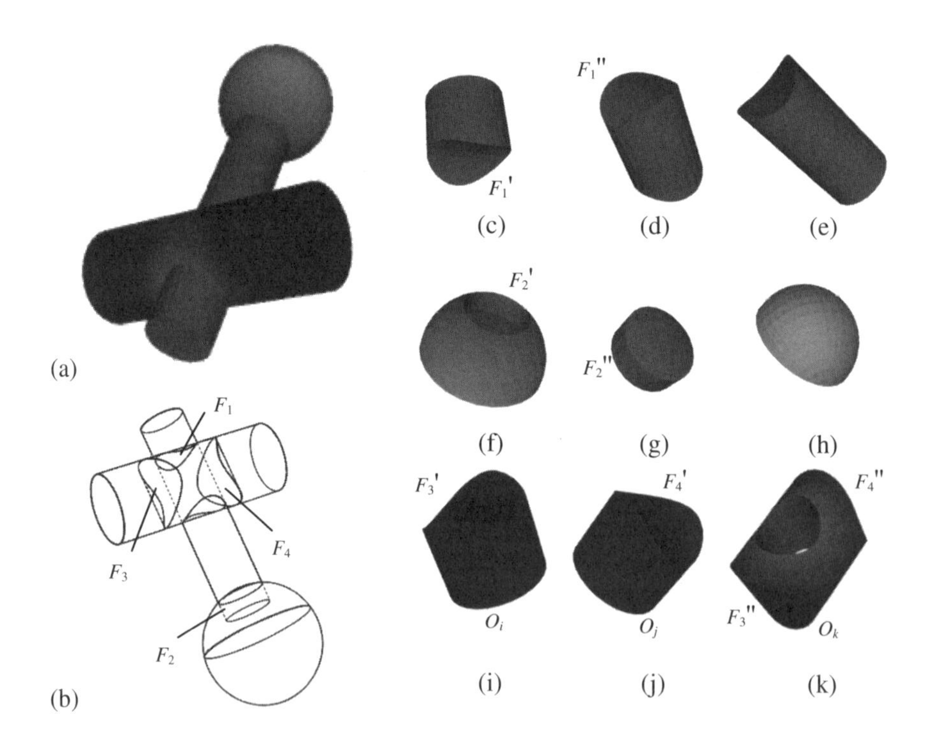

Fig. 5. A heterogeneous object composed of multiple components

To remedy this, a natural idea is to introduce multiple HFT structures to model objects with complex heterogeneities. Using the part-assembly model to represent the object geometries and associating each part with a HFT structure seem to be an intuitive solution. However, the part-assembly model introduces serious data redundancies and inconsistency problems (as will be further elaborated in Section 4). For instance, the faces F_i' and F_i'' (i=1, 2, 3, 4) in Fig. 5 represent exactly the same geometries, however are repetitively kept in separate parts. From the visualization point of view, such data redundancies do not matter much, however, when the model undergoes further manipulations (for instance, local translation or deformations of F_i), inconsistent and self-intersected geometries may occur. This is because F_i' and F_i'' are separately translated or deformed and there is no guarantee that they will be exactly identical, especially when the computation error or other noise sources are also involved.

Using multiple, independent HFT structures to represent the material distribution also suffer from similar problems. For instance, if one of the component's material (e.g. Fig. 5 (j)) is changed to another material, its neighbor component's material distributions will not change accordingly (Fig. 5 (k)), resulting in sharp material transitions and possibly, stress concentrations.

Our solution to this problem is to use a novel Heterogeneous Cellular Representation (HC-Rep) to represent complex heterogeneous objects. A heterogeneous object is defined as a collection of heterogeneous cells, each of which, graphically, resembles the parts shown in Fig. 5 (c) to (k). However, the part-assembly model is substituted with the *non-manifold cellular model* and *extended HFT* structures [22] are utilized to model the material heterogeneities.

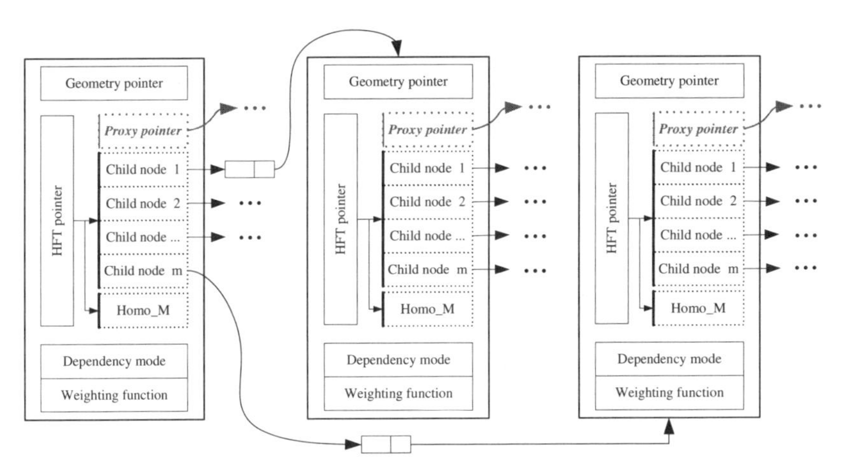

Fig. 6. The extended Heterogeneous Feature Tree structure (eHFT), reprinted from [22] with permission from Elsevier[4]

[4] The figure was reprinted from Computer-Aided Design, 38(5), Kou, X.Y., S.T. Tan, and W.S. Sze, *Modeling complex heterogeneous objects with non-manifold heterogeneous cells*, pp. 457-474, Copyright (2006), with permission from Elsevier.

The non-manifold cellular model uses both the oriented faces (single-sided) and the 'double-sided' boundary faces (co-boundaries) in the computer model. As is shown in Fig. 5 (b), the faces F_i (i=1, 2, 3, 4) which delimit the complex object are modeled as *co-faces* and are *shared* between adjacent cells (rather than explicitly copied to each cell). This naturally solves the aforementioned data consistency problem because the applied transformations will be exerted on the shared entities. The computation error, if exists, has the same impact on both cells and there will be no contradictory geometries (e.g. self-intersections) in the modified computer model.

Based on a similar entity-sharing mechanism, the extended HFT (eHFT) structure is proposed to model the local (cell level) as well as the overall (object level) material distributions. The eHFT is characterized as sharing part of the tree branches with other HFT structures. To enable this HFT sharing, the Proxy-HFT (PHFT) is proposed, as shown in Fig. 6. The PHFT points to an existing heterogeneous feature in the modeling space. The feature that is pointed to by the PHFT pointer is called a *proxy*, and the cell that uses the PHFT is called a *client* [22]. Based on this proxy-client mechanism, the material composition evaluation for a cell can be directly *forwarded* or *delegated* to its proxy features. If a client feature contains a valid proxy HFT pointer as shown in Fig. 6, the material composition for a point inside or on the client's boundary is dynamically determined by calling the proxy feature's material evaluation procedure. Otherwise if the proxy feature pointer is a NULL pointer, then it degenerates to the common HFT representation as previously discussed; and the material evaluation can be executed according to the encoded hierarchical dependencies.

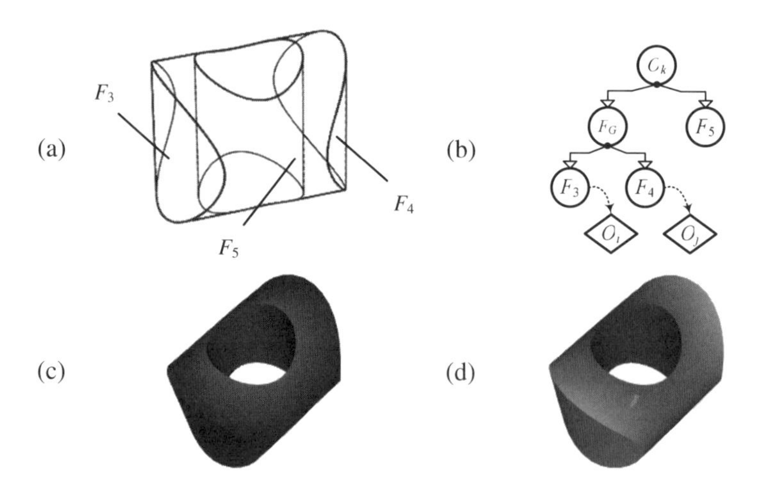

Fig. 7. The eHFT representation for the cell in Fig. 5 (k)

Fig. 7 shows the eHFT representation for the cell in Fig. 5 (k), whose material distribution is defined as a gradation between the face (F_5) and the two co-faces (F_3 and F_4), while the material distributions of F_3 and F_4 are not represented by the usual HFT structures, instead, two proxies (O_i and O_j, see Fig. 5 and Fig. 7) are proposed to 'share' their material distributions with the clients (i.e. F_3 and F_4).

The benefits of sharing the HFT branches are obvious: data redundancy is eliminated; modifications on the material distributions are efficiently updated; and smooth material gradations can be guaranteed. For instance, when the proxy O_i's or O_j's (Fig. 5 (i), (j)) material compositions are changed from "blue" to "green" (see Fig. 7 (c) and (d)), its neighbor O_k's material (Fig. 5 (k)) can be immediately reflected due to the feature sharing mechanism. Note that this auto-update ability allows for local material editions to be properly propagated to the entire heterogeneous object, and this is crucial to avoid abrupt material transitions in FGM object modeling.

3 Finite Element Analysis (FEA) of Heterogeneous Object

The previous section focuses on CAD modeling of heterogeneous objects. Based on the CAD models, other specific modeling tools can be developed to facilitate users to design objects with the *desired* geometries and material distributions. Nevertheless, the word "desired" here mostly refers to visual appearances of the objects. A visually pleasing heterogeneous object may *look right*, however, there is no guarantee that it can *functionally work right*. To assure the designed object can properly work as required, finite element analysis and other numerical approaches can be conducted to evaluate its physical properties or performances.

FEA of heterogeneous objects is a well investigated topic in CAE communities. However, most investigations focus on objects with simple material heterogeneities. For instance, FEA on unidirectionally graded objects account for the majority of case studies in existing literature [2-6]. The primary reason for this phenomenon is not because the finite element method is incapable of handling more complex objects, but possibly due to insufficient support in formulation/representation of complex material heterogeneities. Also note that contemporary CAD modelers (e.g. Solidworks,

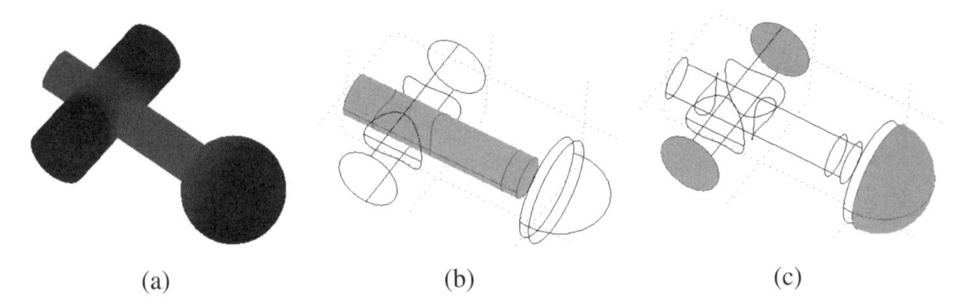

Fig. 8. A steady state thermal conduction analysis. (a) The heterogeneous material distributions of the object under examination; (b) Boundary condition (I), highlighted surfaces are constrained at a temperature of 773.15K; (c) Boundary condition (II), highlighted surfaces are constrained at a temperature of 313.15K.

Pro-Engineer, Unigraphics NX etc.) have dramatically enhanced the communications between CAD and CAE modules; however, the material heterogeneities are not included in the data exchanges and information flows.

In what follows, we present an integrated approach to conduct the CAD modeling and FE analysis on the designed heterogeneous object. A heterogeneous object shown in Fig. 8 (a) is used as an example to demonstrate the proposed scheme. The object is modeled with a Heterogeneous Cellular Representation (HC-Rep) as described earlier. The "red" and "blue" colors are defined to represent the material "Alumina" and "Aluminum" whose thermal conductivities (k) are 27 W/m/K and 155 W/m/K respectively. A steady state thermal conduction analysis is conducted, and the boundary conditions of the object are illustrated in Fig. 8 (b) and (c): the highlighted surfaces in Fig. 8 (b) and (c) are constrained at a temperature of 773.15K and 313.15K respectively.

Fig. 9 outlines the integrated CAD modeling and FEA approach [26].

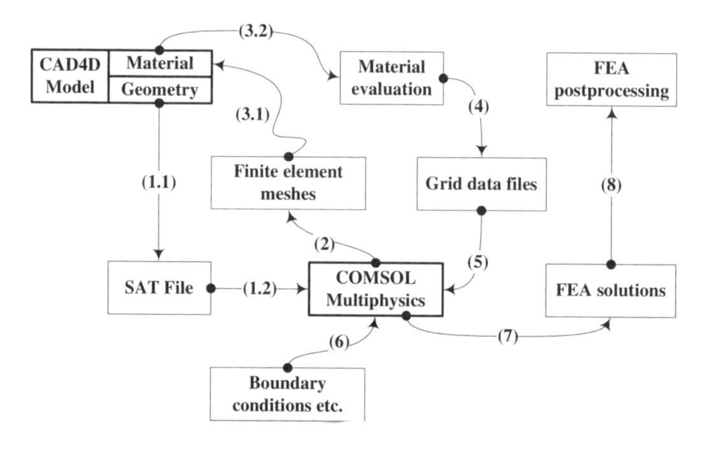

Fig. 9. A flowchart of integrated CAD modeling and FEA of heterogeneous objects, reprinted from [27] with permission from Elsevier[5]

1. The geometric model of the object is first converted into the Standard ACIS Text (SAT) format [28, 29] and then exported to a commercial FEA package COMSOL Multiphysics [30];
2. COMSOL interprets the geometric data and generates finite element meshes based on the object's geometric information (pre-processing);
3. The material compositions of the object sampled at some regular grids are interrogated from the heterogeneous CAD models, relevant material properties (in this example, the thermal conductivity) are then calculated;

[5] The figure was reprinted from Materials & Design. In Press, Corrected Proof, Kou, X.Y. and S.T. Tan, *A systematic approach for Integrated Computer-Aided Design and Finite Element Analysis of Functionally-Graded-Material objects*, Copyright (2007), with permission from Elsevier.

4. The material properties at sampled points are saved in grid data files, following a prescribed format recognizable to COMSOL;
5. COMSOL imports the grid data files and determine the material properties at FE nodes through interpolations (in this example linear interpolation is used);
6. Boundary conditions are defined within COMSOL;
7. COMSOL computes the local stiffness matrix, assembles global stiffness matrix and solves the nodal variables (temperatures);
8. COMSOL performs post processing to generate graphical outputs.

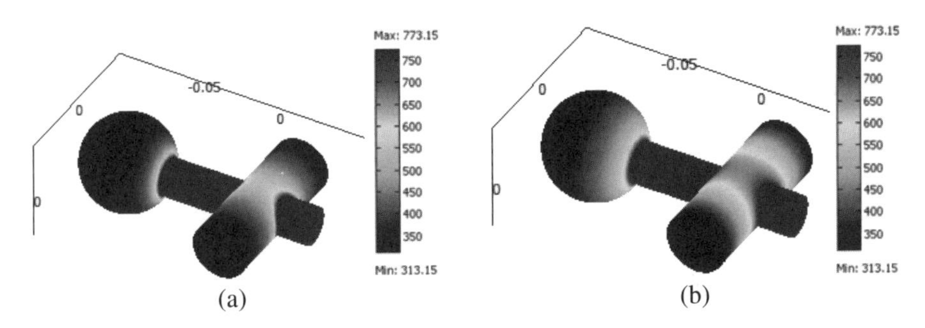

Fig. 10. Results of the steady state thermal conduction analysis. (a) Temperature distribution of an FGM object (gradation between Aluminum k=155 W/m/K and Alumina k=27 W/m/K); (b) Temperature distribution of a homogeneous object (Aluminum), unit in (K).

Fig. 10 (a) shows the obtained temperature distribution of the heterogeneous object. As a comparison, the temperature of a homogeneous object (Aluminum, subject to the same boundary condition) is also provided in Fig. 10 (b). As can be seen the high temperature region of Fig. 10 (a) is smaller than that of Fig. 10 (b), this is because the heterogeneous object uses a less conductive primary material (Alumina) which helps to impede thermal conductions, as compared with the homogeneous object.

For brevity, Fig. 10 only shows the thermal conduction results of the heterogeneous object, however other coupled analysis (for instance, the thermal stress, strained energy density etc.) can be also conducted using similar approaches. A multi-physics based finite element analysis on 2D heterogeneous objects have been reported in our recent paper [27] and interested readers may find more technical details there.

The benefit of this integrated CAD modeling and FE analysis is that objects with complex heterogeneities can be easily analyzed with the finite element methods. Without proper CAD models, however, the complex material distributions can hardly be formulated or interrogated; further material property calculations and physical performance simulations are therefore very difficult to be obtained.

Using the integrated CAD modeling and FEA of heterogeneous object, the users can first design heterogeneous CAD models and then transfer the models to the CAE modules for property simulations. Modifications on the object geometries or material distributions can be carried out if the simulated properties do not satisfy the functional requirements. The modified CAD models can be further delivered to CAE modules to re-validate its efficacies. The above process can be repeated until the users' functional requirements are fulfilled.

4 Rapid Prototyping of Heterogeneous Object

Once the designed heterogeneous object has been validated via numerical simulations, prototypes of the object can be then fabricated to test its physical behaviors under working conditions. In this section, Rapid Prototyping (RP) of heterogeneous objects is discussed and the manufacturing efficiency of objects with complex material heterogeneities is addressed.

4.1 Fundamental Algorithms

Typical processes involved in rapid prototyping of a heterogeneous object includes the following procedures [31]:

1. The object geometry is first sliced by parallel planes and a collection of boundary profiles are obtained, as shown in Fig. 11 (a) and (b);
2. For each slice, the silhouette boundary curves are covered into 2D regions, see Fig. 11 (c);

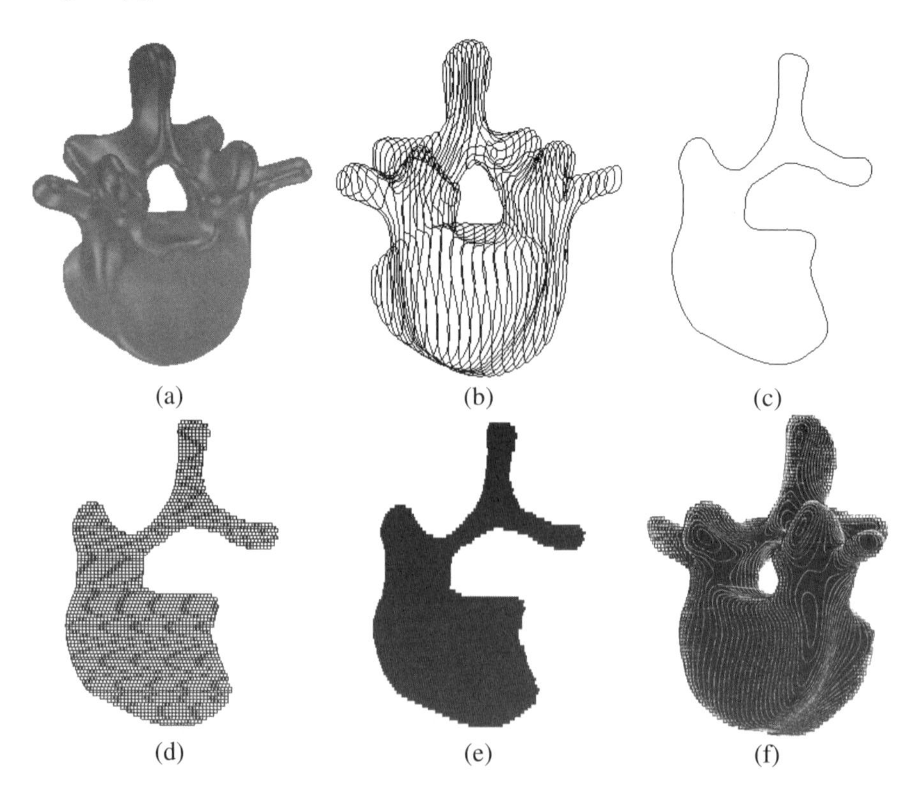

Fig. 11. Rapid prototyping of an example unidirectional FGM object

3. Each 2D region is intersected with scan lines and a collection of 1D solids is obtained;
4. Each scan line is further decomposed into an array of voxels;
5. For each voxel, its material composition is interrogated from the heterogeneous CAD model, as shown in Fig. 11 (d) and (e);
6. The obtained voxel data are used to drive the hardware setup (e.g. nozzles) to selectively deposit materials in a continuous point-wise, line-wise and slice-wise fashion until the object is completely fabricated, as shown in Fig. 11 (f).

4.2 RP Data Generation for Complex Heterogeneous Object

The aforementioned approach works well with objects with simple material distributions (for instance the unidirectional FGM object in Fig. 11), however if the objects under fabrication have complex material heterogeneities (e.g. the one shown in Fig. 5 (a)), simply applying the above discussed algorithms may introduce additional problems. As mentioned in Section 4.2, the part-assembly model is a widely used representation for objects with complex material heterogeneities; however, if the presented procedures in Section 4.1 are used in conjunction with the assembly model, significant robustness and efficiency problems may occur [31].

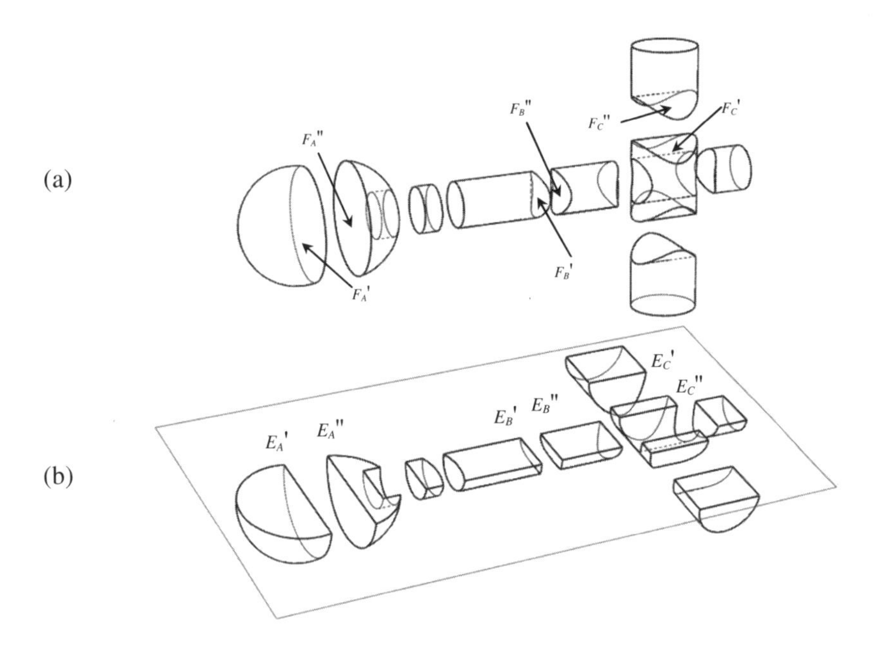

Fig. 12. Planar slicing of a part-assembly model in rapid prototyping of a complex heterogeneous object (a) Subdivided components of the object and redundant faces; (b) Redundant edges generated by repetitive planar slicing

Take the object in Fig. 5 (a) as an example. In the planar slicing stage, to get the boundary profile of the object, the slicing algorithm must be separately applied on all the nine components, as shown in Fig. 12. Note that many faces are repetitively kept in the assembly, for instance F_A', F_A'', F_B', F_B'' and F_C', F_C'' in Fig. 12 (a). All these redundant faces subsequently take part in the boundary slicing process, repetitive plane-face intersections are performed and redundant edges (e.g. E_A', E_A'', E_B', E_B'' and E_C', E_C'' in Fig. 12 (b)) are generated. Similarly in the line scanning process, identical line-edge intersections will be conducted on these redundant edges. Also note that these repetitive plane-face and line-edge intersection computations are ubiquitous in the entire RP data generation process and they are performed in *every* planar slicing and line scanning step. These repetitive and unnecessary boundary intersections therefore, greatly degraded the overall efficiencies.

A careful study into this problem shows that in most cases, these redundant entities (faces and edges) serve as the delimiting boundaries of some "sub-domains" [31], and each of such sub-domain has different material distributions [31]. They are introduced solely for the purpose of point membership classifications and material interrogations [22, 31, 32]. Conceptually, such material delimitation entities should not be included in the geometric operations (such as section slicing and line scanning); but rather, they should be utilized only in the material evaluation process (the Step 5 in Section 0).

To improve the computational efficiency in the RP data generation, these repetitively kept entities should be temporarily excluded from the boundary intersection computations; and only the boundary elements which bound the object geometries should actually participate in the plane-face, line-edge intersections.

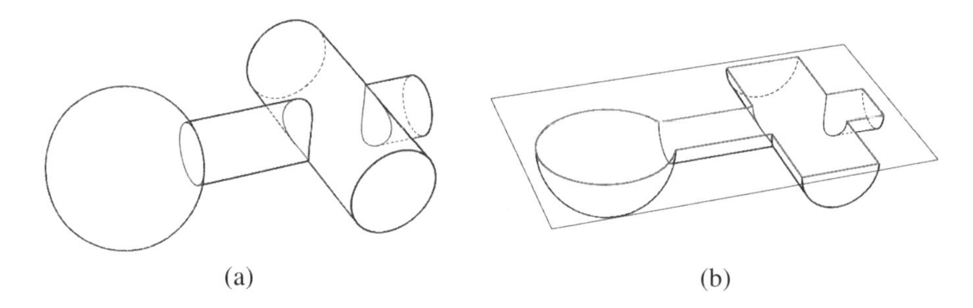

(a) (b)

Fig. 13. Selective boundary slicing in rapid prototyping of a complex heterogeneous object

With the traditional part-assembly model, the redundant entity removal is almost unattainable since all the boundary elements are equally treated. By using the proposed HC-Rep, however, the unnecessary and repetitive boundary-interaction calculations can be avoided and this is accomplished through a selective boundary slicing algorithm [31]. In this algorithm, all the boundary faces are first retrieved from the non-manifold cellular models and kept in a face list; the internal material delimitation boundaries, which share themselves with other boundary elements (i.e. *double-sided*, see also Section 0) are then removed from the list. The remaining faces are subsequently sewn

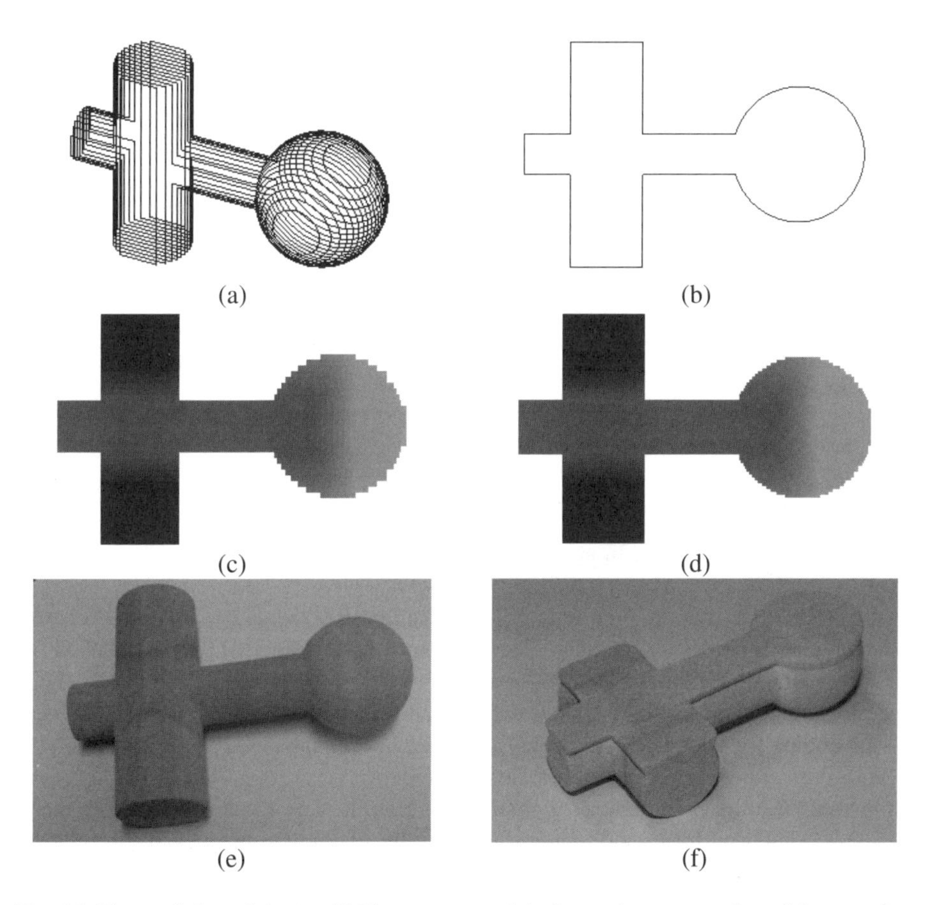

Fig. 14. Planar slicing of the manifold geometry and the layered representation of the complex heterogeneous object. (a) and (b): results of applying the selected planar slicing on manifold geometry; (c) and (d): layered representations of the generated data to be used in the RP process. (e) and (f): Prototypes fabricated with 3D Printer [33].

together to form a manifold solid, as shown in Fig. 13 (a). Note that it is the boundaries of the manifold solid that participate in the actual planar slicing and region covering, as shown in Fig. 13 (b).

Fig. 14 shows the results of applying the selective planar slicing on the *manifold* geometries. The sliced 2D regions are then scanned line by line, and further decomposed into an array of voxels, as described earlier. Based on the presented eHFT structure, the material composition of each voxel is then interrogated from the HC-Rep model.

The final layered representations of the generated data are shown in Fig. 14 (c) and (d), and the effects of different fabrication resolutions are demonstrated. Fig. 14 (e) and (f) show two prototypes fabricated using the Z Corporation's 3D printer [33]; the material distributions of both the outer boundary and internal parts are illustrated.

5 Implementations

The proposed integrated CAX based design approach is partially implemented in our standalone heterogeneous object modeler — CAD4D [34], jointly with COMSOL Multiphysics [30].

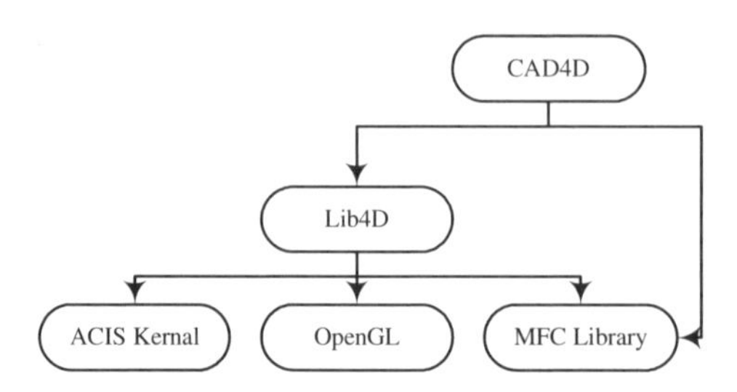

Fig. 15. Software implementation architecture of CAD4D

CAD4D is based upon a reusable object class library *Lib4D* [26, 34] and Microsoft Foundation Class (MFC) libraries. The commercial 3D kernel *ACIS* [28] is used for handling geometric modeling related issues, C++ Standard Template Library (STL) is used to implement container related data structures, and OpenGL is used as the rendering engine for object visualizations, as shown in Fig. 15.

In a typical heterogeneous object design process, the users invoke the CAD modeling tools through graphical user interfaces (e.g. clicking toolbar buttons or menu items). CAD4D performs modeling operations by manipulating relevant object data structures with proper algorithms. For instance, in object constructions, Lib4D object instances are created and appended to the object database (Fig. 16 (b1) and (c1)), visualizations are then updated in response of the user actions as shown in Fig. 16. Fig. 17 shows a snapshot of the proposed CAD4D modeler.

The integrated CAD modeling and finite element analysis are conducted using CAD4D and COMSOL Multiphysics, as detailed in Section 3.

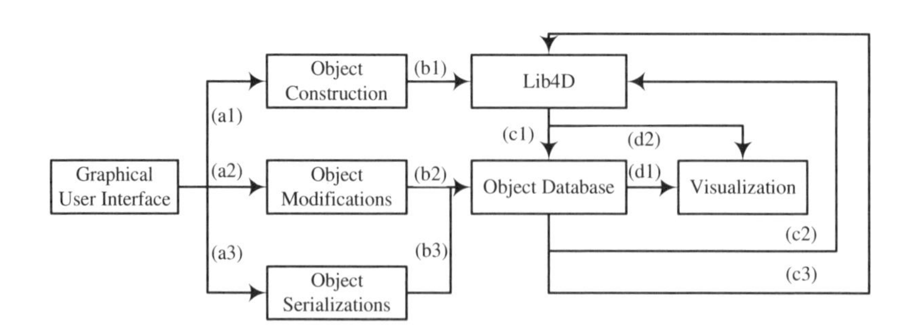

Fig. 16. Typical user interactions and modeling processes using CAD4D modeler

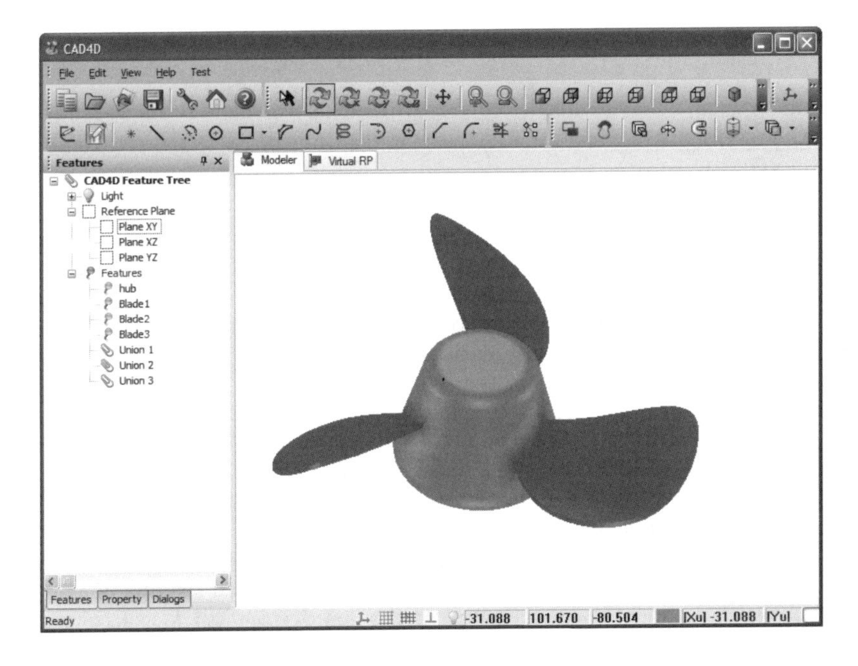

Fig. 17. A Snapshot of the CAD4D modeler

The integrated CAD and CAM of heterogeneous objects are carried out by data communications with CAD4D and Z Corporation's 3D printer, mostly via the VRML data format. A heterogeneous CAD model is designed by CAD4D and validated/verified by COMSOL Multiphysics; once the functional requirement of design is satisfied, the HC-Rep based CAD models are then converted to VRML format for physical fabrications.

6 Conclusions

This article presents an integrated CAX (CAD/CAE/CAM) perspective on heterogeneous object design. Different from many existing approaches which emphasize a particular aspect of the design problem, heterogeneous object design is envisaged as an integral process which combines the CAD modeling, property analysis and physical realization. The article is motivated to present such a perspective, demonstrate the typical design procedures, paradigms and benefits. The emphasis of this article is on the linkage and integration of the CAX tools, rather than each separate topic. The readers may, however, refer to our previous papers [18, 22], [27] and [31] for more technical details on each subject of interest. Flash animations are also available on the first author's website http://web.hku.hk/~kouxy. Interested readers may download them to get a rough idea of the relevant schemes before delving into the technical details.

Acknowledgement

The authors would like to thank the Department of Mechanical Engineering, The University of Hong Kong and the Research Grants Council for supporting this project [HKU7200/04E].

References

1. Cho, W., et al.: Local composition control in solid freeform fabrication. In: Proceedings of the 2002 NSF Design, Service and Manufacturing Grantees and Research Conference, San Juan, Puerto Rico (2002)
2. Huang, J., et al.: Bi-objective optimization design of functionally gradient materials. Materials & Design 23(7), 657–666 (2002)
3. Elishakoff, I., Gentilini, C., Viola, E.: Three-dimensional analysis of an all-round clamped plate made of functionally graded materials. Acta Mechanica 180(1 - 4), 21–36 (2005)
4. Eraslan, A., Akis, T.: On the plane strain and plane stress solutions of functionally graded rotating solid shaft and solid disk problems. Acta Mechanica 181(1 - 2), 43–63 (2006)
5. Cho, J.R., Shin, S.W.: Material composition optimization for heat-resisting FGMs by artificial neural network. Composites Part A: Applied Science and Manufacturing 35(5), 585–594 (2004)
6. Praveen, G.N., Reddy, J.N.: Nonlinear transient thermoelastic analysis of functionally graded ceramic-metal plates. International Journal of Solids and Structures 35(33), 4457–4476 (1998)
7. Cho, J.R., Ha, D.Y.: Optimal tailoring of 2D volume-fraction distributions for heat-resisting functionally graded materials using FDM. Computer Methods in Applied Mechanics and Engineering 191(29-30), 3195–3211 (2002)
8. Hu, Y., Blouin, V.Y., Fadel, G.M.: Design for manufacturing of 3D heterogeneous objects with processing time considerations. In: Proceedings of ASME 2005 Design Engineering Technical Conferences, Long Beach, California USA, September 24-28 (2005)
9. Jackson, T.R.: Analysis of functionally graded material object representation methods, Ph.D. Thesis, Massachusetts Institute of Technology (2000)
10. Pasko, A., et al.: Constructive hypervolume modeling. Graphical Models 63(6), 413 (2001)
11. Adzhiev, V., Kartasheva, E.: Cellular-functional modeling of heterogeneous objects. In: Proceedings of the seventh ACM symposium on Solid modeling and applications, Germany (2002)
12. Hua, J., He, Y., Qin, H.: Multiresolution heterogeneous solid modeling and visualization using trivariate simplex splines. In: Proceedings of the Ninth ACM Symposium on Solid Modeling and Applications, Genova, Italy, June 2004, p. 47 (2004)
13. Martin, W., Cohen, E.: Representation and extraction of volumetric attributes using trivariate splines: a mathematical framework. In: Proceedings of the sixth ACM symposium on Solid modeling and applications, p. 234 (2001)
14. Siu, Y.K.: Modeling and prototyping of heterogeneous solid CAD models, PhD Thesis, Department of Mechanical Engineering, The University of Hong Kong (2003)
15. Samanta, K., Koc, B.: Feature-based design and material blending for free-form heterogeneous object modeling. Computer-Aided Design 37(3), 287 (2005)
16. Biswas, A., Shapiro, V., Tsukanov, I.: Heterogeneous material modeling with distance fields. Computer Aided Geometric Design 21(3), 215–242 (2004)

17. Liu, H., et al.: Methods for feature-based design of heterogeneous solids. Computer-Aided Design 36(12), 1141 (2004)
18. Kou, X.Y., Tan, S.T.: A hierarchical representation for heterogeneous object modeling. Computer-Aided Design 37(3), 307 (2005)
19. Kumar, V., Dutta, D.: Approach to modeling & representation of heterogeneous objects. Journal of Mechanical Design, Transactions of the ASME 120(4), 659–667 (1998)
20. Sun, W., Hu, X.: Reasoning Boolean operation based modeling for heterogeneous objects. Computer-Aided Design 34(6), 481 (2002)
21. Cavalcanti, P.R., Carvalho, P.C.P., Martha, L.F.: Non-manifold modelling: an approach based on spatial subdivision. Computer-Aided Design 29(3), 209 (1997)
22. Kou, X.Y., Tan, S.T., Sze, W.S.: Modeling complex heterogeneous objects with non-manifold heterogeneous cells. Computer-Aided Design 38(5), 457–474 (2006)
23. Chen, M., Tucker, J.V.: Constructive volume geometry. Computer Graphics Forum 19(4), 281–293 (2000)
24. Adzhiev, V., et al.: Hybrid cellular-functional modeling of heterogeneous objects. Journal of Computing and Information Science in Engineering 2(4), 312 (2002)
25. Kou, X.Y., Tan, S.T.: Heterogeneous object modeling: A review. Computer-Aided Design 39(4), 284–301 (2007)
26. Kou, X.Y.: Computer-Aided Design of Heterogeneous Objects, PhD Thesis, The University of Hong Kong (2006)
27. Kou, X.Y., Tan, S.T.: A systematic approach for Integrated Computer-aided design and finite element analysis of Functionally-Graded-Material objects. Materials & Design 28(10), 2549–2565 (2007)
28. http://www.spatial.com/products/acis.html
29. Corney, J.: 3D modeling with the ACIS kernel and toolkit, vol. xix, p. 294. J. Wiley & Sons, Chichester, New York (1997)
30. http://www.comsol.com/
31. Kou, X.Y., Tan, S.T.: Robust and efficient algorithms for rapid prototyping of heterogeneous objects. Technical Report CADTR01/07, Department of Mechanical Engineering, The University of Hong Kong (2007), Available at: http://web.hku.hk/~kouxy/TR0107.pdf
32. Qian, X., Dutta, D.: Heterogeneous object modeling through direct face neighborhood alteration. Computers & Graphics 27(6), 943 (2003)
33. http://www.zcorp.com/
34. Kou, X.Y., Tan, S.T.: An intractive CAD environment for heterogeneous object design. In: Proceedings of ASME 2004 Design Engineering Technical Conferences, Salt Lake City, Utah, USA, September 28-October 2 (2004)

The HybridTree: A Hybrid Constructive Shape Representation for Free-Form Modeling

Rémi Allègre[1,2], Eric Galin[1], Raphaëlle Chaine[1], and Samir Akkouche[1]

[1] LIRIS UMR 5205 CNRS / Université Lyon 1, Villeurbanne, France
{eric.galin,raphaelle.chaine,samir.akkouche}@liris.cnrs.fr
[2] Now at: LSIIT, UMR CNRS 7005 / Université Louis Pasteur, Strasbourg, France
remi.allegre@lsiit.u-strasbg.fr

Abstract. In this paper, we describe a hybrid modeling framework for creating complex 3D objects incrementally. Our system relies on an extended CSG tree that assembles skeletal implicit primitives, triangle meshes and point set models in a coherent fashion: we call this structure the HybridTree. Editing operations are performed by exploiting the complementary abilities of implicit and polygonal mesh surface representations in a complete transparent way for the user. Implicit surfaces are powerful for combining shapes with Boolean and blending operations, while triangle meshes are well-suited for local deformations such as FFD and fast visualization. Our system can handle point sampled geometry through a mesh surface reconstruction algorithm. The HybridTree may be evaluated through four kinds of queries, depending on the implicit or explicit formulation is required: field function and gradient at a given point in space, point membership classification, and polygonization. Every kind of query is achieved automatically in a specific and optimized fashion for every node of the HybridTree.

1 Introduction

For the purpose of modeling complex free-form shapes, a large number of geometric representation have been developed, each with specific properties and limitations. For certain modeling operations, some surface representations are thus more advantageous than others. Our goal is to overcome this kind of restriction by mixing multiple shape representations into a single coherent modeling framework that takes benefit from the complementary advantages of volume and surface models. We focus here on three fundamental representations: implicit surfaces, triangle meshes and point sets.

Implicit surfaces [1,2] are powerful for representing objects of complex geometry and topology. As a volumetric model, They naturally lend themselves for blending [3] and CSG Boolean operations, and can be deformed by space warping techniques [4]. Pasko et al. [5] and later Wyvill et al. [6] have proposed two hierarchical models that incorporate Boolean, blending and warping operations in a unified system. We have contributed to develop the *BlobTree* model [6] that is characterized by a combination of skeletal primitives in a tree data-structure.

A. Pasko, V. Adzhiev, and P. Comninos (Eds.): HOMA, LNCS 4889, pp. 60–89, 2008.

The BlobTree has proved to be an intuitive and effective tool for modeling and animating complex and realistic organic shapes [7]. However, in this framework as in most implicit modeling frameworks, local deformations are difficult to implement and are restrictive. Visualizing complex implicit surfaces is also a computationally expensive task.

In contrast, triangle meshes can be efficiently visualized thanks to common graphic hardware. These surfaces can be edited interactively by a variety of powerful tools, such as free-form deformations [8] or Laplacian editing [9], that provide very intuitive local control over geometry. However, combining surfaces in boundary representation with Boolean operations is a complicated task, which is prone to topological inconsistencies. Polygonal meshes also do not naturally blend themselves together.

Point sampled geometry, that can be obtained from scanning devices, can be efficiently visualized and edited through point-based implicit surface models, such as Moving Least Squares surfaces [10]. This kind of representation is strongly dependent on the sampling density of the input point set and extrapolating reliable topological information can be a hard task. A connectivity structure can be provided by a surface reconstruction process [11].

In this paper, we describe a hybrid shape representation mixing implicit and polygonal mesh representations for incremental modeling of complex 3D shapes. We review the ideas and results that were first introduced in [12,13], and provide additional comments and perspectives. Our model is characterized by a tree data-structure that combines skeletal implicit surfaces, triangle meshes and point set models by means of Boolean, blending and warping operations, including free-form deformations. We call our model the *HybridTree*, which may be seen as a generalization of the BlobTree [6]. We evaluate this structure on-the-fly through three fundamental queries: field function, gradient and point membership classification, and a polygonization process. The originality of our model relies in the evaluation system that dynamically switches from one surface model to the other so as to use the most suitable representation for every type of editing operation. The core of our current implementation is a dual skeletal implicit/triangle mesh representation for every node. Each kind of node in the HybridTree is evaluated automatically in a specific and optimized fashion, depending on the formulation required by each operation. To handle point set models, we rely on an intermediate mesh representation obtained through the dynamic reconstruction technique proposed in [14,15,16], that can also be used for local remeshing purposes.

The remainder of this paper is organized as follows. In Section 2, we provide an overview of related shape modeling frameworks. Section 3 describes the architecture of our system and present how implicit and mesh representations are combined together. We explain in Section 4 how fundamental queries are performed on the HybridTree, and detail our polygonization algorithms in Section 5. Applications to complex shape modeling are discussed in Section 6. Eventually, in Section 7, we conclude and present future work.

2 Related Work

Modeling complex 3D shapes, either from scratch and/or from existing surfaces, e.g. acquired with scanning devices, is an active research area in Geometric Modeling and Computer Graphics. Conversion techniques from one model to the other make it possible for objects in different representations to coexist and interact in the same environment through a unified representation. Recent developments in implicit, mesh and point set modeling have lead to new interesting solutions to tighten the gap between these representations. Combining volumetric and boundary models into a single hybrid model has also received much attention for geometry processing and shape modeling.

Conversion Techniques. Techniques to translate geometry from the implicit to the polygonal mesh representation or from the polygonal mesh to the implicit representation have been extensively studied, but still remain a challenging research field. In the Computer Graphics community, state-of-the-art implicit surface meshing techniques include 3D-space cell decompositions [17,18,19,20], particle systems [21], and surface marching methods [22,23]. Recent work in Computational Geometry focused on how to produce a mesh approximation of an implicit surface with guaranteed topology and geometry [24,25]. A polygonal mesh surface can be converted into an implicit surface either through distance field computation [26], or reconstruction methods [27,28]. Surface reconstruction techniques from point sets may be used as well in this case, depending on the sampling conditions [29,30,31,32,33]. There also exists wide literature on converting point sets into polygonal mesh models, with Computational Geometry techniques [11], and more local approaches [34,35,36].

Implicit Surface Editing. In recent work, discrete implicit representations have been proposed as a general-purpose model for editing complex shapes. The level set framework proposed by Museth et al. [37] provides conversion algorithms from many other surface representations and a wide range of editing tools. This model is memory consuming and the quality of the result is dependent on the gid resolution. The Adaptively Sampled Distance Fields introduced by Frisken et al. [38] rely on a hierarchical structure that provides local control over geometric error. In constrast, the framework in this paper manipulates a continuous implicit representation and preserves existing surfaces when possible. As far as local deformations are concerned, Schmitt et al. [39] have proposed an approach based on specific skeletal elements and field functions to simulate free-form deformations on functionally-defined implicit surfaces. This method is not as intuitive as mesh deformation tools and does not offer as many degrees of freedom.

Mesh Surface Editing. Surface mesh editing has been recently enriched with new techniques based on differential coordinates. The Laplacian representation encodes the location of each vertex relatively to its neighborhood, which provides an approximation of the Laplacian of the underlying surface. Sorkine et al. [9] developed local deformation and blending tools that preserve geometric details. All operations require to solve least-squares systems, which can be achieved at interactive rates for not too dense meshes. A similar approach based on the discrete Poisson equation was introduced by Yu et al. [40]. In both methods, mesh blending requires to solve a vertex matching problem between corresponding mesh boundaries. The meshes that are blended together should have near equal edge length, which may require an initial remeshing process.

Kanai et al. [41] have proposed a method for cutting and pasting mesh parts. After selection of mesh parts of interest, matching boundary vertices are first determined. A registration process is applied between the two boundaries and a B-spline function is used to interpolate vertex locations smoothly. Using this technique, Funkhouser et al. [42] developed a new kind of "data-driven" mesh modeling framework. Starting from a mesh model, the user can select parts to edit thanks to an interactive segmentation algorithm, and then query a mesh database for similar parts. The desired parts can be extracted from the retrieved models and then merged with the base mesh.

Polygonal mesh blending is also closely related to shape interpolation and morphing techniques [43]. A source polygonal mesh and a target one can be locally interpolated so as to achieve local blending effects. Related work in this domain include the work by Alexa [44], in which Laplacian coordinates are linarly interpolated. Xu et al. [45] proceed by non-linear interpolation of gradient fields by solving Poisson equations defined on meshes. This approach involves numerous parameterization and remeshing issues [46,47]. The very limitation of mesh blending based on morphing methods is that the source and target models should have the same topology.

Hybrid Modeling. Depending on the surface representation, some operations cannot be performed easily in a direct way. In some cases, this issue can be addressed with the help of an intermediate representation. Over the past few years, hybrid models have been investigated by several authors for this purpose. In the field of geometry processing, some specific problems may be efficiently solved by combining implicit and parametric representations [48]. For shape modeling, this approach has first attracted attention for mesh deformation and blending. Several implicit models have been used to deform meshes [49,50,51]. Decaudin [52] and Singh and Parent [53] introduced mesh blending techniques based on an intermediate implicit representation. More recently, point-based modeling techniques mixing point set and local implicit surface representations have attracted considerable attention.

Point-Based Modeling. Due to the recent advances of 3D digital acquisition, shape modeling from point-sampled geometry has become very popular for a few years. The Moving Least Squares implicit surface model introduced by Levin [54] has proved abilities for both interactive surface editing [10] and physical simulation [55]. Its major interest is that it can handle digitally acquired surfaces without preliminary surface reconstruction step. Explicit information about topology is not maintained, which has advantages for some operations. Whenever neighborhood information is needed, connectivity relations based on k-neighborhoods are computed on-the-fly using a spatial search data-structure. However, establishing correct connectivity relations is not always a trivial problem, that depends on the sampling distribution of the points on the input surfaces [56,57]. Moreover, the sampling density has to be updated frequently to maintain a coherent surface, which requires surface reconstruction steps.

The variational technique presented in [58] generates an implicit surface from point-sets via an interpolation scheme based on compactly supported radial basis functions. The resulting shapes can be locally controlled in an intuitive way by acting on the constraint points. The evaluation may be performed at interactive rates using an octree data-structure. However, this approach remains computationally demanding when manipulating dense point sampled geometry, and sharp features are difficult to handle in this framework.

3 The HybridTree

The HybridTree model relies on a tree data-structure whose leaves can hold either complex skeletal implicit primitives as described in [59], triangle meshes with manifold topology or point set models. Those models are combined by Boolean, blending and warping operations located at the nodes of the tree. Warping nodes include affine transformations, Barr deformations [60] and free-form deformations [8]. Constructive operations are binary whereas warping operations are unary operations.

Figure 1 shows the HybridTree structure of a winged snake-woman model. The snake-woman model (a) has been entirely built from skeletal implicit primitives. Using Boolean difference, only the body has been conserved, which has been then blended with the Igea point set model (b) so as to obtain the result in (c). The wings of a mesh model of the Victory of Samothrace (d) have been extracted by intersecting the model with a box. The wings and the modified snake-woman model have been finally blended together in (e).

The evaluation of the HybridTree is achieved in an incremental way by recursively traversing the tree data-structure. The architecture of our evaluation system is presented in Figure 2. Each pole corresponds to a geometric representation that provides the set of operations for which it is the most well-suited. Arrows depict the gateways from one model to the other, that correspond to conversion procedures. Starting from an implicit, mesh or point set object, it is first converted on-the-fly into the required representation before applying a given operation. The gateways available in our current implementation are depicted by solid arrows in the diagram. Skeletal implicit surfaces and triangle meshes are currently the core of our system. Every node of the HybridTree can generate both a potential field in space and a triangle mesh. Point sets models are plugged through a surface reconstruction technique that produces a triangle mesh representation. The completeness of the system is thus achieved by transitivity. Conversion from the point set representation to the implicit one could be

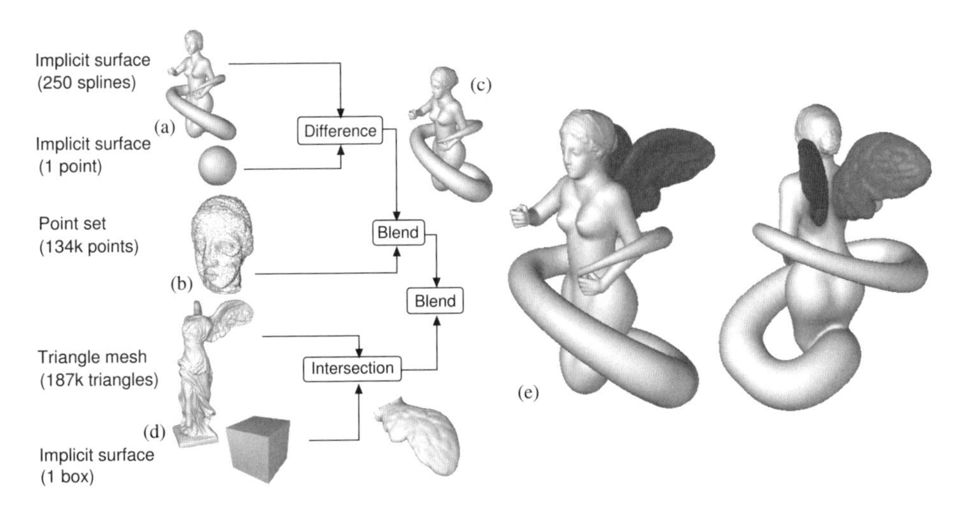

Fig. 1. The HybridTree structure of a winged snake-woman model

performed directly using Moving Least Squares or variational techniques. However, these techniques are not compatible with our skeletal implicit model, that requires the signed distance function to the surface to be reliably evaluable everywhere in space. Depending on the sampling density, converting a triangle mesh into a point set model may require a resampling process, as proposed in [24], that is not currently included in our system. Conversion from a skeletal implicit surface into a point set is achieved using ad hoc algorithms. We convert point set models into triangle meshes using the dynamic convection-driven surface reconstruction technique developed by Allègre et al. in [14,15,16], that offers user control over the level of detail of the resulting mesh.

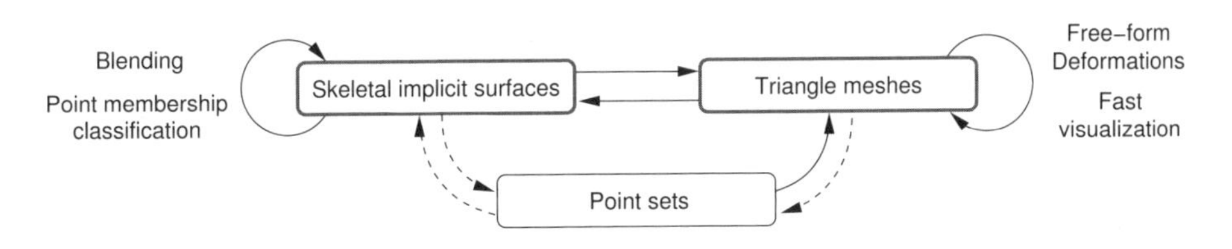

Fig. 2. The HybridTree's evaluation system

The HybridTree is evaluated through three fundamental queries and a polygonization process that are implemented in a specific fashion for each kind of node. Field function queries at a given point in space are performed whenever the implicit formulation is required. We essentially rely on the implicit formulation to achieve blending operations. Gradient queries allow to obtain the exact normal at sample points. Some operations only require to know whether a point lies inside or outside the surface, such as Boolean operations. Instead of evaluating the sign of the field function explicitely, we developed specific methods to perform accelerated point membership queries. The last query type in our framework is an incremental polygonization process that is invoked at a given node if the mesh formulation is needed, for local deformations or visualization. Local results are combined into a coherent fashion by binary operations.

Notations. An implicit surface is mathematically defined by a field function $f : \mathbb{R}^3 \rightarrow \mathbb{R}$ as the points in space that satisfy the equation:

$$S = \{\mathbf{p} \in \mathbb{R}^3 | f(\mathbf{p}) = T\}$$

where T is a threshold level. The field function of a node A will be denoted as f_A, and the corresponding gradient as ∇f_A. We will call $c_A(\mathbf{p})$ the point membership function of A at point $\mathbf{p}$, that can take three different values $\{1, 0, -1\}$ depending on $\mathbf{p}$ respectively lies inside, on, or outside the surface of A. The notation $\mathcal{M}_A$ will refer to the mesh of the surface of A, and the bounding box of the object A will be denoted as $\mathcal{B}_A$.

4 Fundamental Queries

In the following paragraphs, we detail how the field function, gradient and point membership are evaluated for the different kinds of node in the HybridTree.

4.1 Skeletal Implicit Primitives

Skeletal implicit primitives are built around a geometric object called *skeleton*. The field function for a given skeleton is evaluated analytically using the following formulation:

$$f(\mathbf{p}) = g \circ d(\mathbf{p})$$

where $d : \mathbb{R}^3 \rightarrow \mathbb{R}_+$ denotes the Euclidean distance to the skeleton, and $g : \mathbb{R}_+ \rightarrow \mathbb{R}$ refers to the potential field function. The latter is a compactly supported radial basis function that is parameterized by a maximum field value $I \in \mathbb{R}$ reached on the skeleton, and a radius of influence that will be denoted as $R \in \mathbb{R}_+$. The associated region of influence, characterized by non-zero field values, will be denoted as Ω. In our system, we use polynomial potential field functions of the form:

$$g(r) = \begin{cases} I\left(1 - \dfrac{r^2}{R^2}\right)^n, & n \geq 2 \quad \text{if } r \in [0, R] \\ 0 \quad \text{otherwise} \end{cases}$$

The corresponding inverse potential field functions $g^{-1} : \mathbb{R} \rightarrow \mathbb{R}_+$ is defined as follows:

$$g^{-1}(t) = \begin{cases} R\sqrt{1 - \left(\dfrac{t}{I}\right)^{\frac{1}{n}}}, & n \geq 2 \quad \text{if } 0 < t \leq I \text{ or } I \leq t < 0 \\ 0 \quad \text{otherwise} \end{cases}$$

Normals can be obtained directly from the gradient of the field function $\nabla f(\mathbf{p})$ which is evaluated as follows:

$$\nabla f(\mathbf{p}) = g' \circ d(\mathbf{p}) \, \nabla d(\mathbf{p})$$

Since $d(\mathbf{p})$ is the Euclidean distance function, $\nabla d(\mathbf{p})$ is computed as the vector between the orthogonal projection of $\mathbf{p}$ onto the skeleton and $\mathbf{p}$.

The HybridTree implements a wide range of complex skeletal primitives including curve, surface and volume skeletons as described by Barbier et al. in [59]. Figure 3 shows a bottle built using surface of revolution, spline, circle and hollow cylinder skeletons. Every type of primitive implements a specific algorithm that computes the distance $d(\mathbf{p})$ to its skeleton analytically in an optimized fashion. Algorithms become more sophisticated as the complexity of the skeletons increases [61,62].

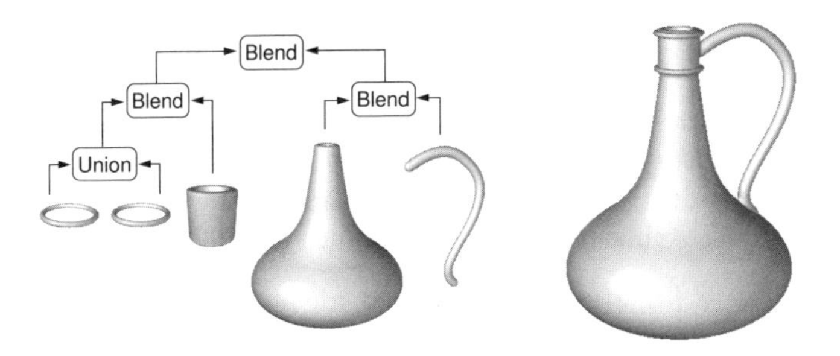

Fig. 3. A bottle model built using complex skeletal implicit primitives

For implicit surfaces, point membership classification is usually obtained through an evaluation of the field function and comparing its value to the threshold level. Since we manipulate skeletal implicit primitives, we do not need to compute the full field function for point membership queries. The T level surface of these primitives may indeed be defined by sweeping a sphere of constant radius $r_T = g^{-1}(T)$ along the boundary of the skeleton. This isosurface exists if and only if $r_T \geq 0$. Therefore, for a given point $\mathbf{p}$ in space, the point membership function $c(\mathbf{p})$ may be defined as follows:

$$c(\mathbf{p}) = \begin{cases} -1 \text{ if } 0 \leq r_T < d(\mathbf{p}) \\ \;\;\,1 \text{ if } 0 < d(\mathbf{p}) < r_T \\ \;\;\,0 \text{ otherwise} \end{cases}$$

The radius r_T is computed once for each primitive. For volume skeletal primitives, the location of query points with respect to the interior of the skeleton is obtained analytically as part of the computation of the distance $d(\mathbf{p})$.

Skeletal implicit primitives are defined only by a few parameters, which yields particularly low storage cost. Complex skeletons are also easier to control than a combination of simple primitives, but distance evaluations are more computationally demanding. A basic acceleration technique consists in pre-computing and caching some results required for several evaluations, e.g. along a ray [59]. When the evaluation of $f(\mathbf{p})$ and $\nabla f(\mathbf{p})$ are both required, common intermediate results are also computed only once. At global scale, every primitive is equiped with a bounding box that allows to save useless evaluations.

4.2 Polygonal Meshes

For a polygonal mesh, the field function is computed using the same formulation as for skeletal implicit primitives. The distance function $d_{\mathcal{M}}$ from a point $\mathbf{p} \in \mathbb{R}^3$ to a triangle mesh $\mathcal{M}$ is defined as the minimal Euclidean distance between $\mathbf{p}$ and any triangle $\mathcal{T}$ of the boundary of $\mathcal{M}$:

$$d_{\mathcal{M}}(\mathbf{p}) = \min_{\mathcal{T} \in \mathcal{M}} d(\mathbf{p}, \mathcal{T})$$

The implicit surface generated by the skeletal mesh for a given threshold T is a rounded surface S which differs for the original mesh $\mathcal{M}$. This surface may be defined by sweeping a sphere of constant radius $r_T = g^{-1}(T)$ along the boundary of $\mathcal{M}$ (Figure 4, left). To make the boundary of $\mathcal{M}$ and the T level surface to correspond independently from the field function parameters, we incorporate the threshold as an offset in a pseudo-distance function which is defined as follows:

$$d(\mathbf{p}) = \begin{cases} d_{\mathcal{M}}(\mathbf{p}) + r_T & \text{if } \mathbf{p} \text{ is outside } \mathcal{M} \\ r_T - d_{\mathcal{M}}(\mathbf{p}) & \text{if } \mathbf{p} \text{ is inside } \mathcal{M} \text{ and } d_{\mathcal{M}}(\mathbf{p}) < r_T \\ 0 & \text{otherwise} \end{cases}$$

Our distance function guarantees that the isosurface and the mesh boundary mesh are the same for any value of T, as shown in Figures 4(right) and 5(right).

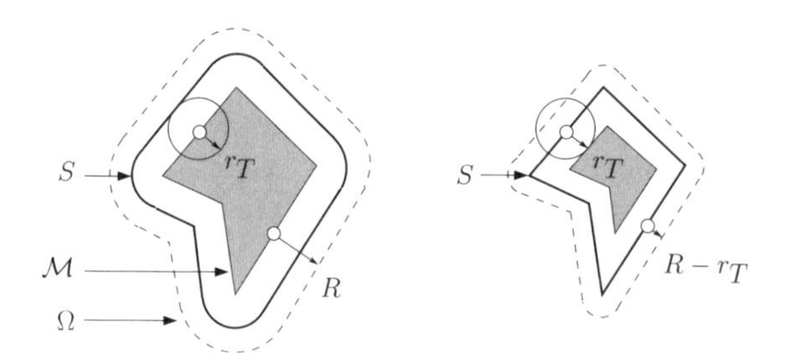

Fig. 4. The distance offset mechanism on a 2D polygon. On the left, the basic distance formula is used. The effect of our distance function is illustrated on the right.

The user keeps control on every parameter of the field function, and can precisely control the range of the blend between two objects. The radius of influence of the mesh, which falls from R to $R - r_T$, is rescaled to $\frac{R}{(1-\frac{r_T}{R})}$ before the evaluation so that the distance offset is hidden.

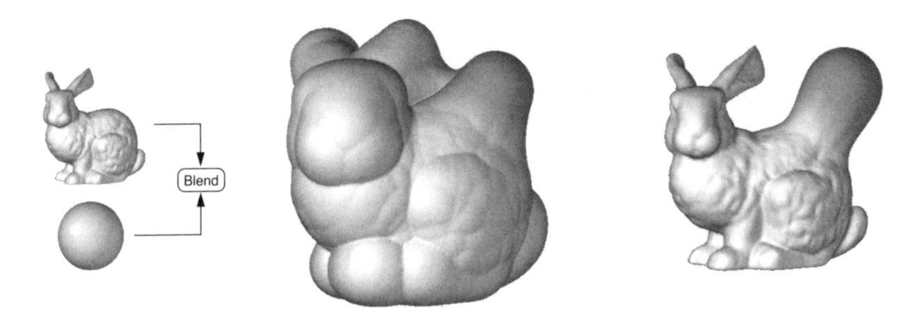

Fig. 5. The Stanford Bunny mesh model (69,674 triangles) blended with an implicit sphere. The result on the left uses the distance $d_{\mathcal{M}}(\mathbf{p})$ whereas the model on the right has been computed using our pseudo-distance function $d(\mathbf{p})$.

The gradient of the field function $\nabla d(\mathbf{p})$ is evaluated as follows:

$$\nabla d(\mathbf{p}) = \begin{cases} \nabla d_{\mathcal{M}}(\mathbf{p}) & \text{if } \mathbf{p} \text{ is outside } \mathcal{M} \\ -\nabla d_{\mathcal{M}}(\mathbf{p}) & \text{if } \mathbf{p} \text{ is inside } \mathcal{M} \text{ and } d_{\mathcal{M}}(\mathbf{p}) < r_T \\ 0 & \text{otherwise} \end{cases}$$

As for implicit primitives, $\nabla d(\mathbf{p})$ is computed as the vector between the orthogonal projection of $\mathbf{p}$ onto the mesh $\mathcal{M}$ and $\mathbf{p}$, that can be obtained as part of the computation of $d(\mathbf{p})$.

Computing the minimum distance between a point $\mathbf{p}$ and all the triangles $\mathcal{T}$ of the mesh $\mathcal{M}$ is computationally expensive. Acceleration techniques have been widely studied, not only in the Computer Graphics community [26]. Our framework implements an algorithm inspired by Johnson and Cohen's lower-upper bound strategy [63]. We rely on a bounding box hierarchy based on a Binary Space Partition tree, which is traversed breadth-first for each point-to-mesh distance query. For each node we compute a lower and an upper bound of the minimum point-to-mesh distance, which yields efficient space-pruning. Moreover, we use the fact that the potential field falls to zero beyond the distance R from the mesh boundary so as to reject more useless point-to-mesh distance computations.

Point membership classification is obtained by computing the number of intersections between a ray and the mesh using the bounding box hierarchy. In order to reduce the number of cells to be traversed, the direction and the orientation of the ray are chosen so that the distance between the query point and the intersection point between the ray and the mesh bounding box is minimized.

4.3 Blending Operations

In our system, we propose two kinds of formulations for blending two surfaces : one global and one local. Let A and B denote two models that blend together. Global blending between two objects is functionally defined as originally proposed by Blinn [3]:

$$f_{A+B} = f_A + f_B$$

The major drawback of this basic approach is the well-known unwanted blending problem [64]. To perform blending with local control, we have implemented a new local blending operation adapting the local blending technique described by Pasko et al. in [65]. This operation has three children. The first two, denoted as A and B, represent the two models that will be partially blended together, whereas the third, denoted as C, represents the region of space where blending will occur. In our system, C is characterized by a potential field that is defined as a union of implicit primitives denoted as C_i. Such a combination makes it possible to build complex blending regions with predictible results. The field function f_C characterizes the blending region and is used to scale the amount of blending between the two sub-trees A and B. The values taken by the field functions f_{C_i} should range between 0 and 1. At a given point in space $\mathbf{p}$, if $f_C(\mathbf{p}) = 0$ then only union occurs, which is the case for any point outside the region of influence of C. In contrast, if $f_C(\mathbf{p}) = 1$, full blending takes place normally. The evaluation of the local blending operation is performed as follows. We first compute the potential field value resulting from the blending of the children nodes $f_{A+B}(\mathbf{p}) = f_A(\mathbf{p}) + f_B(\mathbf{p})$, and the field function value $f_{A\cup B}(\mathbf{p})$. We define the resulting field function as a weighted average:

$$f_{A+B}(\mathbf{p}) = f_C(\mathbf{p})\, f_{A+B}(\mathbf{p}) + (1 - f_C(\mathbf{p}))\, f_{A\cup B}(\mathbf{p})$$

Primitives build from a volume skeleton are very useful to define regions in space where full blending occurs. In Figure 6, the local blending region C is defined as the union of two implicit cylinders.

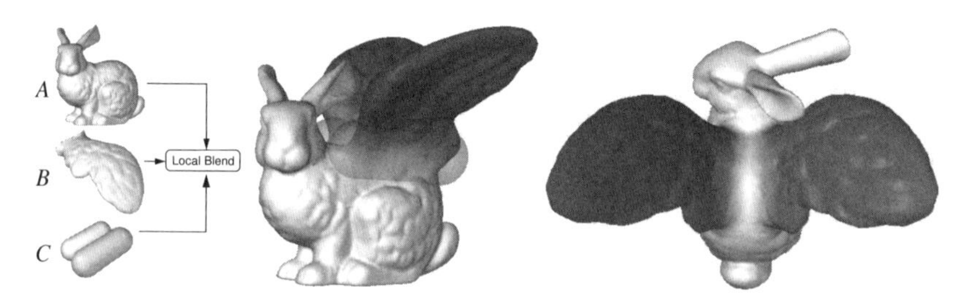

Fig. 6. Local blending. Two implicit cylinders define the blending region between the Bunny and the wing pair.

For both blending techniques, the gradient is obtained by deriving the field functions. For point membership classification, we distinguish *positive* and *negative* potential fields. A model A will be said to generate a positive potential field if one of its primitive is such that $I > 0$. Conversely, a model A will be said to generate a negative potential field if and only if every primitive it consits of is such that $I < 0$.

Let A and B two models that globally blend together such that A and B both generate a positive potential field. Point membership classification is then achieved as follows:

1. If $\mathbf{p}$ is located inside $\mathcal{B}_A \cap \mathcal{B}_B$, then evaluate $v = f_{A+B}(\mathbf{p})$.
   (a) If $v < T$, $\mathbf{p}$ lies inside the surface: return 1.
   (b) Else, if $v > T$, $\mathbf{p}$ lies outside the surface: return -1.
   (c) Else, $\mathbf{p}$ is on the surface: return 0.
2. Else, if $\mathbf{p}$ belongs to $\mathcal{B}_A \setminus \mathcal{B}_B$ (resp. $\mathcal{B}_B \setminus \mathcal{B}_A$), then query point membership on A (resp. B) and return the result.
3. Else, $\mathbf{p}$ lies outside the surface: return -1.

For local blending nodes such that A, B and C generate positive potential fields, the previous algorithm is slightly modified. In Step 1, we consider the bounding box $\mathcal{B}_C$ of the local blending region and evaluate f_{A+B}. In Step 2, we consider the boxes $\mathcal{B}_A \setminus \mathcal{B}_C$ and $\mathcal{B}_B \setminus \mathcal{B}_C$.

For global blending nodes such that A generates a positive potential field and B generates a negative potential field, the previous algorithm is modified as follows: In Step 2, if $\mathbf{p}$ belongs to $\mathcal{B}_B \setminus \mathcal{B}_A$, then $\mathbf{p}$ is outside the surface. If $\mathbf{p}$ belongs to $\mathcal{B}_A \setminus \mathcal{B}_B$, then point membership is queried on A. If both A and B generate negative potential fields, $\mathbf{p}$ lies outside the surface: -1 is returned.

Let us finally consider a local blending nodes such that C generates a negative potential field. If $\mathbf{p}$ is located inside $\mathcal{B}_C$, then $\mathbf{p}$ lies outside the surface: 1 is returned. Otherwise, if $\mathbf{p}$ belongs to $\mathcal{B}_A \setminus \mathcal{B}_C$ (resp. $\mathcal{B}_B \setminus \mathcal{B}_C$), then point membership is queried on A (resp. B) and the result is returned.

4.4 Boolean Operations

The min and max functions prescribed in [6] for union and intersection produce gradient discontinuities in the potential function. This results in visible unwanted normal discontinuities on the surface. We have adapted R-Functions with C^n continuity prescribed in [5] to our model by incorporating the threshold value as an offset in the previous equations. A weighting coefficient appears so as to guarantee that the resulting field function has a compact support. We have:

$$f_{A \cup B} = T + \frac{1}{2-\sqrt{2}} \left[(f_A - T) + (f_B - T) + \sqrt{(f_A - T)^2 + (f_B - T)^2} \right]$$

$$f_{A \cap B} = T + \frac{1}{2+\sqrt{2}} \left[(f_A - T) + (f_B - T) - \sqrt{(f_A - T)^2 + (f_B - T)^2} \right]$$

Although min and max functions on the one hand, and R-Functions on the other hand produce different potential fields in space, both representations produce the same implicit surface if the Boolean nodes are located at the top of the tree structure. In this case, the computation of the min and max is computationally inexpensive compared to R-Functions. In contrast, we use the modified R-Function equations to create a continuously differentiable potential field if blending nodes are located above Boolean operations in the HybridTree. Our system automatically adapts the function used to evaluate Boolean operations depending on the context during the evaluation.

For Boolean operations defined using the min and max functions, the gradient at a given point in space is the gradient of the minimum or maximum contributing field function between the two input models. Using the R-Function formulation, the gradient is simply derived from the field functions.

Point membership classification is obtained through the standard operations of Boolean algebra. For instance, the point membership classification for a difference operation between two models A and B is performed as follows:

1. If $c_A(\mathbf{p}) = 1$, then:
   (a) If $c_B(\mathbf{p}) = -1$, then return 1.
   (b) Else, if $c_B(\mathbf{p}) = 0$, then return 0.
   (c) Else, return -1.
2. Else, if $c_A(\mathbf{p}) = 0$ and $c_B(\mathbf{p}) \in \{0, 1\}$ then return 0.
3. Else, return -1.

4.5 Warping Operations

In our system, the shape of a surface can be distorted by locally warping space. Our warping operations include affine transformations and Barr's twist, taper and bend deformations [60]. We also handle free-form deformations [8], denoted as FFD, so as to perform local deformations. Throughout the following paragraphs, A will denote the child object of a warping node, w a space transformation that maps $\mathbb{R}^3$ into $\mathbb{R}^3$, and w^{-1} the corresponding inverse transformation.

Barr Deformations. When the implicit formulation is required, twist, taper and bend deformations are applied as warp functions. The resulting field function is defined using the inverse warp function as follows:

$$f_w(\mathbf{p}) = f_A \circ w^{-1}(\mathbf{p})$$

The gradient of the field function may be evaluated as:

$$\nabla f_w(\mathbf{p}) = J_{w^{-1}}^{\top}(\mathbf{p}) \times \nabla f_A \circ w^{-1}(\mathbf{p})$$

where $J_{w^{-1}}^{\top}(\mathbf{p})$ denotes the transpose Jacobian matrix of the inverse warp function w^{-1} at $\mathbf{p}$. For these deformations, the closed form expressions of w^{-1} and $J_{w^{-1}}^{\top}(\mathbf{p})$ can be easily computed. The detailed equations can be found in [60] (it should be noted that there are some mistakes in the formulas related to the inverse bend transformation).

Point membership classification is achieved by computing the location $w^{-1}(\mathbf{p})$ of the query point $\mathbf{p}$ in the unwarped space and then querying point membership on A as follows:

$$c_w(\mathbf{p}) = c_A \circ w^{-1}(\mathbf{p})$$

Free-Form Deformations. Free-form deformations have been first introduce by Sederberg and Parry in [8], and then have been extended by several authors [66,67,68]. Applying local deformations in the implicit formulation is not straightforward as there is no easy way of computing an analytical formulation for w^{-1}. In our framework, the field function for FFD operations is evaluated using an intermediate mesh representation. The algorithm proceeds as follows:

1. Generate the mesh $\mathcal{M}_A$ of A.
2. Deform $\mathcal{M}_A$ by applying w to its vertices.
3. Evaluate the field function using the distance to the deformed mesh $\mathcal{M}_A$.

FFD nodes hold a mesh representation of their own resulting surface using this method. Subsequent field function, gradient and point membership queries are performed directly on this mesh without further recursion down to the subtree. If A is a single primitive, the field function is evaluated using the potential field function of A. If the local deformation extends to several primitives, their local blending properties are replaced by a new potential field function that is associated with the computed mesh. Therefore, these nodes should be located at the lowest levels of the tree in order to preserve these local properties if they are involved in further operations.

Affine Transformations. Affine transformations can be applied to implicit surfaces as warp functions. The resulting field function can be defined using the inverse transformation in the same way as for Barr deformations. However, this requires to evaluate w^{-1} for every queries on the subtree. Benefitting from the distributivity of affine transformations over Boolean and blending operations, Fox et al. [69] prescribed to remove affine transformation nodes and directly integrate them into the parameters of the primitives. In their method, the process is blocked by warping nodes. In our system, we have extended the algorithm to our local blending nodes and optimized it so that Euclidean similarities, including rigid transformations plus uniform scaling, can be cast through Barr and FFD nodes either. For these three operations, affine transformations are transmitted to both the arguments and space parameters of the operation.

5 Polygonizing the HybridTree

The resulting surface of a HybridTree may be triangulated using standard implicit surface meshing techniques, thanks to the previously defined field functions. However, these techniques rely on many evaluations of the potential field function, which is computationally demanding in the general case. In particular, sampling the field function of a mesh primitive is an expensive task that is clearly unprofitable if the mesh surface only interacts locally with other primitives.

For efficient meshing, we developed an incremental approach that preserves existing mesh surfaces as much as possible and optimizes the mesh generation for every kind of node using specific meshing algorithms. Local meshes are merged together at blending and Boolean nodes to form the resulting mesh surface. The following paragraphs detail our meshing methods for skeletal implicit primitives and for the different editing operations.

5.1 Skeletal Implicit Primitives

In our system, every implicit primitive automatically generates its mesh representation very efficiently for a target level of detail. For every kind of skeleton, we developed a specific and optimized meshing procedure that outputs a manifold mesh characterized by an almost uniform sampling distribution and regular connectivity (Figure 7).

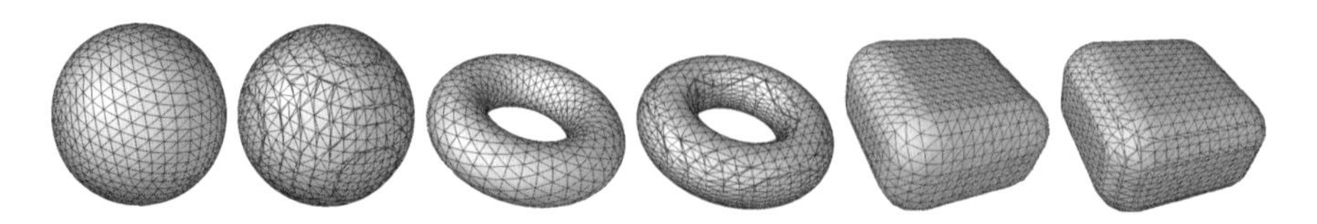

Fig. 7. Meshes obtained by direct meshing (left) vs. Marching Cubes results (right) for identical target edge length. From left to right: point, circle and box skeletons.

As mentioned previously, the T level surface of a skeletal implicit primitive can be described by sweeping of a sphere of constant radius $r_T = g^{-1}(T)$ along the skeleton. This isosurface corresponds to a $2d$-manifold and will be polygonized if and only if $r_T > 0$. For most primitives, this surface can be defined as a patchwork of simple surface pieces such as portions of sphere or cylinder, planes, disks, which facilitates direct meshing. After having identified the different components for a given primitive, taking symmetries into account, we sample each part iteratively while establishing consistent connectivity relations. The level of detail is fixed by the choice of a maximum edge length.

In comparison to standard implicit surface triangulation algorithms like the Marching Cubes algorithm [17], timings demonstrate that our direct meshing approach accelerates the polygonization process up to 200 times and produces up to 30% fewer, better shaped triangles. Figure 7 shows the meshes of point, circle and box primitives produced by our algorithms which compares favorably to the Marching Cubes mesh outputs for the same precision.

5.2 Blending Operations

We generate the mesh at blending nodes using the implicit representation of the surface. The computation of each sample point on the isosurface is an expensive task that requires several field function evaluations. To save computational time and preserve existing mesh surfaces, we restrict the computation of new sample points to blending regions. Wherever two objects do not overlap very much, we observed that it is better to generate the meshes of these objects before combining them rather than computing the overall mesh from scratch.

In the following paragraphs, we describe our meshing algorithms for global and local blending nodes. For the global case, we will first assume that the objects that blend together are associated with positive potential fields, which is the most common situation. The algorithm for blending nodes involving negative potential fields will be exposed subsequently.

Global Blending

We present two kinds of algorithms that differ in the way they process a HybridTree. The first *node-based* algorithm recursively traverses the tree data-structure, every node generating the mesh of its own subtree. The second algorithm, called *primitive-based*, is more independent from the tree structure and focuses on the interactions between primitives.

Node-Based Algorithm. Let A and B to objects that globally blend together into an object C. We suppose that A and B generate positive potential fields. If A and B are only partially blended, then we create the mesh of C after the meshes $\mathcal{M}_A$ and $\mathcal{M}_B$. We use the Marching Cubes algorithm to generate the mesh in the blending region. Otherwise, the whole mesh of C is generated by applying the Marching Cubes algorithm from scratch. In both cases, Marching Cubes sample points are computed by evaluating the local field function $f_{A+B}(\mathbf{p})$.

To determine the most favorable approach, we estimate how much the models A and B overlap, i.e. which proportion of volume each one shares in the blending region. Between A and B, blending occurs in the regions of space where f_A and f_B are both positive. These blending regions are enclosed in the intersection of $\mathcal{B}_A$ and $\mathcal{B}_B$. We introduce a ratio ρ called *overlapping ratio* such that $0 \leq \rho \leq 1$, which is computed as follows:

$$\rho = \frac{V_{A \cap B}}{\min(V_A, V_B)}$$

where V_A, V_B and $V_{A \cap B}$ denote the volume of the bounding boxes $\mathcal{B}_A$, $\mathcal{B}_B$ and $\mathcal{B}_A \cap \mathcal{B}_B$ respectively. Let $0 \leq \rho_0 \leq 1$ denote a fixed threshold. Without loss of generality, assume $V_A \leq V_B$. Our algorithm then proceeds as follows:

1. If $\rho > \rho_0$, then:
   (a) If $\frac{V_A}{V_B} \leq \rho_0$, then apply the Marching Cubes algorithm in the box that bounds $\mathcal{B}_A \cup \mathcal{B}_B$.
   (b) Else:
      - Create the mesh $\mathcal{M}_B$ of B.
      - Remove the triangles of $\mathcal{M}_B$ that have at least one vertex $\mathbf{p}_i$ such that $\mathbf{p}_i \in \mathcal{B}_A$.
      - Apply the Marching Cubes algorithm in $\mathcal{B}_A$.
2. Else:
   (a) Create the meshes $\mathcal{M}_A$ and $\mathcal{M}_B$ of A and B respectively.
   (b) Remove the triangles of $\mathcal{M}_A$ and $\mathcal{M}_B$ that have at least one vertex $\mathbf{p}_i$ such that $\mathbf{p}_i \in \mathcal{B}_A \cap \mathcal{B}_B$.
   (c) Apply the Marching Cubes algorithm in $\mathcal{B}_A \cap \mathcal{B}_B$.

We invoke a crack fixing algorithm after each local Marching Cubes meshing process in order to bridge the narrow gap between boundary triangles and output a closed manifold mesh.

Before launching the polygonization process, we evaluate the ratio ρ for terminal blending nodes and propagate the information up the tree structure. We cluster consecutive blending nodes whose child nodes strongly overlap along the same branch so as to treat them in one single Marching Cubes pass at the highest possible level of the branch, out of efficiency. The clustering process along a branch stops as soon as a non-blending node is encountered.

The user can provide a value for ρ_0 or directly specify which method should be used for each blending operation involved in a tree. We choose $\rho_0 = 0.5$ as a default threshold, which appears as a good guess in most cases. In some situations however, the overlapping volume between two objects can be significantly smaller than the overlapping volume between their bounding boxes. The estimation of the amount of blend could be refined by subdividing the bounding boxes and considering overlapping sub-volumes.

Based on the analysis of the bounding box hierarchy, this approach is particularly simple and systematic. However, its performance clearly depends on the structuration of the HybridTree. Some blending mesh parts generated from different blending subtrees may be locally destroyed and remeshed several times due to a non optimal binary tree structure. This limitation could be compensated by a restructuration of the tree, but this kind of optimization is known to be computationally expensive [59], and would not improve every configurations, as illustrated by the model in Figure 8. Here, the blending node would be globally polygonized using Marching Cubes although blending only occurs locally.

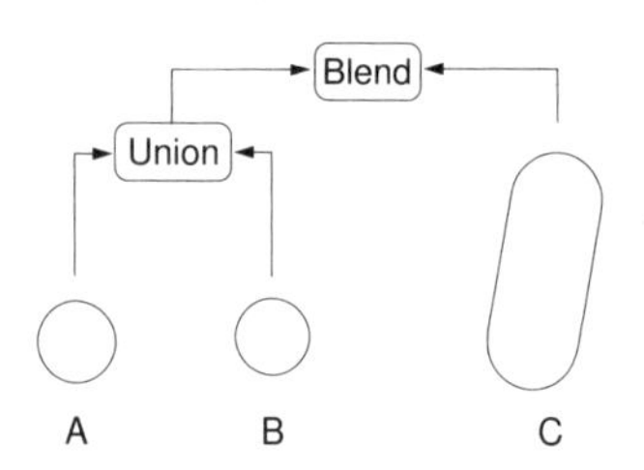

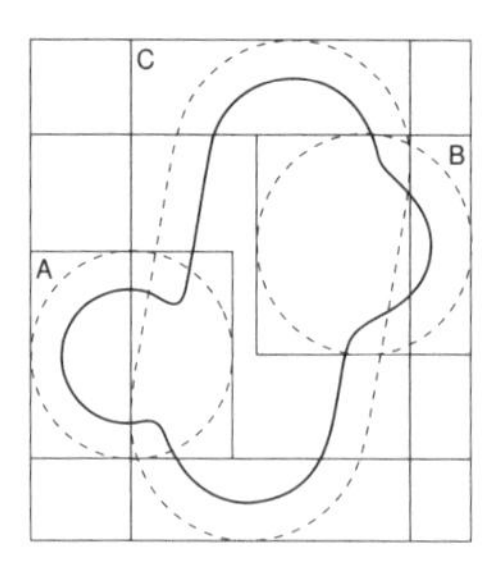

Fig. 8. A HybridTree configuration for which our node-based polygonization technique for blending nodes is not optimal

The difficulty is in fact intrinsic to the binary tree representation that cannot explicitely describe the interactions that occur between more than two primitives. To cope with non optimal HybridTree structures, we propose to extend our first algorithm in the spirit of the space decomposition approach introduced by Fox et al. in [69] that focuses on how primitives interact in space rather than on the global tree structure.

Primitive-Based Algorithm We distinguish two kinds of regions of space: the regions $\mathcal{R}_P$ that are influenced by a single primitive (Figure 9(a), light gray), and the regions $\mathcal{R}_+$ where blending occurs (Figure 9(a), dark grey). Our goal is to polygonize primitives using direct meshing in regions $\mathcal{R}_P$ and to apply the Marching Cubes algorithm in a single pass in regions $\mathcal{R}_+$. For this purpose, the HybridTree is embedded in a regular grid and we rely on a modified Marching Cubes algorithm to compute sample points in the bounding boxes of the blending regions.

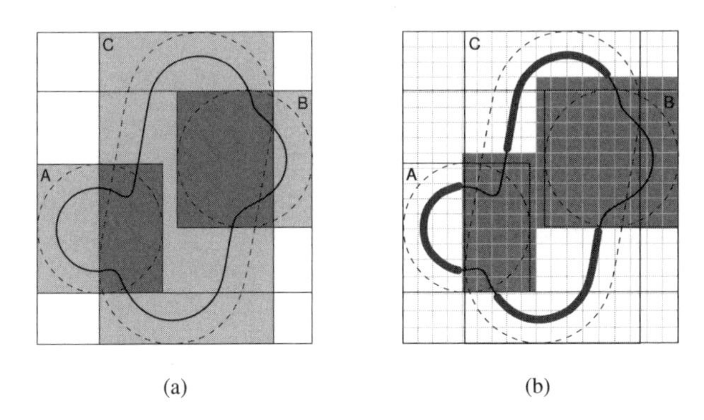

Fig. 9. Our primitive-based optimized local meshing approach

Let $\{P_j\}, j = 1..n$ denote the set of primitives involved in a tree, with bounding boxes $\mathcal{B}_j$. We define the overlapping ratio ρ_i for a primitive P_i as follows :

$$\rho_i = \frac{V_O(\mathcal{B}_i)}{V(\mathcal{B}_i)}$$

where $V(\mathcal{B}_i)$ denotes the volume of $\mathcal{B}_i$ and V_O is the amount of volume of $\mathcal{B}_i$ that is shared with the bounding boxes of the other primitives, i.e.:

$$V_O(\mathcal{B}_i) = V\left(\bigcup_{j\neq i} \mathcal{B}_i \cap \mathcal{B}_j\right)$$

where $V\left(\cup_{k=1}^{n} \mathcal{B}_k\right)$ is computed using the inclusion-exclusion formula:

$$V\left(\bigcup_{k=1}^{n} \mathcal{B}_k\right) = \sum_{l=1}^{n}(-1)^{l+1} \sum_{m_1<m_2<\ldots<m_l} V\left(\mathcal{B}_{m_1} \cap \mathcal{B}_{m_2} \cap \ldots \cap \mathcal{B}_{m_l}\right)$$

Let $\mathcal{G}$ denote a regular grid in which the HybridTree is embedded. All the grid cells are first initialized to "0". We proceed as follows for all primitives P_i:

1. Compute the set $\{\mathcal{B}_i \cap \mathcal{B}_j\}$, $j \neq i$, $\mathcal{B}_i \cap \mathcal{B}_j \neq \emptyset$, such that the first common ancestor A of P_i and P_j is a blending node and P_j is not a right descendent of a difference operation located below A.

2. Align $\mathcal{B}_i$ and every box $\mathcal{B}_i \cap \mathcal{B}_j$, $j \neq i$ on the grid $\mathcal{G}$.
3. Compute ρ_i.
4. If $\rho_i < \rho_0$ then:
   (a) Create the mesh $\mathcal{M}_i$ of P_i.
   (b) Remove the triangles of $\mathcal{M}_i$ that have at least one vertex $\mathbf{p}_i$ such that $\mathbf{p}_i \in \cup_{j \neq i} \mathcal{B}_j$.
   (c) Mark "1" all cells of $\mathcal{G}$ that lie in $\cup_{j \neq i} \mathcal{B}_j$.
5. Else, mark "1" all cells of $\mathcal{G}$ that lie in $\mathcal{B}_i$.

Our modified Marching Cubes algorithm then computes sample points along the edges of "1" cells and triangulates these cells.

Using this method, more implicit primitive mesh parts are obtained by direct meshing and more existing mesh parts can be preserved. In Figure 9(b), the regions that are polygonized using the direct meshing strategy are depicted with a bold contour. Figure 10 shows mesh outputs of the restored Igea model from Figure 14 for three different meshing approaches. In this model, several implicit primitives are locally blended with the reconstructed mesh model of the Igea point set, yielding a configuration that is similar to the one in Figure 8. The polygonization results were obtained by applying the global Marching Cubes algorithm (top-left), then our node-based algorithm (top-center), and finally our primitive-based algorithm (bottom-left). Computational timings are reported in Table 1, that also shows the number of new sample points computed on the surface through the Marching Cubes technique. For this model, our primitive-based algorithm is two times faster than our node-based algorithm.

In general, the local results of the Marching Cubes algorithm is overly dense with respect to the local geometry. This may be improved using some mesh simplification technique. Another interesting approach is to carry out the whole meshing process, not only for input point sets, in the dynamic surface reconstruction framework by Allègre et al. [14,15,16]. Starting from a reconstructed surface, such as the Igea head model, together with the set of sample points generated by the Marching Cubes algorithm in blending regions, the reconstruction local update capability of this framework makes its possible to directly obtain a mesh whose sampling density is adapted to the local geometry, which saves some connectivity computations and avoids a crack-fixing step. This technique for local meshing can be also applied to polygonal meshes as well as to meshed implicit primives in input of the HybridTree, provided the facets are embedded in the Delaunay triangulation of their vertices. The bottom-right image in Figure 8 shows a remeshed version of the result of the primitive-based algorithm obtained through this dynamic reconstruction approach (note that a slightly finer sampling was chosen for the initial reconstruction of the Igea model in this picture). Local reconstruction update with on-the-fly subsampling can be performed at a rate of 1500 points per second, which means an interactive rate in most cases. A complete integration of the HybridTree framework with the dynamic surface reconstruction framework is currently under development.

The computation of the intersection boxes between every pair of primitives is achieved in time $O(n^2)$, where n denotes the number of primitives. For a given

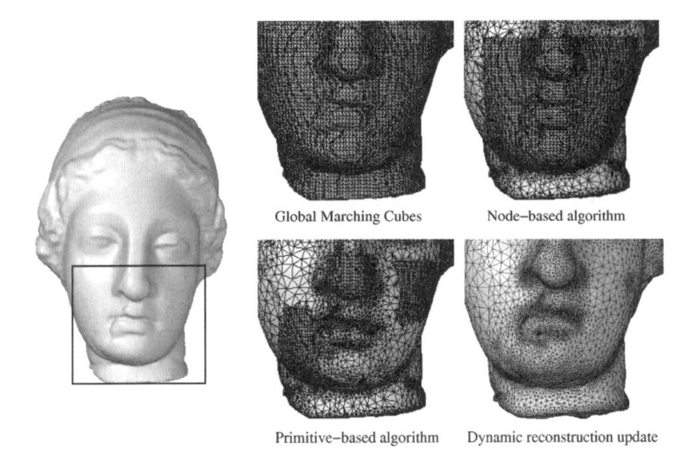

Fig. 10. Several polygonization results for the restored Igea model from Figure 14

Table 1. Polygonization timings (in seconds), number of new sample points computed by the Marching Cubes technique and number of triangles for global Marching Cubes, node-based and primitive-based meshing (computations performed on a Pentium IV 3.0GHz - 1GB workstation)

Method	Polyg. time	#Sample points M.C.	#Triangles
Global M.C.	1,533.02	54,420	108,836
Node-based	19.82	8,299	61,980
Primitive-based	9.67	2,592	58,458

primitive, the overlapping ratio is evaluated in time $O(n^2)$ in the worst case. As a consequence, the performance of our primitive-based algorithm may decline over a set of primitives that are all tightly blended together. However, in practice, the number of primitives that effectively contribute to the final potential field at a given point in space is generally small compared to the overall number of primitives involved in a particular model.

Our primitive-based method also requires to store a grid of size m^3 with only 1 bit per cell. For a grid with 300^3 cells, which was the maximum in our tests, this represents less than 3.5 megabytes of main memory. The time for traversing the set of cells is negligible against the polygonization process. If more precision is needed or if memory is a critical resource, then the node-based approach may be more profitable, or an adaptive grid could be used.

Negative Blending. Suppose that A generates a positive potential field and B a negative one. In this case, our algorithm proceeds as follows:

1. Create the mesh $\mathcal{M}_A$ of A.
2. Remove the triangles of $\mathcal{M}_A$ with at least one vertex $\mathbf{p}_i$ such that $\mathbf{p}_i \in \mathcal{B}_B$.

3. Apply the Marching Cubes algorithm in $\mathcal{B}_B$ and invoke the crack fixing algorithm to close the gap.

Local Blending. Here we suppose that A and B generate positive potential fields, which is not required for C. The polygonization local blending nodes is achieved as follows:

1. Create the meshes $\mathcal{M}_A$ and $\mathcal{M}_B$ of A and B respectively.
2. Compute the mesh $\mathcal{M}_D$ of the union D between A and B.
3. Remove the triangles of $\mathcal{M}_D$ that have at least one vertex $\mathbf{p}_i$ such that $\mathbf{p}_i \in \mathcal{B}_A \cap \mathcal{B}_C$ or $\mathbf{p}_i \in \mathcal{B}_B \cap \mathcal{B}_C$.
4. Apply the Marching Cubes algorithm in $\mathcal{B}_R$ and invoke a crack fixing algorithm to close the gap.

5.3 Boolean Operations

Computing the mesh resulting from Boolean operations is achieved as performed by standard B-Rep modelers. Our approach takes advantage of the dual implicit/mesh representation of the HybridTree. We rely on the implicit representation of the child nodes to perform point membership classification efficiently. The algorithm may be written as follows for any of the union, intersection or difference operations:

1. Create the meshes $\mathcal{M}_A$ and $\mathcal{M}_B$ of A and B respectively.
2. If $\mathcal{B}_A$ and $\mathcal{B}_B$ overlap, then compute the resulting mesh surface using the point membership function of A and B for point membership classification.

To determine whether two triangles overlap and clip them properly, we use the fast and robust triangle-triangle overlap test proposed by Guigue and Devillers [70].

5.4 Warping Operations

We first create the mesh $\mathcal{M}_A$ of the child node A. Then the deformation is applied to the mesh $\mathcal{M}_A$ by simply changing the coordinates of the vertices of the mesh $\mathbf{p}_i$ into $w(\mathbf{p}_i)$ so as to obtain the deformed mesh. Translation, rotation and uniform scaling preserve the aspect ratio of the triangles, whereas non uniform scaling or twisting, tapering and bending may stretch the triangles into flat triangles. In those cases, we apply a simple local remeshing process based on edge collapse and vertex insertion to get better shaped triangles. Another interesting approach could be to exploit the dynamic surface reconstruction framework as for blending operators.

6 Results and Discussion

In this section, we present some complex models created by combining and deforming skeletal implicit models built from hundreds of implicit primitives, and meshes and point sets with tens of thousands of elements. Table 2 reports the timings for polygonizing the final models (in seconds), as well as the overall number of triangles. The given preprocessing timings take into account the time taken to build the bounding box hierarchy for mesh models and the initialization of the Marching Cubes grid when the second local meshing algorithm is used for blending nodes. These timings do not include the time needed to reconstruct a mesh model from an input point set when involved in our hybrid models. For the Igea point set that we used, the time taken to produce a triangle mesh was 29 seconds with the latest implementation of the dynamic geometric convection framework. Measures were performed on a Pentium IV 3.0GHz - 1GB RAM workstation.

Table 2. Preprocessing and polygonization timings (in seconds) and number of triangles for polygonizing several complex hybrid models

Figure	Preproc. time	Polyg. time	#Triangles
1	4.63	63.85	171,562
6	1.72	48.16	105,467
11	0	14.02	121,271
12	1.97	21.24	94,862
13	6.85	56.35	269,698
14	1.58	9.12	58,458

6.1 Free-Form Modeling

The Winged Snake-Woman. Figure 1 shows blending and Boolean operations applied to implicit and mesh input models. The original snake-woman (Figure 1(a)) is an implicit model built from 250 spline implicit primitives blended together, which is stored in our own library of models. The body has been first blended with a mesh of the Igea model (62,323 triangles) that was automatically reconstructed from a dense point set, and with the wings of the Victory of Samothrace (16,340 triangles). The mesh creation process first invokes the polygonization of the implicit snake-woman model. The Marching Cubes algorithm is used as all implicit primitives are overlapping much. The resulting mesh consists of 121,524 triangles, and took 6 seconds to generate. The head has been removed using Boolean difference with an implicit sphere primitive, and the body has been blended with the Igea model using our local meshing method. The wings were extracted from the Victory of Samothrace mesh model by intersecting the original model with an implicit box. The wings and the modified snake-woman model have finally been blended together using the local meshing method.

The bowl. The bowl in Figure 11 has been created using blending operations. The interior of the Igea model has been carved using a negative potential field generated by a cylinder implicit primitive. Handles built from two implicit circle primitives have then been added using our local blending operation.

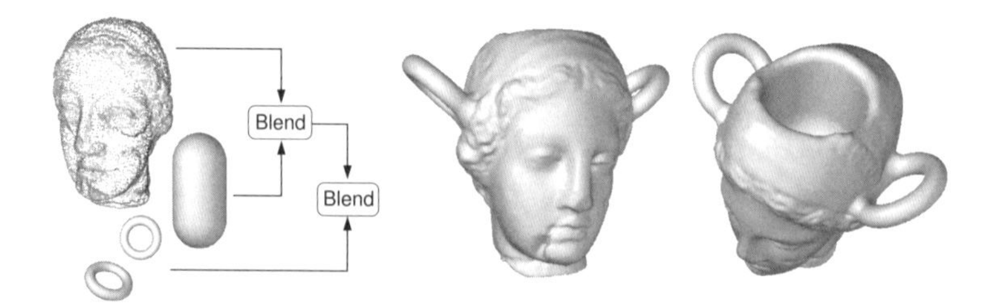

Fig. 11. A bowl created from the Igea model

The bottle. The bottle in Figure 12 has been create from the implicit bottle model of Figure 3 that incorporates 5 complex skeletal implicit primitives. We first applied our Free-Form Deformation tool, which necessitates the polygonization of the bottle model. Additionally, 12 holes have been created using Boolean differences with implicit spheres.

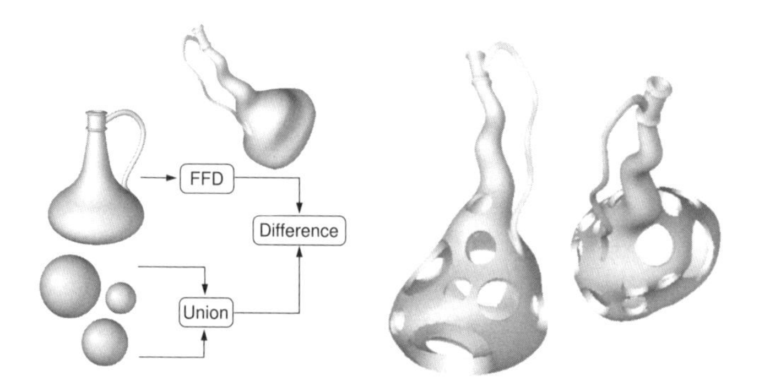

Fig. 12. Bottle with holes

The Victory Figure 13 shows a statue model based on the Victory of Samothrace mesh model (187,072 triangles), that has no head and no arms. We picked up the arms of the original snake-woman implicit model, that consist of 18 implicit spline primitives each, and we have blended them locally with the Victory of Samothrace mesh model using implicit spheres located at each shoulder. We have blended the resulting surface locally with the Igea head using an implicit cylinder placed around the neck. We have finally completed our custom Victory model by adding a shepherd's crook in the right hand.

6.2 Virtual Restoration of Artwork

Our model is well-suited for modeling complex shapes either from existing models or from scratch. It could also be advantagely used for the purpose of digital

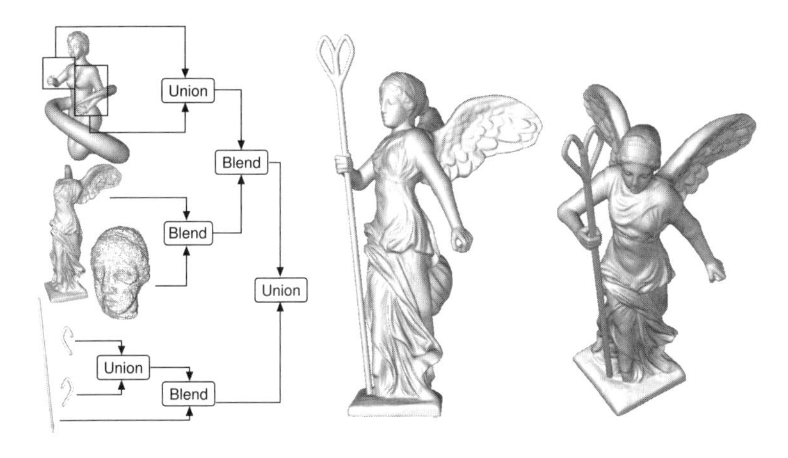

Fig. 13. A Victory model

preservation of cultural heritage artwork, which has become a very challenging research domain. Our HybridTree structure can be efficiently used to simulate restoration or natural phenomena [71] effects on digitized pieces of artwork and it naturally maintains the history of every operation, which is useful for archiving purposes.

Figure 14 shows a virtual restoration process on the Igea model using the HybridTree. We were interested in filling in the ridges on the right of the chin and on the left cheek, and restoring the nose, exactly as a specialist could do. We used our blending tools to simulate cementing in a very intuitive and realistic way. We have placed implicit spline primitives along each ridge and one implicit point primitive at the tip of the nose, and we have blended them with an Igea mesh model. We have built an independent subtree for the set of primitives of the chin and the another for the nose. The former has been polygonized using the Marching Cubes algorithm, as the primitives overlap much. Then, the resulting mesh has been blended with the Igea model using the local method. The same approach has been used for the nose.

6.3 Discussion

Performance Our system can handle complex implicit primitives and polygonal meshes of up to 25,000 triangles at interactive rates. Free-form deformations as well as local blending may be performed interactively for not too fine resolutions. Boolean operations combining small objects compared to the overall size of the final surface may also be performed at interactive rates.

The conversion step between triangles meshes and implicit surfaces is a critical limiting factor regarding computational performance. The computation of the potential field function generated by a mesh at a given point in space remains computationally expensive, despite our acceleration technique. Experiments demonstrate that a field function query performed on a complex mesh

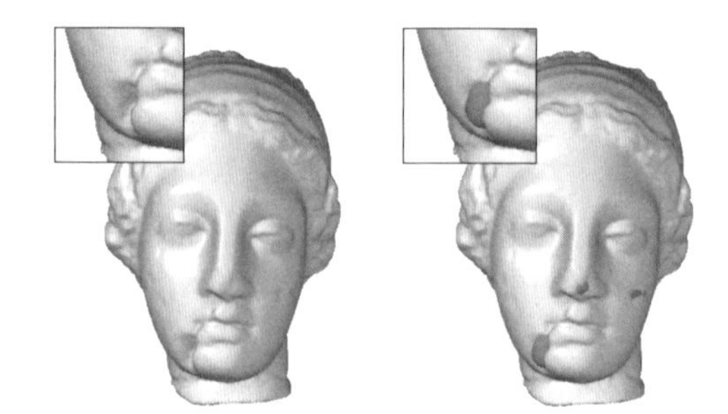
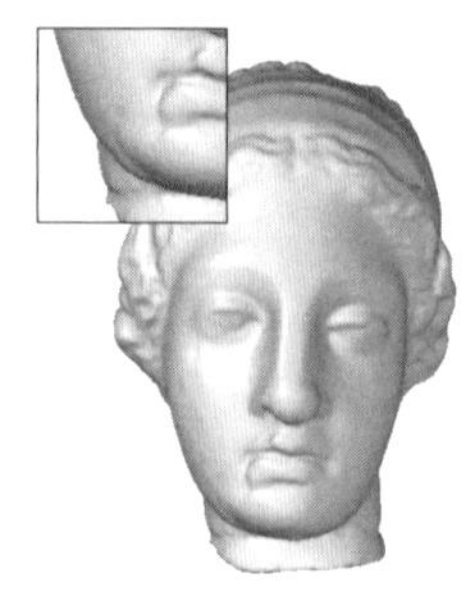

Fig. 14. Virtual restoration stages of the Igea Greek artifact. The initial reconstructed Igea mesh is on the left. In the center image, material has been added to fill in some cavities. On the right is shown the partially restored model.

can have a cost in time that is up to several hundreds times the cost of the same evaluation performed on a point primitive.

For interactive shape design or animation, the evaluation could be accelerated by sampling the potential field on a regular grid and caching computed field values [72]. An approximation of the surface can be retrieved by tricubic interpolation. This approach involves an increased cost in memory and possible loss of geometric and topological information.

Storage. The HybridTree data-structure significantly reduces the amount of memory needed for storing complex models. Contrary to Level Set [37] or Adaptive Distance Fields models [38], we do not store any voxel grid or octree, which saves memory. The use of complex implicit skeletal primitives enables us to design complex shapes with a very compact representation. The snake-woman model represented in Figure 1 was created by blending a few hundred spline skeletal primitives together. The corresponding HybridTree representation takes less than 64 kilobytes in memory.

Shape control. The ability to combine mesh models and skeletal implicit surfaces in a coherent framework not only extends the range of models that can be created but also permits us to have a tight control when editing our models.

The implicit surface representation enables blending of meshes of arbitrary topology and geometry. This compares favorably with other specific mesh blending methods such as [53] or [41] that impose some geometric or topological restrictions. Moreover, our local blending technique provides fine control on the way shapes blend together. The designer may simply tune the radius of influence for mesh or implicit primitives so as to control the geometry of the blend with other objects. The implicit representation also provides means of creating negative blending between shapes, which is useful for simulating carvings. Eventually, our Free-Form Deformation tool enables intuitive, non restrictive local deformations on hybrid models.

7 Conclusion and Future Work

In this paper, we have presented a hybrid constructive shape representation. Our model combines skeletal implicit surfaces, triangle meshes and point set models in a coherent framework. The HybridTree's evaluation system is designed to exploit the complementary advantages of these geometric models. The core of our current system is based on a dual skeletal implicit/triangle mesh representation. Editing operations are performed in the most suitable representation in a totally transparent way for the user. The mesh representation is useful for fast visualization and free-form deformations, and the implicit one lend themselves for Boolean, and local and global blending. The HybridTree is evaluated through field function, gradient, point membership classification and polygonization queries that are optimized for every kind of node.

The HybridTree should be considered as an open model, capable of integrating many other representations through gateways with available surface models. The most well-suited representation for an operation could then be chosen according to finer criteria, such as the representation that provides the fastest result or the best quality possible result. Other criteria could be progressivity or point-of-view in a scene, in order to avoid some useless computations.

The HybridTree model still deserves many improvement to develop its potential and bring it to maturity. Our implementation currently has only a declarative interface. It would be interesting to develop an interactive modeling interface. That would yield new research issues, such as bringing unicity in the representation of the geometric information and achieving reversibility in the evaluation process. This also arises the question of animating HybridTree models. The current representation can already support keyframing animation, provided the same tree structure is maintained all along an animation. Our local meshing strategy can be a benefit for interactive visualization of animations, and could be improved for this particular purpose, in order to discard useless computations. If the structure is likely to change, e.g. for metamorphosis, then a new evaluation strategy would be required, taking matching problems into account [7].

We will investigate the automatic management of levels of detail in the HybridTree. We think that it should be possible to combine skeletal implicit primitives with levels of detail as presented in [73] with multiresolution meshes and subdivision surfaces. Moreover, levels of detail in the HybridTree could be exploited to improve the locality of conversions, e.g. with blending operations, by making it possible to determine the affected regions more precisely than when just using bounding boxes. In the near future, we also plan to extend the integration of the point set representation into the HybridTree. This representation, that avoids the management of connectivity relations, could be interesting for interactive visualization [74].

The evaluation system of the HybridTree could benefit from an integration with the dynamic surface reconstruction framework developed by Allègre et al. in [14,15,16] for every meshing step between two subtrees. This could make it possible to directly produce meshes of hybrid models in a seamless fashion, and with sampling density adapted the local geometry. Another appealing perspective of this integration would be to develop tools for locally repairing defective reconstructed surfaces from point sets with missing or noisy data using an intermediate implicit representation [75,76].

References

1. Bajaj, C., Blinn, J., Bloomenthal, J., Cani-Gascuel, M.P., Rockwood, A., Wyvill, B., Wyvill, G.: Introduction to Implicit Surfaces. Morgan Kaufmann, San Francisco (1997)
2. Velho, L., Gomes, J., Figueiredo, L.H.: Implicit Objects in Computer Graphics. Springer, New York (2002)
3. Blinn, J.F.: A generalization of algebraic surface drawing. ACM Transactions on Graphics 1(3), 235–256 (1982)
4. Crespin, B., Blanc, C., Schlick, C.: Implicit Sweep Objects. In: Proc. Eurographics, vol. 15, pp. 165–174 (1996)
5. Pasko, A., Adzhizev, V., Sourin, A., Savchenko, V.: Function Representation in Geometric Modeling: Concepts, Implementation and Applications. The Visual Computer 11(8), 429–446 (1995)
6. Wyvill, B., Galin, E., Guy, A.: Extending The CSG Tree. Warping, Blending and Boolean Operations in an Implicit Surface Modeling System. Computer Graphics Forum 18(2), 149–158 (1999)
7. Barbier, A., Galin, E., Akkouche, S.: Controlled Metamorphosis of Animated Objects. In: Proc. Shape Modeling International, pp. 184–196 (2003)
8. Sederberg, T.W., Parry, S.R.: Free-Form Deformation of Solid Geometric Models. In: Proc. SIGGRAPH, pp. 151–160 (1986)
9. Sorkine, O., Cohen-Or, D., Lipman, Y., Alexa, M., Rössl, C., Seidel, H.-P.: Laplacian Surface Editing. In: Proc. Eurographics/ACM SIGGRAPH Symposium on Geometry Processing, pp. 179–188 (2004)
10. Pauly, M., Keiser, R., Kobbelt, L.P., Gross, M.: Shape Modeling with Point-Sampled Geometry. ACM Transactions on Graphics 22(3), 641–650 (2003)
11. Cazals, F., Giesen, J.: Delaunay Triangulation based Surface Reconstruction: Ideas and Algorithms. In: Boissonnat, J.D., Teillaud, M. (eds.) Effective Computational Geometry of Curves and Surfaces (2006)
12. Allègre, R., Barbier, A., Galin, E., Akkouche, S.: A hybrid shape representation for free-form modeling. In: Proc. Shape Modeling International, pp. 7–18 (2004)
13. Allègre, R., Galin, E., Chaine, R., Akkouche, S.: The hybridtree: Mixing skeletal implicit surfaces, triangle meshes and point sets in a free-form modeling system. Graphical Models 68(1), 42–64 (2006) (SMI 2004 special issue)
14. Allègre, R., Chaine, R., Akkouche, S.: Convection-Driven Dynamic Surface Reconstruction. In: Proc. Shape Modeling International, pp. 33–42. IEEE Computer Society Press, Los Alamitos (2005)
15. Allègre, R., Chaine, R., Akkouche, S.: A Dynamic Surface Reconstruction Framework for Large Unstructured Point Sets. In: Proc. IEEE/Eurographics Symposium on Point-Based Graphics, pp. 17–26 (2006)
16. Allègre, R., Chaine, R., Akkouche, S.: A flexible framework for surface reconstruction from large point sets. Computers & Graphics 31(2), 190–204 (2007)
17. Lorensen, W.E., Cline, H.E.: Marching Cubes: A high Resolution 3D surface reconstruction algorithm. Computer Graphics (Proc. SIGGRAPH) 21(4), 163–169 (1987)

18. Kobbelt, L.P., Botsch, M., Schwanecke, U., Seidel, H.P.: Feature Sensitive Surface Extraction from Volume Data. In: Proc. SIGGRAPH, pp. 57–66 (2001)
19. Ju, T., Losasso, F., Schaefer, S., Warren, J.: Dual contouring of hermite data. ACM Transactions on Graphics (Proc. SIGGRAPH) 21(3), 339–346 (2002)
20. Ho, C.C., Wu, F.C., Chen, B.Y., Chuang, Y.Y., Ouhyoung, M.: Cubical marching squares: Adaptive feature preserving surface extraction from volume data. Computer Graphics Forum (Proc. Eurographics) 24(3), 537–545 (2005)
21. Witkin, A.P., Heckbert, P.S.: Using particles to sample and control implicit surfaces. Computer Graphics 28(2), 269–277 (1994)
22. Hilton, A., Stoddart, A.J., Illingworth, J., Windeatt, T.: Marching Triangles: Range image fusion for complex object modelling. In: IEEE International Conference on Image Processing, pp. 381–384 (1996)
23. Akkouche, S., Galin, E.: Adaptive Implicit Surface Polygonization using Marching Triangles. Computer Graphics Forum 20(2), 67–80 (2001)
24. Boissonnat, J.D., Oudot, S.: Provably Good Surface Sampling and Approximation. In: Proc. Symposium on Geometry Processing, pp. 9–18 (2003)
25. Boissonnat, J.D., Cohen-Steiner, D., Vegter, G.: Isotopic Implicit Surface Meshing. In: Proc. ACM Symposium on Theory of Computing, pp. 301–309 (2004)
26. Guéziec, A.: Meshsweeper: Dynamic Point-to-Polygonal-Mesh Distance and Applications. IEEE Transactions on Visualization and Computer Graphics 7(1), 47–61 (2001)
27. Turk, G., O'Brien, J.F.: Modelling with Implicit Surfaces that Interpolate. ACM Transactions on Graphics 21(4), 855–873 (2002)
28. Shen, C., O'Brien, J.F., Shewchuk, J.R.: Interpolating and Approximating Implicit Surfaces from Polygon Soup. ACM Transactions on Graphics 23(3), 896–904 (2004)
29. Hoppe, H., DeRose, T., Duchamp, T., McDonald, J., Stuetzle, W.: Surface Reconstruction from Unorganized Points. Computer Graphics (Proc. SIGGRAPH) 26(2), 71–78 (1992)
30. Carr, J.C., Beatson, R.K., Cherrie, J.B., Mitchell, T.J., Fright, W.R., McCallum, B.C., Evans, T.R.: Reconstruction and Representation of 3D Objects with Radial Basis Functions. In: Proc. SIGGRAPH, pp. 67–76 (2001)
31. Ohtake, Y., Belyaev, A., Alexa, M., Turk, G., Seidel, H.P.: Multi-level Partition of Unity Implicits. ACM Transactions on Graphics 22(3), 463–470 (2003)
32. Kazhdan, M.: Reconstruction of solid models from oriented point sets. In: Proc. Symposium on Geometry Processing, pp. 73–82 (2005)
33. Hornung, A., Kobbelt, L.: Robust reconstruction of watertight 3d models from non-uniformly sampled point clouds without normal information. In: Proc. Symposium on Geometry Processing 2006 (to appear)
34. Ohtake, Y., Belyaev, A.G., Seidel, H.P.: An integrating approach to meshing scattered point data. In: Proc. Symposium on Solid and Physical Modeling, pp. 61–69 (2005)
35. Sharf, A., Lewiner, T., Shamir, A., Kobbelt, L., Cohen-Or, D.: Competing fronts for coarse-to-fine surface reconstruction. In: Proc. Eurographics 2006 (to appear)
36. Boubekeur, T., Heidrich, W., Granier, X., Schlick, C.: Volume-surface trees. In: Proc. Eurographics, pp. 399–406 (2006)
37. Museth, K., Breen, D.E., Whitacker, R.T., Barr, A.H.: Level Set Surface Editing Operators. ACM Transactions on Graphics 21(3), 330–338 (2002)
38. Frisken, S.F., Perry, R.N., Rockwood, A.P., Jones, T.R.: Adaptively Sampled Distance Fields: A General Representation of Shape for Computer Graphics. In: Proc. SIGGRAPH, pp. 249–254 (2000)

39. Schmitt, B., Pasko, A., Schlick, C.: Shape-Driven Deformations of Functionally Defined Heterogeneous Volumetric Objects. In: Proc. ACM Graphite 2003, pp. 127–134 (2003)
40. Yu, Y., Zhou, K., Xu, D., Shi, X., Bao, H., Guo, B., Shum, H.Y.: Mesh Editing with Poisson-based Gradient Field Manipulation. ACM Transactions on Graphics (Proc. SIGGRAPH) 23(3), 644–651 (2004)
41. Kanai, T., Suzuki, H., Mintani, J., Kimura, F.: Interactive Mesh Fusion Based on Local 3D Metamorphosis. In: Proc. Graphics Interface 1999, pp. 148–156 (1999)
42. Funkhouser, T., Kazhdan, M., Shilane, P., Min, P., Kiefer, W., Tal, A., Rusinkiewicz, S., Dobkin, D.: Modeling by Example. ACM Transactions on Graphics (Proc. SIGGRAPH) 23(3), 652–663 (2004)
43. Alexa, M.: Recent Advances in Mesh Morphing. Computer Graphics Forum 21(2), 173–196 (2002)
44. Alexa, M.: Differential Coordinates for Local Mesh Morphing and Deformation. The Visual Computer 19(2–3), 105–114 (2003)
45. Xu, D., Zhang, H., Wang, Q., Bao, H.: Poisson Shape Interpolation. In: Proc. ACM Symposium on Solid and Physical Modeling, pp. 267–274 (2005)
46. Kraevoy, V., Sheffer, A.: Cross-Parameterization and Compatible Remeshing of 3D Models. ACM Transactions on Graphics (Proc. SIGGRAPH) 23(3), 861–869 (2004)
47. Schreiner, J., Asirvatham, A., Praun, E., Hoppe, H.: Inter-Surface Mapping. ACM Transactions on Graphics (Proc. SIGGRAPH) 23(3), 870–877 (2004)
48. Kobbelt, L.P., Botsch, M.: Freeform Shape Representations for Efficient Geometry Processing. In: Proc. Shape Modeling International, pp. 111–115 (2003)
49. Singh, K., Parent, R.: Implicit Surface Based Deformations of Polyhedral Objects. In: Proc. Implicit Surfaces (1995)
50. Crespin, B.: Implicit Free-Form Deformations. In: Proc. Implicit Surfaces, pp. 17–23 (1999)
51. Decaudin, P.: Geometric Deformation by Merging a 3D-Object with a Simple Shape. In: Proc. Graphics Interface, pp. 55–60 (1996)
52. Decaudin, P., Gagalowicz, A.: Fusion of 3D Shapes. In: Proc. Computer Animation and Simulation, pp. 1–14 (1994)
53. Singh, K., Parent, R.: Joining Polyhedral Objects using Implicitly Defined Surfaces. The Visual Computer 17(7), 415–428 (2001)
54. Levin, D.: Mesh-Independent Surface Interpolation. Geometric Modeling for Scientific Visualization (2003)
55. Mueller, M., Keiser, R., Nealen, A., Pauly, M., Gross, M., Alexa, M.: Point-Based Animation of Elastic, Plastic, and Melting Objects. In: Proc. ACM Siggraph/Eurographics Symposium on Computer Animation, pp. 141–151 (2004)
56. Pauly, M., Mitra, N., Guibas, L.: Uncertainty and Variability in Point Cloud Surface Data. In: Proc. Symposium on Point-Based Graphics, pp. 77–84 (2004)
57. Kolluri, R.: Provably Good Moving Least Squares. In: Proc. ACM-SIAM Symposium on Discrete Algorithms, pp. 1008–1018 (2005)
58. Reuter, P., Tobor, I., Schlick, C., Dedieu, S.: Point-based Modelling and Rendering using Radial Basis Functions. In: Proc. ACM Graphite 2003, pp. 111–118 (2003)
59. Barbier, A., Galin, E., Akkouche, S.: Complex Skeletal Implicit Surfaces with Levels of Detail. Journal of WSCG 12(1), 35–42 (2004)
60. Barr, A.H.: Global and Local Deformations of Solid Primitives. Proc. SIGGRAPH 18(3), 21–30 (1984)
61. Schneider, P., Eberly, D.H.: Geometric Tools for Computer Graphics. Morgan Kaufman Series in Computer Graphics and Geometric Modeling (2002)

62. Barbier, A., Galin, E.: Fast distance computation between a point and cylinders, cones, line swept spheres and cone-spheres. Journal of Graphics Tools 9(2), 31–39 (2004)
63. Johnson, D., Cohen, E.: A Framework for Efficient Minimum Distance Computation. In: Proc. Conf. Robotics and Automation, pp. 3678–3683 (1998)
64. Cani, M.P., Desbrun, M.: Animation of Deformable Models using Implicit Surfaces. IEEE Transactions on Visualization and Computer Graphics 3(1), 39–50 (1997)
65. Pasko, G., Pasko, A., Ikeda, M., Kunii, T.: Bounded Blending Operations. In: Proc. Shape Modeling International, pp. 95–104 (2002)
66. Coquillart, S.: Extended Free-Form Deformation: A Sculpturing Tool for 3D Geometric Modeling. In: Proc. SIGGRAPH, pp. 187–196 (1990)
67. Borrel, P., Bechmann, D.: Deformation of n-dimensional objects. In: Proc. Solid Modeling and Applications, pp. 351–369 (1991)
68. MacCracken, R., Joy, K.I.: Free-form Deformations with Lattices of Arbitrary Topology. In: Proc. SIGGRAPH, pp. 181–188 (1996)
69. Fox, M., Galbraith, C., Wyvill, B.: Efficient Implementation of the Blobtree for Rendering Purposes. In: Proc. Shape Modeling International, pp. 306–314 (2001)
70. Guigue, P., Devillers, O.: Fast and Robust Triangle-Triangle Overlap Test Using Orientation Predicates. Journal of Graphics Tools 8(1), 25–32 (2003)
71. Martinet, A., Galin, E., Desbenoit, B., Akkouche, S.: Procedural Modeling of Cracks and Fractures. In: Proc. Shape Modeling International, pp. 346–349 (2004)
72. Schmidt, R., wyvill, B., Galin, E.: Interactive Implicit Modeling With Hierarchical Spatial Caching. In: Proc. Shape Modeling International, pp. 104–113 (2005)
73. Angelidis, A., Cani, M.P.: Adaptive Implicit Modeling using Subdivision Curves and Surfaces as Skeletons. In: Proc. Solid Modeling and Applications, pp. 45–52 (2002)
74. Galin, E., Allègre, R., Akkouche, S.: A fast particle system framework for interactive implicit modeling. In: Proc. Shape Modeling International, pp. 215–221 (2006)
75. Davis, J., Marschner, S.R., Garr, M., Levoy, M.: Filling holes in complex surfaces using volumetric diffusion. In: First International Symposium on 3D Data Processing, Visualization, and Transmission, June 19-21 (2002)
76. Weyrich, T., Pauly, M., Keiser, R., Heinzle, S., Scandella, S., Gross, M.: Post-processing of scanned 3D surface data. In: Symposium on Point-Based Graphics (2004)

Modelling Function-Based Mixed-Dimensional Objects with Attributes

Benjamin Schmitt[1], Alexander Pasko[2], Valery Adzhiev[2], Galina Pasko[3], and Christophe Schlick[4]

[1] Digital Media Professionals, Hosei Research Institute Tokyo, Japan
hfschmitt@gmail.com
[2] Bournemouth University, United Kingdom
(apasko,vadzhiev)@bournemouth.ac.uk
[3] European University of Lefke, Turkish Republic of Northern Cyprus
gip@tokyo.com
[4] LaBRI, Laboratoire Bordelais de Recherches Informatiques Talence, France
schlick@labri.u-bordeaux.fr

Abstract. The implicit complex model allows for the representation of heterogeneous objects as multidimensional point sets with multiple attributes, where elements of different dimensions and with attributes of different nature are combined together in a topological structure. In this chapter, we present in detail the underlying function-based model for a certain class of cells to be used among several others in the implicit complex model. Each cell is dimensionally homogeneous and defined by a real vector-function.

We provide a brief survey of different modelling techniques related to point sets with attributes. It spans such different areas as solid modelling, scalar fields, volume models, and material modelling. Then, on the basis of this survey we formulate requirements to a more general model.

In the presented generalizing constructive hypervolume model, point set geometry and attributes are represented independently using real-valued scalar functions with their underlying tree data structures. While 3D and higher dimensional entities have been widely studied, we present function-based definitions of lower dimensional entities, such as surface patches and curve segments, with a corresponding trimming technique.

A high level language supporting modelling function-based cells and attributes is described and illustrative examples are provided.

1 Introduction

The implicit complexes presented in this volume [22] allow for modeling heterogeneous objects as multidimensional point sets with multiple attributes, where elements (cells) of different dimensions and with associated different attributes are combined together in a single topological complex. In this survey, we present in detail the underlying model for a certain class of cells, where each cell is dimensionally homogeneous, has a number of attributes assigned at each point,

A. Pasko, V. Adzhiev, and P. Comninos (Eds.): HOMA, LNCS 4889, pp. 90–117, 2008.

and with both geometry and attributes defined by real functions of point coordinates.

We consider cells as point sets in geometric spaces of arbitrary dimension. A point set is a geometric model of a real or abstract object under consideration. An attribute can be defined as a mathematical model of an object property of arbitrary nature defined at any point of the point set. For example, to model a mechanical part with varying internal material distribution one can introduce a three-dimensional solid as a point set and a real-valued scalar function to represent material density as an attribute. Application areas of such models include fabrication of objects with multiple materials and varying material distribution [26,25]; simulations for the analysis of physical fields distribution over geometric areas [31]; modeling and analysis of geological structures [18]; biological modeling and medical examination [17]; and volume graphics [23].

In general, multidimensional point sets with an arbitrary number of attributes of different mathematical nature (scalar, vector, tensor, etc.) can be introduced in various ways depending on the application. Following [35], we present in Section 2 a brief survey of different modeling techniques related to point sets with attributes. This survey spans such areas as solid modeling, heterogeneous objects modeling, scalar fields or "implicit surface" modeling, and volume graphics.

The function representation (FRep) [36] is used as the basic model for both point set geometry and attributes. With this model briefly described in Section 3, the point set and its attributes are represented independently by real functions. Each function can be associated with a tree structure and is evaluated by a tree traversing procedure. This reflects the constructive nature of the symmetric approach to modeling geometry and the associated attributes. FRep provides a rich system of primitives, operations and relations for modeling both geometry and attributes. On the base of recent works in these areas we describe in Section 4 a constructive hypervolume model using vector functions.

Lower dimensional entities can also be modeled using FRep. For example, the geometric domain of FRep in 3D space includes solids with non-manifold boundaries and lower dimensional entities (surfaces, curves, points) defined by zero value of the function. The lower dimensional objects in 3D space can be defined by real functions as follows:

- definition of a surface patch requires a trimming operation implemented as intersection between an "implicit" surface and a trimming 3D solid;
- a curve can be defined as the intersection of two surfaces;
- a point can be defined as the intersection of three surfaces, a curve and a surface, or directly as $d(x, y, z)$, where d is a negative distance to the given point.

Section 5 of this document presents surface and curve modeling using a trimming operation. Finally, Section 6 is devoted to the implementation of the constructive hypervolume model in the form of a special modeling language and its supporting software tools.

2 Models of Point Sets with Attributes

There are several interrelated directions in the research on modeling point sets with attributes. In this section, we provide a brief survey of these directions and discuss the following aspects of different approaches: model of a point set, point set dimensionality, types of attributes, attribute model, and operations on attributes.

Computer graphics, solid modeling, and volume modeling (based on the binary spatial occupancy enumeration [44]) in their early stages dealt with 3D and higher dimensional homogeneous objects of different dimensions (points, curves, surfaces, solids). Attributes could be assigned only to the entire object, but not to its components. No operations on attributes were provided. The first attempt to represent heterogeneous objects was texture mapping in computer graphics with its essential limitation that attributes can be assigned to the object surface, but not to points throughout its volume. The following models attempt to represent a volumetric distribution of attributes with different levels of generality.

Scalar fields. Real functions of three variables (also called scalar fields) defined for point coordinates in 3D space can be interpreted as defining functions of some isosurfaces ("implicit" surfaces [15]). On the other hand, real functions can define volumetric attributes such as material density to model amorphous and gaseous phenomena [14]. Special operations simulating noise and turbulence are applied to the attributes.

Heterogeneous volumes (discrete fields). As an extension of the homogeneous volume models with binary encoding of voxel occupancy, integer or real scalar values can be given in the nodes of a regular or a non-regular space grid of a heterogeneous volume model (voxel array). This model is close to the scalar fields and can be considered a discrete field. Processing of scalar values given at a discrete set of points requires some approximation procedure [30]. The scalar values can represent the geometry of a point set (e.g., by a density field [55] or a distance field [40,21,20]), object's color, and other attributes. Operations on non-geometric attributes include approximation of scattered data [30], different kinds of filtering, and other application specific operations.

Multi-material solids. The next step towards modeling heterogeneous objects was the introduction of solids composed of multiple materials. A systematic approach to multi-material solid modeling was proposed in [26]. A 3D solid is subdivided into components made of unique materials. A non-manifold Boundary Representation (BRep) is used to model such objects. Each component is homogeneous inside and has an assigned index of material. Regularized set-theoretic operations are applied to the solid components. Corresponding operations on material indices are introduced on the basis of the resulting material selection for each pair of materials and for each set-theoretic operation. A similar approach was adopted in Svlis [8] which is an elaborated Constructive Solid Geometry (CSG) system.

Constructive Volume Geometry. The Constructive Volume Geometry (CVG) [11,12] defines a spatial object as a tuple of scalar fields in 3D space. Special attention is paid to the first field in the tuple, which is an opacity field specifying the visibility of every point in space. This has obvious limitations for defining the objects geometry which has to be independent of any visual characteristics. Other visual (photometric) attributes can be included in the model: color, ambient, diffuse, and specular reflection parameters. Several operations are introduced for the opacity field (union, intersection, difference, blending, etc.) together with corresponding operations for other attributes. Discrete fields can be used in the tuple along with some interpolation procedure.

Hypervolumes. The term hypervolume was introduced in [4] to denote a discrete scalar field embedded in n-dimensional (nD) space. The hypervolume is defined by an nD regular grid with scalar values given at the grid nodes. A 3D volume changing in time is a typical example of a hypervolume in the 4D space. The authors described a projection operation of such a hypervolume to a 2D point set with color attributes used for visualization. A hierarchical representation of nD discrete scalar fields in the form of tree structures was proposed.

Object model. A general object model [25] was designed to include all the characteristics and attributes of an object. Geometry is considered the most fundamental attribute of an object. All other attributes are described as a function of geometry. A 3D point set (so-called r-set in E^3) is represented by its decomposition (atlas) into a finite set of closed 3-cells. The authors proposed to use BRep scheme to model individual cells and the entire point set. Each point of the point set is mapped to its corresponding attribute, which can be a vector or a tensor. The model of attributes is a collection of functions mapping the object geometry to several attributes. This is a generalization of the multi-material solid model [26] discussed above. Basic operations on attributes include vector sum and product with scalar, union, intersection and complement specialized for specific attributes as abstraction of the material combining operation in [26].

Continuous attributes modeling. Recently, a particular attention has been paid to heterogeneous object modeling, where an object has a number of non-uniformly distributed attributes assigned at each point and varying in space. These attributes may be continuous or piecewise continuous and are of different nature such as material density, stress or other physical fields distribution.

In the work [25] mentioned above, a more general model is proposed where the attributes are defined by a collection of functions, which map the object geometry to several attributes. Such a mathematical model is known as a fiber bundle, with the geometrical model playing the role of the base space. Several other works are using the same model, extending it in various directions [5,10]. However, as noticed in [6], such a model does not really offer concrete computational solutions.

Discrete volumetric representations using voxel arrays or scattered data points can be extended to support continuous attributes with some special approximation procedures [31]. The inherent drawback of these models is the difficulty to directly describe the material distribution without using data acquisition devices (therefore it is supposed that the object to be modeled already exists).

A continuous volumetric representation was proposed in [43], where a B-spline volume is used to model the object geometry, whereas the attributes are modeled by means of diffusion. This model seems to suffer from the lack of flexibility of the geometry model limited to volume splines.

Biswas et al. [6] are interested in the representation and control of material distributions using some intuitive parameters related to the geometry of the solid and/or its material features for mesh-free modeling. They propose to use the distance functions from material features (point-sets of any dimension with known material properties) as these parameters. It appears from the existing literature that the (Euclidean) distance, or functions of the distance are the most common types of material functions considered in the area of the functionally graded material modeling [19,53,28,1]. The authors of [6] also prove that this approach is theoretically complete as it can represent all material functions.

Discussion. Historically, separate treatment of geometry and attributes was introduced in computer graphics for rendering textured surfaces. Voxel arrays in volume graphics can be considered as attribute models with the default geometry represented by a bounding box. The next step of models development was to combine geometric and attribute representations in a single model. In solid modeling, this was done for multi-material solids [26] with the material indices assigned to different geometric regions. Then, this approach was generalized in the object model [25] covering arbitrary geometry and multiple attributes of different mathematical types (scalars, vectors, tensors) defined at each point. Only 3D geometry is considered in the object model with the boundary representation being the primary geometric model. The object model does not include voxel arrays or scalar fields for modeling geometry.

In volume modeling, CVG [11,12] was the first model combining geometry and attributes in a systematic manner. The model is presented as an algebra of spatial objects with operations available for both geometry and attributes. The model allows for utilizing both voxel arrays and continuous scalar fields. The use the opacity field in the CVG model to "implicitly define the visible geometry of an object" is somewhat controversial, because in reality, the shape of an object does not necessarily predefine its photometric characteristics and vice versa. We believe it is important that a point set and its visual and physical characteristics are represented independently.

Note that CVG has been originated in volume graphics and is mainly aimed to providing more flexible object and scene definitions in volume rendering. The idea of hypervolumes reflected the importance of modeling and visualization of time-dependent volumetric objects. On the other hand, the object model introduced in the area of solid modeling is oriented towards the mechanical design and rapid prototyping applications. Functionally graded materials modeling and fabrication is one of the active research areas in CAD/CAM [1]. All the above mentioned areas of research exist separately and the motivation of our work is to introduce a universal model that can be suitable for all application areas of heterogeneous objects modeling.

3 Function Representation

As it can be seen from the previous section, scalar fields and constructive operations are useful components in modeling point sets and attributes. In this section, we discuss their further integration in the framework of the function representation. The function representation (FRep) was introduced in [37,36] as a uniform representation of multidimensional geometric objects. FRep is formulated as an algebraic system including sets of objects, operations and relations on them. An object (point set) in multidimensional space is defined by a continuous real-valued function of point coordinates $F(X)$. The points with $F(X) \geq 0$ belong to the object, and the points with $F(X) < 0$ are outside of the object.

The idea of representing the entire object by a single function has been used in modeling implicit surfaces. In this sense, FRep generalizes implicit surface modeling by combining them with the constructive modeling approach. The complex object is defined by starting from simple (primitive) ones and applying a sequence of constructive operations to them. Thus, FRep generalizes Constructive Solid Geometry (CSG) by providing a single function for a complex constructive solid. The geometric domain of FRep in 3D space includes solids with non-manifold boundaries and lower dimensional entities (surfaces, curves, points) defined by zero value of the function. The main distinctive characteristic of FRep is that the real-valued function defining the point set is evaluated at the given point by a procedure traversing a tree structure with primitives in the leaves and operations in the nodes of the tree. This construction tree is the generalization of the one used in CSG.

A primitive can be defined by an equation or by a "black box" procedure converting point coordinates into the function value. Solids bounded by algebraic surfaces, skeleton-based implicit surfaces, and convolution surfaces, as well as procedural objects (such as solid noise), and voxel objects can be used as primitives (leaves of the construction tree). In the case of a voxel object (discrete field), it should be converted to a continuous real function, for example, by applying the trilinear or higher-order interpolation.

Many operations such as set-theoretic, blending, non-linear deformations, metamorphosis, sweeping, hypertexturing, and others, have been formulated for this representation in such a manner that they yield continuous real-valued functions as output [36,48], thus guaranteeing the closure property of the representation. As it was mentioned in the previous section, the application of min/max functions for set-theoretic operations results in C^1 discontinuity of the resulting function. On the other hand, R-functions originally introduced in [46] provide C^k continuity for the functions exactly defining the set-theoretic operations. Because of this property, blending, deformations, metamorphosis and other geometric operations can be formulated using algebraic operations applied to defining functions of complex constructive objects [36,48]. More details on specific primitives and operations can be found in [33].

Relations defined on a set of objects are used to formulate some operations and to appropriately process FRep objects in applications. Such basic relations as point membership and the collision (interpenetration) relation are also important for point sets with attributes as long as they are used not only for visualization but also for modeling purposes.

We can state that FRep satisfies the following requirements to the basic model of a point set with attributes: constructive type of model, usage of continuous and discrete scalar fields, and dimensionality independence. In this survey, we describe applications of FRep for modeling point sets geometry, space partitions for attributes, and lower dimensional trimmed objects. In contrast to the approaches described in the previous section, uniform treatment of objects of different dimensions provides basis for modeling time-dependent and multidimensional point sets with attributes.

4 Constructive Hypervolume Modeling

Based on the survey presented in the previous sections, we can formulate the requirements for a general model of point sets with attributes:

- Independent representation of the point set and its attributes;
- Coverage of time-dependent and other multidimensional point sets;
- Uniform treatment of point set geometry, photometric, physical, and other attributes of an arbitrary nature;
- Constructive modeling of both point set geometry and attributes using primitives, operations, and relations;
- The ability to model geometry and attributes using real-valued functions (scalar fields).

In this section we discuss a general model of constructive hypervolumes introduced in [34,35] to satisfy the requirements listed above. Extending the FRep formal model introduced in [36], let us describe a general hypervolume model as a triple (O, Φ, W), where O is a set of hypervolume objects, Φ is a set of hypervolume operations, and W is a set of relations for the set of objects. Mathematically, the triple can be treated as an algebraic system. Here we give an outline of the formal framework to be further elaborated elsewhere.

4.1 Objects

A hypervolume object can be expressed as a tuple, $o = (G, A_1, \ldots, A_k)$, where G is a multidimensional point set and A_i is an attribute. In 3D, a point set G can be defined using any existing representational schemes for solids: Brep, CSG, spatial partitioning, generative models, ray implementation, and others (see section 2.1). In the multidimensional case, one can apply multidimensional extensions of CSG or Brep [56,16], or originally multidimensional models such as the generative model [54] and the FRep [36]. Here we introduce a specific "FRep" representation of the hypervolume object that can be expressed as:

$$o = (G, A_1, \ldots, A_k) : (F(X), S_1(X), \ldots, S_k(X)) \quad (1)$$

where :

- $X = (x_1, \ldots, x_n)$ is a point in n-dimensional Euclidian space E^n,
- $F : X \rightarrow \Re$ is a real-valued defining function of point coordinates to represent point sets G. Therefore, F is at least a C^0 continuous function, which is positive inside the point set, negative outside, and has a zero value on its boundary.
- $S_i : X \rightarrow \Re$ is a real-valued scalar function representing an attribute A_i that is not necessarily continuous.

4.2 Constructive Tree Data Structure

We call the introduced representation *a constructive hypervolume model* to emphasise the underlying constructive process while modelling functionally based multidimensional point sets with attributes. As it was described in [36], formally specified in [33], and recalled in the previous section, the main distinctive feature of FRep is that the real-valued function F defining the point set is associated with a tree structure that serves as its underlying representation. The function F is evaluated at the given point by a procedure traversing the tree structure with primitives in the leaves and operations in the nodes of the tree.

As to the constructive hypervolume model, its underlying representation can be defined in a similar way by introducing a set of tree structures. Along with the tree corresponding to a function F defining the point set, there are constructive trees associated with functions $\{S_j\}$ defining attributes and reflecting the construction logic of the attribute definition. Two main types of elements of the set O are considered: basic hypervolume objects (primitives) and complex hypervolume objects. A hypervolume primitive is a specific instance of a function chosen from a finite set of possible types. A complex hypervolume object is the result of operations on primitives. The tree structure with hypervolume primitives in the leaves and hypervolume operations in the nodes of the tree provides the computational scheme for complex hypervolume objects. Some nodes, including root nodes corresponding to the whole complex object, can refer to the hypervolume relations.

The function S_j is evaluated at the given point by a tree traversing procedure. Thus, symmetry in treating the point set and its attributes can be achieved in accordance with the constructive nature of the definition and the underlying representation. A formal description for the traversing procedure for the FRep constructive tree [37,33] is easily adaptable to hypervolume constructive trees.

The constructive tree is similar to one used in CSG, and is created during the object construction process. In contrast to classical CSG, the sets of primitives and operations are not fixed and can easily be extended without redesigning the modelling system, and all operations are applicable on any level of the tree.

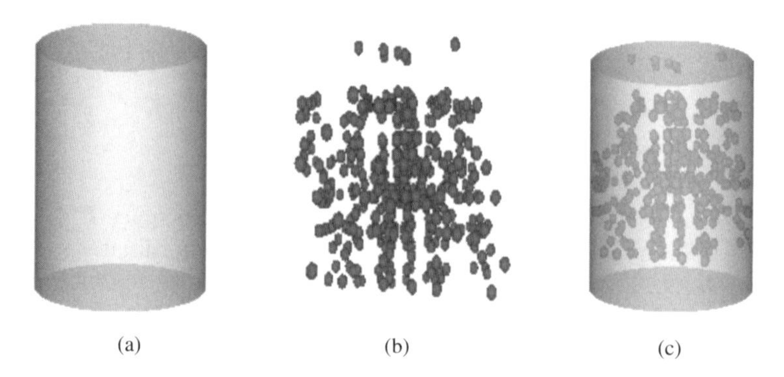

Fig. 1. Density of a composite material rendered using greyscale. (a) The cylinder corresponds to the matrix with a constant density. This is the geometrical tree. (b) Reinforcement material composed of microspheres with constant density. (c) Visualisation of the density of the composite material.

As to the geometric constituent, solids bounded by algebraic surfaces, skeleton-based implicit surfaces and convolution surfaces, as well as procedural objects (such as solid noise), swept, and discrete field objects can be used as primitives. Let us mention in particular that the framework is general enough to embrace multidimensional discrete field (voxel) objects represented as "hybrid volumes" [2] that can also be treated as primitives.

Many operations that have been formulated for FRep in such a manner that they in turn yield continuous real-valued functions as their output [36,48] can be generalised to produce more specific hypervolume operations. Of course, there can be introduced a much more application-specific operations over attributes that can hardly be sensibly applied to the geometry.

4.3 Heterogeneous Material Modeling

Heterogeneous objects are omnipresent around us. We consider two simple examples here that are direct applications of the constructive hypervolume model.

Composite materials are widely used in the industry. They are composed of several elementary materials providing properties that each single element does not have. Usually, such materials can be decomposed in two parts called matrix and reinforcement materials. The reinforcement material confers a skeleton to the composite material, and the matrix makes an envelope. In Fig. 1, we show an example of composite material, and focus on a single attribute A, the material density. There are two steps in making this model: description of the geometry and description of the attribute. The geometry, i.e., the matrix, is defined as a cylinder $F(X)$ shown in Fig. 1a. The reinforcement material is defined as microspheres, corresponding to a function F_s that defines their location in space. It is defined as a FRep tree with several spheres in the leaves and set-theoretic unions in the nodes. A constant density corresponds to each material.

To visualise the resulting object, the density value is mapped to a greyscale colour. Then, for every given point X, a first tree traversing procedure is applied to the geometrical tree $F(X)$. When $F(x) \geq 0$, another tree traversing procedure is applied to $F_s(X)$ to determine the density value. In the case where $F_s(X)$ is positive, the resulting density is the density of the microspheres, otherwise, the density of the cylinder is returned. The resulting composite material is shown in Fig 1c.

The second example shows another heterogeneous object, where a sheathed electric cable is modeled. The sheath is made of plastic and three different cables are embedded inside. One of them has the same orientation as the sheath, and the two other round it up. Each cable is composed of a gainer made of different plastics too, and it has copper inside. The three cables are then surrounded with a twisted pair made of another material. We model this object as a constructive hypervolume. The geometry is defined as a single cylinder, using a function F_{geom}, and the attributes a represented by a material index vector. The constructive hypervolume object is defined as $o = (F_{geom}, A)$. One needs then to define the spatial occupation of each material. Different constructive trees are built for this purpose, i.e., three for the different kinds of plastics corresponding to the material indices A_1, A_2 and A_3, one tree for the copper index A_4, and another tree for the material of the twisted pair, corresponding to the index A_5. The material of the embedding sheath is the default attribute, and does not require an additional tree. Figure 2 shows these tree structures.

5 Lower Dimensional Objects Modelling Using Trimming

When one considers a heterogeneous object, the heterogeneity does not include only material distribution. An object can be dimensionally heterogeneous (mixed-dimensional). It means that an object can be composed of several parts of different dimensions, i.e., points, curves, surfaces and solids. A mathematical model of heterogeneous objects based on cellular complexes and functional representations combned into implicit complexes is described elsewhere in this volume. This model provides reliable mathematical operations for combining heterogeneous cells, either explicitly defined (BRep, parametric curves, wireframes, point lists) or implicitly defined (implicit surfaces, FRep).

In some cases, it is easier and more useful to include cells of lower dimensions (1 or 2) instead of a 3D cell in a model. For instance, for numerical simulations, heavy calculations can not be performed directly on 3D cells, and a good approximation of the result can be obtained while using 2D cells instead, i.e. 2D surfaces embedded in a 3D space. Similarly, 1D cells can be used for some numerical simplifications. Such cells can be defined either explicitly, or implicitly.

The definition of explicit cells is well known, and includes polygonal meshes, parametric curves and surfaces. The definition of implicit cells is based on the FRep model. To define a cell of dimension 2, one can use an implicit surface patch. Such patches can be defined in various ways. We consider the use of implicit surfaces and FRep models as a framework for modeling cells of lower dimension using a trimming technique [38,39,50].

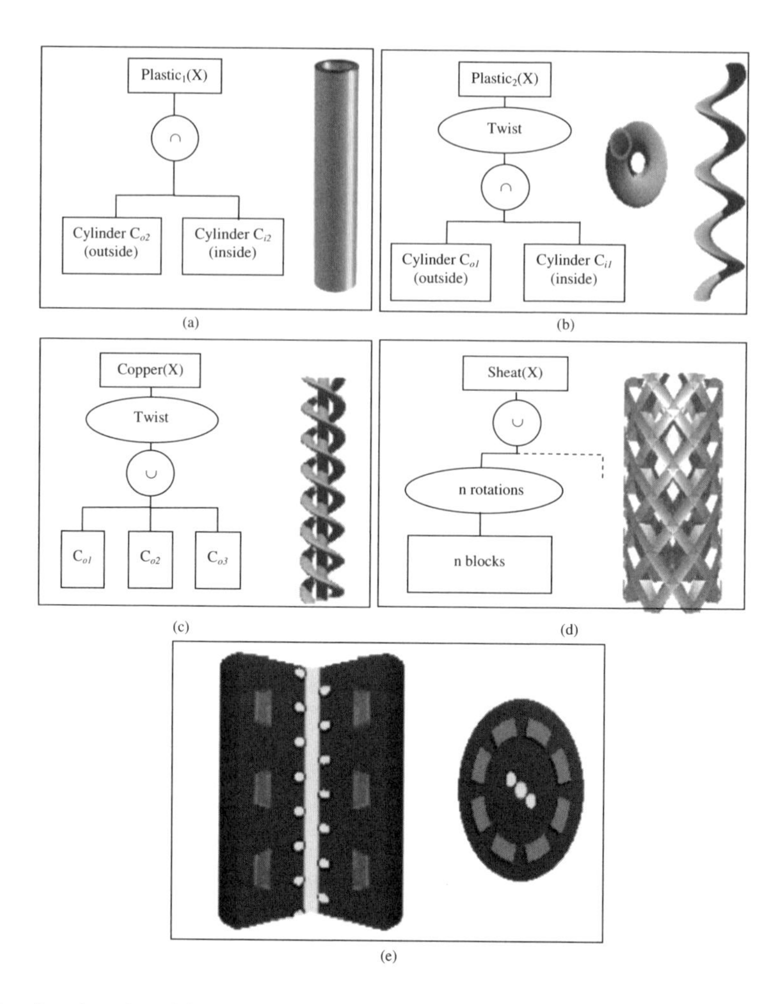

Fig. 2. An electric cable with a metallic wire. The geometry is defined as a cylinder. Constructive trees to determine the location of each material, corresponding to the indices A_1 to A_5 are shown respectively in (a) to (d). Each material is mapped to a grey color. The heterogeneous object is shown in (e), with a cut according to a quadrant (left), and a top view (right).

5.1 Trimmed Surfaces

Trimmed surfaces are now used as a standard tool for modeling complex objects in various areas such as computer animation [13] or CAD modeling. Several software systems (Maya, Lightwave, Catia) have specific tools to define and to render trimmed surfaces.

The most popular technique to define a trimmed surface is to introduce a base parametric surface and to specify the trimming area by a closed parametric curve. The orientation of the trimming curve determines its inner and outer parts. The trimming techniques using NURBS and other parametric objects usually require a re-parameterization step in order to obtain a correct visual result [24].

Although the trimming technique based on parametric curves is very popular, it has some severe limitations. For instance, the trimming curve cannot self-intersect, and the trimming area is a simple hole. It means that for every hole one wants to model, he/she needs to define the corresponding trimming curve on the surface. This may be a costly operation and requires tedious work.

Furthermore, it is well known that set-theoretic operations on BRep models and on parametric surfaces suffer from the lack of robustness, and unwanted holes and cracks often appear when performing trimming operations. An alternative way to define a trimmed surface is to use a trimming solid instead of a trimming curve. In this case, the trimmed surface is defined as an intersection (difference) of the surface with the trimming solid. The trimming solid can be defined using the FRep model by a real valued function. The idea of using a trimming solid was proposed in [45], then applied and extended in [38,39,50].

Hereafter, we give a description of the mathematical formulation of a trimmed implicit object followed by illustrative examples. By the term *trimmed implicit objects*, we denote trimmed implicit surfaces and trimmed implicit curves.

5.2 Trimmed Implicit Objects Definition

In FRep, any object is defined by the inequality $f \geq 0$. To include a surface in the FRep model, we can define it as $F \geq 0$, where $F = -f^2$ takes zero value on the surface only and negative values at all other points of space.

A trimmed implicit surface is defined as the intersection of a carrier surface, F_c, and a trimming solid f_t. The carrier surface is defined as $F_c = -f_c^2$, where f_c is a standard FRep object (sphere, convolution surface, or any complex constructive object). The trimmed surface is then functionally defined as $F = F_c \& f_t$ or

$$F = -f_c^2 \& f_t \tag{2}$$

where & stands for the set-theoretic intersection operation, and can be defined either using the min function, or any other R-functions corresponding to the intersection operation [47,36].

A trimmed implicit surface is defined as an intersection of a surface with a solid. In a similar way, one can define a trimmed curve as the intersection of two surfaces. Such a curve can be functionally defined as $F = F_c \& F_t$ or

$$F = -f_c^2 \& (-f_t^2) \tag{3}$$

where f_t is a defining function of the trimming solid and $F_t = -f_t^2$ is the FRep definition of its surface. The resulting trimmed surface or curve properties can be derived from the underlying FRep model. For instance, a normal vector of the trimmed surface and of the trimmed curve is defined as a gradient of the carrier surface.

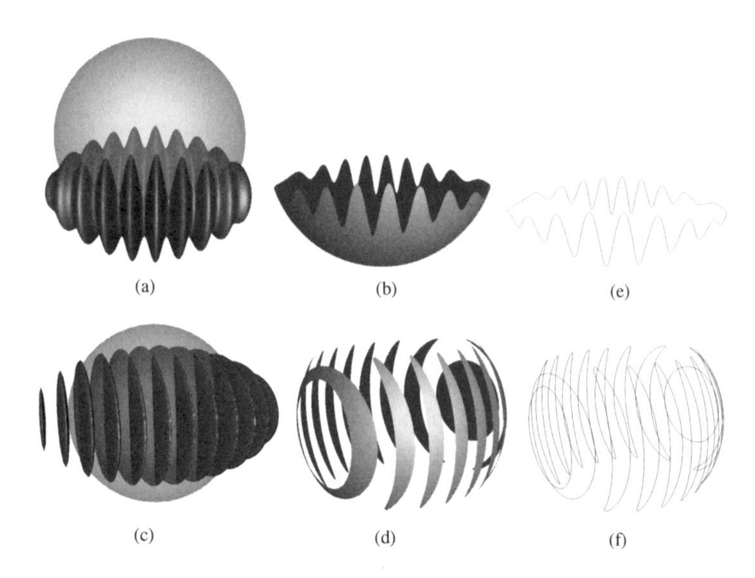

Fig. 3. Concept of trimming implicit surfaces. (a) and (c): Carrier surface (transparent) and trimming object (opaque) in different relative positions. (b) and (d): Trimming the surface. (e) and (f): Trimmed curve.

5.3 Case Study: Trimmed Implicit Objects

In the following, we provide examples of trimmed implicit objects to illustrate the mathematical definition given in the previous subsection.

Trimmed implicit surface. Let us consider an example of a trimmed implicit object shown in Fig. 3. In this example, the carrier surface is a sphere described using a defining function for a solid ball to illustrate the general approach:

$$\begin{cases} f_{ball}(x,y,z) &= R^2 - x^2 - y^2 - z^2 \\ F_{sphere}(x,y,z) &= -f_{ball}^2(x,y,z) \\ G_{sphere}(x,y,z) &= \{(x,y,z)/F_{sphere}(x,y,z) \geq 0\} \end{cases} \tag{4}$$

where R is radius of the sphere. The point set G_{sphere} is the set of points that belong only to the sphere surface shown in Fig. 3a. The trimming solid, shown in Fig. 3a (fully opaque), is defined using a real valued function as an ellipsoid combined with a sine function with an amplitude α as follows:

$$\begin{cases} f_{ell}(x,y,z) &= 1 - \left(\frac{x}{a}\right)^2 - \left(\frac{y}{b}\right)^2 - \left(\frac{z}{c}\right)^2 + \alpha sin(x) \\ G_{ell}(x,y,z) &= \{(x,y,z)/f_{ell}(x,y,z) \geq 0\} \end{cases} \tag{5}$$

The trimmed surface is then a point set G_{trim} defined as the intersection of the point sets G_{sphere} and G_{ell}:

$$G_{trim} = \{(x,y,z)/F_{sphere} \geq 0, f_{ell} \geq 0\} \tag{6}$$

which can be expressed equivalently in a functional form $F_{trim} = F_{sphere} \& f_{ell}$, where F_{trim} is a defining function of the trimmed implicit surface:

$$F_{trim} = -f_{ball}^2 \& f_{ell} \tag{7}$$

where & is an R-function for intersection. The result of the trimming operation is shown in Fig. 3b. With this definition, nothing prevents one from model a trimmed implicit surface with disjoint components. Figure 3c shows the carrier surface and the trimming solid in different relative positions, and Fig. 3d shows a trimmed surface composed of several disjoint parts. To obtain this surface, the amplitude of the sine function has been increased, and the center of the ellipsoid is placed at the center of the sphere.

Trimmed implicit curve. Hereafter, we give an example of curve segments obtained by trimming an implicit surface using another implicit surface. This example follows the previous one, where a trimmed implicit surface was defined. The carrier surface is still a sphere defined by F_{sphere}. To obtain a trimmed curve instead of the trimmed surface, the ellipsoid is replaced by its corresponding surface, and the trimming point set G_{ell} is then defined as follows:

$$\begin{cases} F_{ell} & = -f_{ell}^2(x, y, z) \\ G_{ell}(x, y, z) & = \{(x, y, z)/F_{ell}(x, y, z) \geq 0\} \end{cases} \tag{8}$$

The trimmed curve is defined in a similar way: a point set G_{trim} is defined as the intersection between objects G_{sphere} and G_{ell}:

$$G_{trim} = \{(x, y, z)/F_{sphere} \geq 0, F_{ell} \geq 0\} \tag{9}$$

and is described in the functional form as $F_{trim} = F_{sphere} \& F_{ell}$ or

$$F_{trim} = (-f_{sphere}^2) \& (-f_{ell}^2) \tag{10}$$

The result of the trimming operation is shown in Fig. 3e, and Fig. 3f shows trimmed curves with disconnected components when the amplitude of the sine function is increased and the center of ellipsoid coincides with the center of the sphere.

As one could notice, there are practical issues with rendering trimmed implicit objects using the above definitions, because the defining function does not change its sign at the object points as it happens with traditional implicits. The details of the polygonization and ray-tracing algorithms for trimmed implicit objects can be found in [39,50].

Modeling a hubcap using FRep and trimmed implicit objects. In this example, we model an object typical for CAD applications. We chose to model a hubcap, as it is an object that has been used in several other works on trimming (see [27], for example). When considering CAD applications in general, it appears that for a given object, one needs to perform different calculations, measurements, numerical simulations and other evaluations for the model. One important problem is that for a given type of calculations, one representation of the object is preferable to another. For instance, for calculations linked to internal material distribution, the CAD object should be defined using a mathematical model that can handle heterogeneous objects. Similarly, for heavy numerical calculations, such as pressure measurement or heat transfer, a surface-based model is preferable (as a simplification of the solid object).

In this sense, the use of the cellular-functional model is justified. We propose a dual representation of the hubcap in this example. In Fig. 4a, we define a hubcap using an FRep solid model, without any holes. The constructive tree of this solid is composed of two tori, a cylindrical shell and a block object, deformed by non-linear space mapping (the cap of the hubcap). To create holes, we used five convolution triangles and five cylinders. In the FRep model, holes are obtained by subtracting these primitives from the base object using set-theoretic intersections. The result is shown in Fig. 4b. In Fig. 4c, the cap of the hubcap is modeled using a trimming operation. To define the carrier surface, we used its FRep model.

The only modeling task was to add a few primitives to the constructive tree of the trimming solid in order to obtain the desired trimmed surface. In Figs. 4(d,e), we show different rendering of the trimmed surface, a ray-traced image and a polygonal model respectively. The trimmed curves for this object are shown in Fig. 4f. Note that to render Figs. 4(a,b,c), we used a high quality ray tracing software, PovRay [41]. In order to render the trimmed surface shown in Fig. 4c using this tool, we first polygonalized this surface and exported the triangle mesh into PovRay meshing format. Details of the rendering algorithms can be found in [50,39].

6 Implementation

6.1 Language for Constructive Hypervolume Modeling

HyperFun [42] has been developed as a high-level specialized language for the parameterized description of functionally based multidimensional geometric models. While being minimalist and suitable for easy mastering, it supports all main notions of FRep. The current version of the language that is publicly available [42] only allows for the description of geometry. Here, we describe a new version that allows us deal with the constructive hypervolume model of any degree of generality.

A model in the HyperFun language can contain the specification of several hypervolume objects parameterized by input arrays of point coordinates x_i and numerical parameters a_i whose values are to be passed from outside the object. Each object is defined by a function describing its geometry accompanied, if necessary, by a set of scalar functions s_i representing its attributes. Note the following feature that allows for increasing flexibility while dealing with attributes: values of scalar functions s_i not only can be defined and calculated within the HyperFun object definition but can be passed from the outside the object to be utilized or modified within the program describing the object.

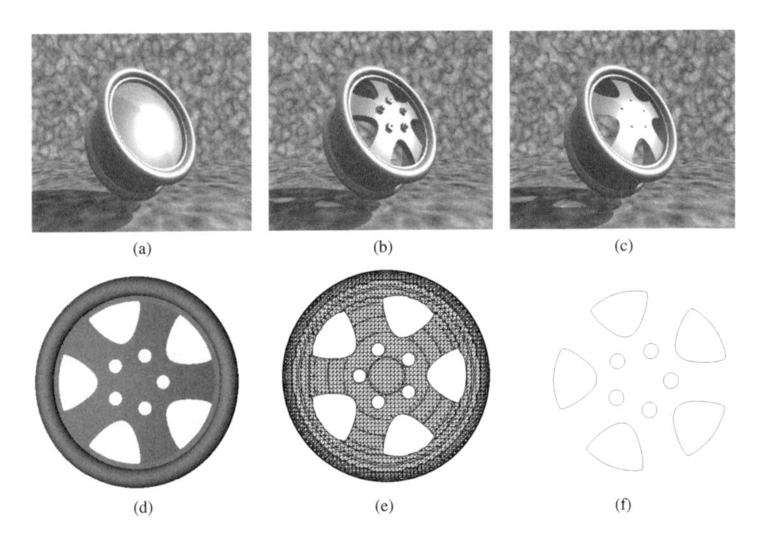

Fig. 4. Modeling a hubcap using FRep and a trimmed implicit surface: (a) carrier object; (b) FRep solid object; (c) Frep solid and trimmed implicit surface (cap of the hubcap); (d) ray traced trimmed implicit surface; (e) polygonalized trimmed implicit surface; (f) trimmed implicit curve

The functions defined in HyperFun are actually symbolic embodiments of the corresponding trees whose structure reflects constructive logic of building both the object's geometry and its attributes. Not only primitives (that can be library functions and local variables defined by algebraic expressions with an appropriate semantics), but other objects can also be the leaves of the tree. At the language level, this means that references to objects that have already been specified can be present in functional expressions. The functions describing geometry and attributes can be built in a step by step manner using assignment statements with introducing local variables and arrays. Conditional selection ('if-then-else') and iterative ('while-loop') structures are also available. Functional expressions are built using conventional arithmetic and relational operators by utilizing standard mathematical functions ('exp', 'log', 'sqrt', 'sin', 'cos', etc.). The distinctive feature of HyperFun is the support of fundamental set-theoretic operations by special built-in operators with the reserved symbols ('|' - union, '&' - intersection, '\' - subtraction, '~' - negation, '@' - Cartesian product).

In principle, the language is self-contained and allows users to build objects from scratch, without using any pre-defined primitives. However, its expressive power is increased by the availability of the system "FRep library" that is easily extendable and can be adapted to a particular application domain and can even be customized for needs of a particular user. The current FRep library version in general use contains the most common primitives and transformations of a quite broad spectrum.

Thus, there are functions implementing conventional CSG primitives (block, sphere, cylinder, cone, torus) as well as their more general counterparts (ellipsoid, superellipsoid, elliptic cylinder, elliptic cone). Another group of the library primitives implements popular implicits: blobby object [7], soft object[57], metaballs [32]), and convolution objects [29] with skeletons of different types (points, line segments, arcs, triangles, curve, and mesh). Primitives derived from parametric functions (cubic spline [52] and Bézier objects [51]) have also been included into the library. As to the transformations, one can mention rotation, scaling, translation, twisting, stretching, tapering, blending union/intersection as well as some more general operations such as non-linear space mapping driven by arbitrary control points.

Taking into account that texturing often requires non-trivial mathematical skills and specialist knowledge (e.g., in color theory), some useful library functions have been developed. These functions allow for creating different texture patterns, such as using Gardner solid noise, wave-like (based on trigonometric functions), checkerboard-like, periodic concentric circles, etc. An important group of library functions deals with color attributes implementing a number of modes for color union and blending. Details of available set of functions for texturing can be found in the following sections.

6.2 HyperFun Software Tools

Application software deals with HyperFun models through using either a built-in interpreter or HyperFun-to-C/HyperFun-to-Java compilers and utilities of the HyperFun API. The latter way concerned with intermediate generation of C/Java code ensures more efficient function evaluation but is much more demanding for developers of application software in a multi-platform environment. All case studies presented in this paper have been developed with a help of software tools with a built-in interpreter.

The HyperFun interpreter has been implemented as a small set of functions in ANCI-C. It is quite easy to integrate them into the application software since the developer needs to deal with only two C-functions. 'Parse' function performs syntax analysis in accordance with the language grammar and semantic rules. For each object described in the HyperFun program, the function generates an internal representation that is actually a collection of the tree structures optimized for subsequent efficient evaluation. If there are any errors in the program, the function outputs a list containing the location and details of each error found.

Another interpreter function ('Calc') is called every time when there is a need to evaluate the defining function and attributes at a given point in the modeling space and for the given external numerical parameters. Externally defined values for attribute scalar functions can be passed too. The object's internal representation serves as an input parameter for 'Calc' function that returns both the value of the "geometric" function and a set of values for "attribute" scalar functions - all evaluated at the given point.

The formal specification of the internal representation and of the function evaluation procedure was given in [37,33]. Note, that the function 'Parse' is invoked just once while processing the HyperFun program; in a way, the internal representation can be treated as "byte-code" and can serve as a protocol for data exchange between system components In fact, these two procedures constitute an application programming interface (API) that is quite simple to utilize.

Software tools for HyperFun creation and processing are being developed in an open source project manner by the international team of developers. Some of them are currently available for free download at the Web site [42]: HyperFun Polygonizer for the surface mesh generation with VRML/STL output, HyperFun plug-in to POVRay [41], which makes it possible to generate high quality photorealistic images on an ordinary PC; HyperFun Java applet executable in a Web browser [9]. The latest release of the HyperFun Polygonalizer also includes options to generate trimmed implicit objects, based on the rendering algorithms proposed in [50,39].

Conceptually, we strive to separate the modeling in multidimensional space with abstract coordinate variables $x_1, \ldots, x_n$ from the subsequent interpretation of the model in "real world" terms (that can be, in particular, a visualization). The concept of multimedia types [3] is exploited here. A special mapping with giving each coordinate an interpretation has been established by default. For instance, 'x', 'y', 'z' types can correspond to Cartesian coordinates; 't' - to "dynamic" coordinate representing continuous values that can be linearly or non-linearly mapped onto physical time; 'u' and 'v' - to 2D "spreadsheet" coordinates, etc. -more details on the "spreadsheet" concept and modeling in multidimensional space can be found in [35].

HyperFun tools have special features allowing users to implement this mapping procedure. With introducing a set of scalar functions for representing object attributes, one can propose a similar methodology. This means that within a HyperFun program, the object's attributes are considered as abstract real-valued functions; as to their actual meaning, it can be determined later - by an appropriate application program. Such a technology allows us to introduce "generic" objects with subsequent generation of their different instances. For example, the same attribute can be treated (without any change in HyperFun program) as color, or as transparency, or as density, or as temperature, depending on circumstances and available application software features. Moreover, it is possible to assign simultaneously a few multimedia types to the same attribute. However, if the user considers it appropriate, it is possible to fix the attribute's meaning as early as on the modeling stage (this is the case for this paper's examples).

6.3 The Hyperfun Library Functions for Texturing

One direct application of the constructive hypervolume model is related to texturing. As a matter of fact, abstract attributes of the model are either defined directly as photometric or texturing attributes at each point in space, or are mapped afterwards to photometric attributes for visualization purposes. This mapping, and more generally the definition of such attributes, is called *constructive texturing*. More details and study of this technique can be found in [49].

To help the user, a set of predefined library functions is available in HyperFun. Taking into account that texturing often requires non-trivial mathematical

Table 1. Examples of utility functions in the HyperFun library for attributes

Function Name	Parameters	Short Description
hfA_Floor	a_f Input Value	Returns the integer part of a
hfA_NoiseG	a_{a_3} Point Coordinates b_f Phase c_f Frequency	Returns Gardner's solid noise
hfA_Turbulence	a_{a_3} Point Coordinates b_f Frequency	Returns Perlin's turbulence function

skills and specialist knowledge (e.g., in color theory), we have been developing the library functions that can facilitate creating constructive hypervolume texturing models. There are three main groups of useful functions. The first group includes functions that are applicable to different attributes irrespective of their specifics. Such functions include noise functions, sinusoidal functions, and linear interpolation. Some other service functions are also provided, such as conversion from one color space to another (RGB to HSV for instance). The table 1 provides more detailed examples of available functions. In this table, a function name is given first, then parameters and a short description of this function. Parameters of the function are alphabetically ordered, and the subscript indicates if this parameter, a for instance, if either a single float value a_f or an array of floating values a_{a_n} of size n. The prefix *hfA_* indicates that the function is related to Attributes.

The second group includes more specialized functions, where different attribute patterns can be defined, such as concentric circles or a brick wall pattern. Three functions are given in the table 2, and a complete list and detailed usage of these functions are available on the HyperFun website. A usage example of one of the functions, *hfA_Crackles*, can be found in the next section, where the HyperFun code for a heterogeneous object is given. Parameters of this function are point coordinates a_{a_3}, resulting array of attributes b_{a_n}, float value controlling the noise behavior of the output pattern, and two arrays defining respectively a set of attributes e_{a_n} corresponding to a set of intervals d_{a_n}. The resulting color pattern is computed as follows. Given an input point coordinate, a noise function is evaluated; then the obtained value is used to determine which interval it belongs to, and the corresponding set of attributes is then returned.

The last group of functions for texturing represents basic operations such as set-theoretic and others. In general, these operations transfer an input array of attribute values to an output array according to the function value. The table 3 gives an outline of some available functions.

Table 2. Examples of specialized attribute library functions in HyperFun

Function Name	Parameters	Short Description
hfA_CheckerBoard	a_{a_3} Point Coordinates b_{a_3} Brick Size c_{a_3} Mortar Size d_{a_n} Output attribute array e_{a_n} Attribute array for blocks f_{a_n} Attribute array for the mortar	Defines a wall pattern including blocks (bricks) and in-between intervals (mortar). If the input point coordinate belongs to a block, the attributes corresponding to the block are returned, otherwise the attributes corresponding to the mortar are returned.
hfA_LookUpMap	a_f Input value b_{a_n} Output attribute array c_{a_n} Mapping array d_{a_n} Array of attribute arrays	Returns the attributes corresponding to the interval a_f belongs to. The mapping array is a set of floats defining intervals, each interval corresponding to a set of attributes defined by the parameter d_{a_n}.
hfA_Crackles	a_{a_3} Point Coordinates b_{a_n} Output attribute array c_f Frequency d_{a_n} Mapping array e_{a_n} Array attribute array	Defines a crackle pattern. A noise value is computed using the point coordinates. This value is then used in a similar way as in the function hfA_LookUpMap: d_{a_n} defines a set of intervals, e_{a_n} defines a set of attributes for each interval, and the noise value is used to determine which interval to consider for the current noise value.

Table 3. Examples of attribute library functions related to set-theoretic operations

Function Name	Parameters	Short Description
hfA_SetAttributes	a_f Input Value b_{a_n} Output attribute array c_{a_n} Input attribute array	If and only if the input value is positive, the input attribute array is copied to the output attribute array
hfA_Union	a_f Function value f_1 b_f Function value f_2 c_{a_n} Output attribute array d_{a_n} Input attribute array (f_1) e_{a_n} Input attribute array (f_2) f_f Union operation for attributes	Returns the R-function value for the set-theoretic union $f_1 \| f_2$. Depending on the union operation for attributes and the value f_1 and f_2, the output array attribute is set to different values depending on d_{a_n} and e_{a_n}

Set-theoretic operations require special attention. The built-in function *hfA_Union*, for instance, calculates union of attributes. The semantics of the function is shown in table 3. The parameters f_1 and f_2 are two function values, and d_{a_n} and e_{a_n} are two arrays of attributes corresponding respectively to objects f_1 and f_2. The returned value f is the value corresponding to the set-theoretic union operation. The array c_{a_n} contains the output attributes. Values of the attributes are set depending on the result of the union of f_1 and f_2. If f_1 is positive and f_2 is negative, then c_{a_n} is equal to d_{a_n}, and vice-versa for f_2 and e_{a_n}. In the case when both f_1 and f_2 are positive, then attribute values contained in c_{a_n} are defined as a combination of d_{a_n} and e_{a_n}. The usage and meaning of the union of attributes may not be straightforward and often is application dependent. The last input parameter of the function f_f is a selection flag, which serves for the selection of different types of attribute unions. Its zero value, for instance, gives priority to the attribute of the input function f_1, the value of 1 gives the priority to f_2's attribute, the value of 2 adds each individual attributes, and the value of 3 takes the minimal of attribute values. Although several predefined operations are available, nothing prevents one from writing their own union operation in the HyperFun code. More details on union of attributes can be found in [35]. The example provided in the next section uses this function.

Note that, although the above examples use three color attributes as RGB values, the available functions are not restricted to this size of the attribute array and can be extended to any arbitrary size. In [49], the number of photometric attributes are equal to 12, corresponding to ambient, diffuse, specular and shininess attributes.

6.4 Example of a Heterogeneous Object Model in HyperFun

An example of a HyperFun model of a heterogeneous object is shown in Fig. 5 and explained in this section. We consider modeling a volumetric multi-layer geological structure. Heterogeneous objects in geo-sciences usually consist of multiple layers of different materials with cavities, wells, and other irregularities. We present here a simplified example of a constructive volumetric model of such a geological object. The corresponding HyperFun code is given at the end of this section. The basic geometric model is described by a single function $F_{geom}(X) \geq 0$, where X is a vector of 3D point coordinates. In the HyperFun code, the F_{geom} is called *my_model*:

$$F_{geom} = F_{relief} \& F_{bbox} \& (-F_{cavity}) \& F_{cut} \tag{11}$$

F_{relief} defines a solid bounded by the top curvilinear surface and the bottom plane, corresponding to *tlayer5* in the HyperFun code, F_{bbox} is a function defining a bounding box for the model, F_{cavity} is a model of cavities made using an algebraic sum between the functions of an ellipsoid and solid noise, and F_{cut} serves for producing a zigzag cut of the full object. The symbol & stands for the R-function defining set-theoretic intersection between two functionally defined solids, and the symbol | stands for the set-theoretic union. Note that an R-function defining set-theoretic difference between two function A and B can be defined as $A\&(-B)$. In HyperFun, the set-theoretic difference is directly expressed by the symbol '\'.

For the model of the "relief" solid, we used the following expression:

$$F_{relief} = (f_{relief}(x, y) - z)\&z \tag{12}$$

where $z = f_{relief}(x, y)$ defines the top curvilinear surface of the object, and z value of the bottom plane is zero. The curvilinear surface is defined as a combination of *sin* functions (note *land5* in the source code).

The five material layers shown in Fig. 5 using different grey scales are presented in the attribute model by the space partition different from the basic geometric model. For the i-th layer, the defining function is

$$F_i = (f_{i+1}(x, y) - z)\&(-f_i(x, y) + z) \tag{13}$$

where $i = 1 \ldots 4$, $f_5 = f_{relief}$, and $z = f_i(x, y)$ defines the top surface of the layer. In the simplest case of the homogeneous material distribution inside the layer, the single material attribute can be defined as

$$A = \begin{cases} M_i \ F_{geom} \geq 0, F_i \geq 0 \\ \theta \quad F_{geom} < 0 \end{cases} \tag{14}$$

where M_i is a material index, and θ stands for the undefined value, equal to zero in the HyperFun code example. The equation 14 is expressed in the code example by the following block:

```
if(model>-0.001) then
...Calculation of the partition corresponding to each layer
...definition of attributes
endif;
```

For instance, when the first partition corresponding to the layer 1 is defined, a built-in function from the HyperFun attribute library is called to set the attributes corresponding to this layer (contained in the array `c1`). Similarly, partitions and attributes for other layers are processed. The attributes corresponding to the current point coordinate $x[3]$ are actually set at the end of this block while using the built-in function `hfA_Union`. Union of the first and second partition is performed, using their respective function values and attribute arrays. The result is the returned function value corresponding to the geometric set-theoretic union, and the output attributes are contained in the array `clayer`. The result of this union operation is then successively used with other partitions, and for the last union, the output array of attributes is finally copied to the 's' array, which is an input/output parameter of the entire function defined in this HyperFun model. The source code of the HyperFun model is given hereafter:

```
my_model(x[3], a[1],s[3])    {
--Declarations of various arrays
array xtt[3], llc[3];
array delta[3], xp[3], center[3];
array map[5], colors[12];
array clayer[3],cedges[3];
array c1[3],c2[3],c3[3],c4[3],c5[3];

map = [0.0,0.4,0.5,0.8,1.0]; s = [0.0,0.0,0.0];
clayer = [0.0,0.0,0.0];      cedges = [0.0,0.0,0.0];
c1 = [0.0,0.0,0.0];          c2 = [0.0,0.0,0.0];
c3 = [0.0,0.0,0.0];          c4 = [0.0,0.0,0.0];
c5 = [0.0,0.0,0.0];

xp[1]= -x[1]+20; xp[2]= x[2];  xp[3]= x[3];
xt   = xp[1];    yt   = xp[2]; zt   = xp[3];

llc = [0,0,0];
bbox = hfblock(xp,llc,20,20,15);
f1 = zt;

--layer5: Top layer and geometry
land5 = 0.4*(sin(xt/1.2)+sin(yt/1.5));
f6 = 12 + land5;
tlayer5 = (f6-zt)&f1;

-- cut of the geometry to visualize inside the
--geological model
fc1 = yt-2;
fc2 = (yt-2)-(xt-5);
fc3 = yt-8;
cut = (fc1 & fc2) | fc3;

--cavity
tmp = hfNoiseG(x,1,1,1);
center = [7,5,3.5];
cavity = hfEllipsoid(xp,center,6,3,2)+tmp;

-Final geometrical model
model = ((tlayer5) \ cavity) & bbox & cut;

if(model>-0.001) then
    ------layer1-----------------------------------------------------
          --Partition
          land1=0.4*(sin(xt/2)+sin(yt/3));
          f2 = 3 + land1;
          tlayer1 = (f2-zt)&zt;
          layer1 = tlayer1;
          --Attributes
          colors=[0.0,0.8,0.6, 0.1,0.8,0.4, 0.2,0.9,0.4,0.0,0.8,0.6];
```

```
        tmp = hfA_Crackles(xp,c1,0.5,colors,map);
  ------layer2----------------------------------------------------
        --Partition
        land2 = 0.4*(sin(xt/1.9)+sin(yt/2.5));
        f3 = 5 + land2;
        tlayer2 = (f3-zt)&zt;
        layer2 = tlayer2\layer1;
        --Attributes
        tmp = hfA_NoiseG(xp,4.0,4.0);
        c2[1] = 0.2+tmp;
        c2[2] = 0.4;
  ------layer3----------------------------------------------------
        --Partition
        land3 = 0.4*(sin(xt/1.7)+sin(yt/2.2));
        f4 = 7 + land3;
        tlayer3 = (f4-zt)&zt;
        layer3 = tlayer3\tlayer2;
        --Attributes
        colors=[0.2,1.0,0.4, 0.4,0.9,0.4, 0.2,0.9,0.2, 0.4,0.9,0.4];
        tmp = hfA_Crackles(xp,c3,0.1,colors,map);
  ------layer4----------------------------------------------------
        --Partition
        land4 = 0.4*(sin(xt/1.4)+sin(yt/1.8));
        f5 = 9 + land4;
        tlayer4 = (f5-zt)&zt;
        layer4 = tlayer4\tlayer3;
        --Attributes
        tmp = hfA_NoiseG(xp,1.0,1.0);
        c4[2] = 0.4;
        c4[1] = (1.0+sin(tmp*10*xp[1]) )/2.0;
        c4[3]= 0.1;
  ------layer5----------------------------------------------------
        --Partition
        layer5 = tlayer5\tlayer4;
        --Attributes
        tmp = hfA_NoiseG(xp,5.0,5.0);
        c5[1] = 1.0-0.7*tmp;
        c5[2] = 1.0-0.7*tmp;
        c5[3] = 0.2;
    --Final Union of attribute of each layer--------------------------
    --to determine the current attribute for the current input
    --point coordinate
       tmp = hfA_union(layer1,layer2,clayer,c1,c2,0);
       tmp = hfA_union(tmp,layer3,clayer,clayer,c3,0);
       tmp = hfA_union(tmp,layer4,clayer,clayer,c4,0);
       layers = hfA_union(tmp,layer5,s,clayer,c5,0);
endif;
my_model = model;
}
```

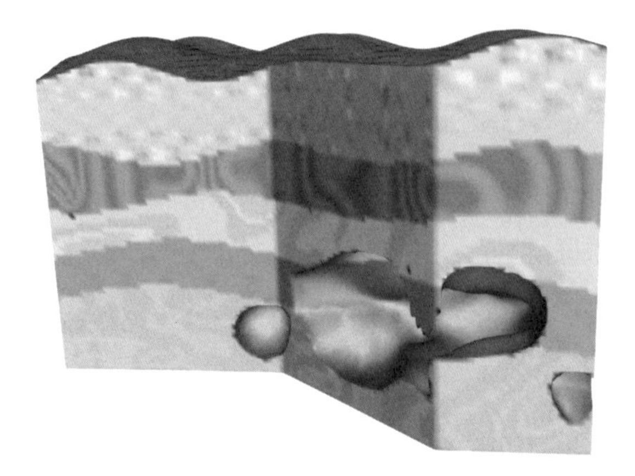

Fig. 5. Volumetric multi-layer geological structure modeled using HyperFun

7 Conclusion

In this survey paper, we presented the development of the function-based models suitable for heterogeneous objects modeling. The function representation was the foundation for this development providing constructive multidimensional models defined by real functions. The constructive hypervolume framework allowed for adding pointwise attributes modelled in a similar way as object geometry. Attributes represent different properties of real or abstract objects defined at each point of the object. The function representation (FRep) is used as the basic model for point set geometry, and attributes are modelled independently using real-valued scalar functions of several variables. Geometry and attributes are modelled constructively in a step-by-step manner. This is reflected in the underlying representation in the form of the constructive trees. Each real function defining geometry or an attribute is evaluated at the given point by a procedure traversing the corresponding constructive tree data structure.

Straighforward application of the proposed constructive hypervolume model is modeling solid objects with internal material distribution. We aslo presented lower dimensional entities such as surfaces and curves. The proposed means for defining such cells relies on the usage of a trimming operation, where surfaces are defined as the intersection of a carrier surface functionally defined with a carrier solid and curves defined as the intersection of two functionally defined surfaces.

Solids, surfaces and curves defined using the constructive hypervolume model can serve as elementary cells in the theoretical framework related to the implicit complexes also presented in this volme. Each entity is dimensionally homogeneous and defines a point set with multiple attributes. In the implicit complex model, a unified framework is proposed to combine cells of different dimensions, as well as cells defined by other means, resuling in a specific cellular complex.

References

1. Pasko, A., Shapiro, V.: Heterogeneous object models and their applications. Computer-Aided Design 37(3) (2005) (Special issue)
2. Adzhiev, V., Kazakov, M., Pasko, A., Savchenko, V.: Hybrid system architecture for volume modelling. Computers And Graphics 24(1), 67–78 (2000)
3. Adzhiev, V., Ossipov, A., Pasko, A.: Multidimensional shape modeling in multimedia applications. In: Karmouch, A. (ed.) Multimedia Modeling 1999, pp. 39–60. World Scientific, Singapore (1999)
4. Bajaj, C., Pascucci, V., Rabbiolo, G., Schikore, D.: Hypervolume visualization: a challenge in simplicity. In: IEEE Symposium on Volume Visualization, ACM SIGGRAPH, pp. 95–102 (1998)
5. Bhashyam, S., Shin, K.H., Dutta, D.: An integrated cad system for design of heterogeneous objects. Rapid Prototyping Journal 6(2), 119–135 (2000)
6. Biswas, A., Shapiro, V., Tsukanov, I.: Heterogeneous material modeling with distance fields. Comput. Aided Geom. Des. 21(3), 215–242 (2004)
7. Blinn, J.: A generalization of algebraic surface drawing. ACM Transactions on Graphics 1(3), 235–256 (1982)
8. Bowyer, A.: Svlis: Introduction and user manual. In: Information Geometers, UK, p. 128 (1995)
9. Cartwright, R., Adzhiev, V., Pasko, A., Goto, Y., Kunii, T.L.: Web-based shape modeling with hyperfun. IEEE Computer Graphics and Applications 25(2), 60–69 (2005)
10. Chen, K., Feng, X.: Computer-aided design method for the components made of heterogeneous materials. Computer-aided design 35(5), 453–466 (2003)
11. Chen, M., Tucker, J.: Constructive volume geometry. Technical Report CS-TR-98-19, University of Wales Swansea, UK, p. 36 (1998)
12. Chen, M., Tucker, J.: Constructive Volume Geometry. Computer Graphics Forum 19(4), 281–293 (2000)
13. De Rose, T.D., Kass, M., Truong, T.: Subdivision surfaces in character animation. In: Proceedings of SIGGRAPH 1998, pp. 85–94 (July 1998)
14. Ebert, D., et al.: Texturing and modelling: a procedural approach. AP Professional, San Diego (1998)
15. Bloomenthal, J., et al.: Introduction to implicit surfaces. In: Proceedings of SIGGRAPH 1984, Computer Graphics, Morgan Kaufmann, San Francisco (1997)
16. Gomes, A., Middleditch, A., Reade, C.: A mathematical model for boundary representations of n-dimensional geometric objects. In: Bronsvoort, W., Anderson, D. (eds.) Fifth Symposium on Solid Modelling and Applications, pp. 270–277. ACM Press, New York (1999)
17. Hohne, K.H., Fuchs, H., Pizer, S.: 3d imaging in medicine: Algorithms, systems, applications. NATO Advanced Science Institutes Series, Series F, Computer and Systems Science 60 (1990)
18. Houlding, S.: 3d geoscience modelling - computer techniques for geological characterization. In: SIGGRAPH 1991, Computer Graphics Proceedings (1994)
19. Jackson, T.R.: Analysis of functionally graded material object representation methods. PhD thesis, MIT, Ocean Engineering Department (2000)
20. Jones, M.: The production of volume data from triangular meshes using voxelization. Computer Graphics Forum 15(5) (1996)
21. Jones, M., Chen, M.: A new approach to the construction of surfaces from contour data. Computer Graphics Forum 13(3) (1994)

22. Kartasheva, E., Adzhiev, V., Comninos, P., Fryazinov, O., Pasko, A.: Implicit complexes framework for heterogeneous objects modelling. Heteroheneous Objects Modeling and Applications (this volume)
23. Kaufman, A., Cohen, D., Yagel, R.: Volume graphics. IEEE Computer 26(7) (1993)
24. Kumar, S., Manocha, D.: Efficient rendering of trimmed NURBS surfaces. Computer-Aided Design Journal 27(7), 509–521 (1995)
25. Kumar, V., Burns, D., Dutta, D., Hoffmann, C.: A framework for object modeling. Computer-Aided Design 31(9), 541–546 (1999)
26. Kumar, V., Dutta, D.: An approach to modeling multi-material objects. Fourth Symposium on Solid Modeling and Applications, ACM SIGGRAPH, 336–345 (1997)
27. Litke, N., Levin, A., Schroder, P.: Trimming for subdivision surfaces. Computer Aided Geometric Design 18(5), 463–481 (2001)
28. Liu, H., Maekawa, T., Patrikalakis, N.M., Sachs, E.M., Cho, W.: Methods for feature-based design of heterogeneous solids. Computer Aided Design 36(12), 1141–1159 (2004)
29. McCormack, J., Sherstyuk, A.: Creating and rendering convolution surfaces. Computer Graphics Forum 17(2), 113–120 (1998)
30. Nielson, G.: Scattered data modelling. IEEE Computer Graphics and Applications 13(1), 60–70 (1993)
31. Nielson, G.: Volume modelling. In: Chen, M., Kaufman, A., Yagel, R. (eds.) Volume Graphics, pp. 29–48. Springer, Heidelberg (2000)
32. Nishimura., H., Hirai, M., Kawai, T., Kawata, T., Shirakawa, I., Omura, K.: Object modelling by distributed function and a method of image generation (in Japanese). Transactions of IECE of Japan J68-D(4), 718–728 (1985)
33. Pasko, A., Adzhiev, V.: Function-based shape modeling: mathematical framework and specialized language. In: Winkler, F. (ed.) ADG 2002. LNCS (LNAI), vol. 2930, pp. 132–160. Springer, Heidelberg (2004)
34. Pasko, A., Adzhiev, V., Schmitt, B.: Constructive hypervolume modelling. Technical Report TR-NCCA-2001-01, National Centre for Computer Animation, Bournemouth University, UK, ISBN 1-85899-123-4, p. 34 (2001) URL: http://wwwcis.k.hosei.ac.jp/~F-rep/BTR001.pdf
35. Pasko, A., Adzhiev, V., Schmitt, B., Schlick, C.: Constructive hypervolume modelling. Graphical Models 63, 413–442 (2002) (Special issue on volume modeling)
36. Pasko, A., Adzhiev, V., Sourin, A., Savchenko, V.: Function representation in geometric modelling: concept, implementation and applications. The Visual Computer 11(8), 429–446 (1995)
37. Pasko, A., Savchenko, V., Adzhiev, V., Sourin, A.: Multidimensional geometric modelling and visualization based on the function representation of objects. Technical Report 93-1-008 (1993)
38. Pasko, A.A.: On escher's spirals - polygonization of 2-manifolds with boundaries. In: Bloomenthal, J., Saupe, D. (eds.) Implicit Surfaces 1998, Eurographics/ACM SIGGRAPH Workshop, Seattle, USA, June 15-16, 1998, pp. 77–80. University of Washington (1998) ISSN 1024-0861
39. Pasko, G., Pasko, A.: Trimming implicit surfaces. The Visual Computer 20(7), 437–447 (2004)
40. Payne, B., Toga, A.: Distance field manipulation of surface models. IEEE Computer Graphics and Applications 12(1), 65–71 (1992)
41. Pov-Ray,: The Persistance of Vision., http://www.povray.org/
42. HyperFun Project. Language and Software for FRep Modelling, http://www.hyperfun.org

43. Qian, X., Dutta, D.: Physics based b-spline heterogeneous object modeling. In: DETC and Computers and Information in Engineering Conference, ASME (September 2001)
44. Requicha, A.: epresentations for rigid solids: theory, methods, and systems. ACM Computing Surveys 12(4), 437–464 (1980)
45. Rossignac, J.: Csg formulation for identifying and for trimming faces of csg models. In: CSG'f96 Set-theoretic Solid Modeling: Techniques and Applications, Information Geometers, UK, pp. 1–14 (1996)
46. Rvachev, V.L.: On the analytical description of some geometric objects. Reports of Ukrainian Academy of Sciences 153(4), 765–767 (1963)
47. Rvachev, V.L.: Theory of r-functions and some applications (in Russian). Naukova Dumka, Kiev (1987)
48. Savchenko, V., Pasko, A.: Transformation of functionnaly defined shapes by exented space mapping. The Visual Computer 14(5/6), 257–270 (1998)
49. Schmitt, B., Pasko, A., Adzhiev, V., Schlick, C.: Constructive texturing based on hypervolume modeling. The Journal of Visualization and Computer Animation 12, 297–310 (2001)
50. Schmitt, B., Pasko, A., Pasko, G., Kunii, T.: Rendering trimmed implicit surfaces and curves. In: AFRIGRAPH 2004 Proceedings, 3-rd International Conference on Computer Graphics, Virtual Reality, Visualization and Interaction in Africa, Stellenbosch, Cape Town, South Africa, November 2004, vol. 12, pp. 7–13 (2004)
51. Schmitt, B., Pasko, A., Savchenko, V.: Extended space mapping with bézier patches and volumes. In: Hughes, J., Schlick, C. (eds.) Implicit Surfaces 1999, Eurographics/ACM SIGGRAPH Workshop, September 1999, pp. 25–31 (1999)
52. Schmitt, B., Pasko, A., Schlick, C.: Constructive modelling of FRep solids using spline volumes. In: Anderson, D., Lee, K. (eds.) Sixth ACM Symposium on Solid Modeling and Applications, pp. 321–322. ACM Press, New York (2001)
53. Siu, Y.K., Tan, S.T.: Modeling the material grading and structures of heterogeneous objects for layered manufacturing. Computer-Aided Design 34, 705–716 (2002)
54. Snyder, J.: Generative Modelling for Computer Graphics and CAD. Academic Press, London (1992)
55. Udupa, K., Odhner, D.: Fast visualization, manipulation, and analysis of binary volumetric objects. IEEE Computer Graphics and Applications 11(6), 53–62 (1991)
56. Wise, K., Bowyer, A.: Using csg models in many dimensions to map where things can and cannot go. In: CSG 96 Set-theoretic Solid Modelling: Techniques and Applications, Information Geometers, UK, pp. 359–376 (1996)
57. Wyvill, G., McPheeters, C., Wyvill, B.: Data structure for soft objects. The Visual Computer 2(4), 227–234 (1986)

SARDF: Signed Approximate Real Distance Functions in Heterogeneous Objects Modeling

Pierre-Alain Fayolle[1], Alexander Pasko[2], and Benjamin Schmitt[3]

[1] Université d'Orléans, France
p.fayolle@free.fr
[2] Bournemouth University, United Kingdom
apasko@bournemouth.ac.uk
[3] Digital Media Professionals, Hosei Research Institute Tokyo, Japan
hfschmitt@gmail.com

Abstract. Distribution of material density and other properties of heterogeneous objects can be parametrized by the Euclidean distance function from the object boundary or from special material features. For objects constructed using geometric primitives and set-theoretic operations, an approximation of the distance function can be obtained in a constructive manner by applying special compositing operations to the distance functions of primitives. We describe such operations based on a smooth approximation of min/max functions and prove their C^1 continuity. These operations on distance functions are called SARDF operations for Signed Approximate Distance Functions. We illustrate their applications by 2D and 3D objects models with heterogeneous material distribution.

1 Introduction

Modeling spatial objects, their properties and relations, has important applications in various engineering fields. Spatial objects refer here to curves and surfaces as well as to three-dimensional volumetric objects with heterogeneous internal properties (such as material, color, density, and others). Fields of applications of modeling and visualization of volumetric objects include: medicine, scientific visualization, physical analysis and simulation, mechanical engineering, and others. Several mathematical models have been developed to construct such volumetric objects; all of them have strengths and weaknesses depending on the field of application. The use of distance-based scalar fields is one of the possible methods for naturally defining the geometry of these objects and their internal properties.

The signed Euclidean distance function defines a solid by giving at each point in space the shortest distance between the current point and any point belonging to the surface of the solid. The sign is used to distinguish between object interior and exterior. Starting from expressions of the Euclidean distance for primitive objects (plane, sphere, torus), it is possible to construct more complex objects by applying set-theoretic operations (union, intersection, difference) to them. In

A. Pasko, V. Adzhiev, and P. Comninos (Eds.): HOMA, LNCS 4889, pp. 118–141, 2008.

the theory of R-functions [22], these set-theoretic operations are expressed by real-valued functions, and applying them to primitives gives a functional expression for the final function defining the complex object. Constructive modeling is an elegant way to represent volumetric objects using a tree data structure with operations in the nodes and primitives in the leaves. The tree, called a constructive tree, keeps information about the structure, construction operations and semantics of the objects.

Distance-based scalar fields present the advantage to naturally define a volume and to simplify the modeling task. The Euclidean distance gives additional advantages such as a physical meaning which provides a natural parameter to define constraints in modeling point sets and their attributes. In constructive modeling, the R-functions do not provide a good Euclidean distance approximation for the resulting function, whereas min/max (another type of functional expressions for the set-theoretic operations) keep a better approximation but add points of C^1 discontinuity for the resulting function, which impacts some modeling operations and applications. We present in the following, functional expressions for the set-theoretic operations that keep a reasonable approximation of the Euclidean distance but also provide smoothness for the resulting function, a useful property for many applications. We call such functions SARDF for Signed Approximate Real Distance Functions.

In this paper, we present our work on volumetric modeling with the following objectives:

- The definition of new functional expressions for the set-theoretic operations intersection, union, and difference, which provide good approximation of the Euclidean distance and smoothness of the resulting function.
- The introduction of a constructive distance-based modeling framework for volumetric objects based on the introduced set-theoretic operations, and other operations and primitives defined by an approximation of the Euclidean distance function.
- Using this constructive framework to allow parameterization and control of object attributes by the distance.

2 Previous Works

The Euclidean distance from a point $\mathbf{p}$ to a set S is the minimum distance, using the Euclidean norm, between $\mathbf{p}$ and any point of S. The signed Euclidean distance function is a concise and powerful way of describing object geometry. We discuss here the existing methods used to construct Euclidean distance functions to define volumetric objects.

Constructive modeling allows to create models of complex solids by combining together simple solids with operations. It can be implemented using the theory of R-functions [20,22,26,16] or min/max functions [24,18]. We review the existing methods and algorithms for constructive modeling volumetric objects with a focus on the quality of approximation of the Euclidean distance and the smoothness of the resulting function.

The Signed Euclidean Distance Function. The signed Euclidean distance function from a point $\mathbf{p} \in \mathbb{R}^n$ to a $(n-1)$ closed orientable manifold M, embedded in $\mathbb{R}^n$, is defined by: $d : \mathbb{R}^n \to \mathbb{R}$, $d(\mathbf{p}) = \epsilon|\mathbf{p} - \mathbf{c}|$, where ϵ is ± 1 corresponding to the orientation of M, $\mathbf{c}$ is the closest point on M to $\mathbf{p}$, and $|.|$ denotes the Euclidean norm. Two conventions exist for the sign of the distance: the outward normal can point in the positive direction or in the negative direction. In this paper, we adopt the convention that the outward normal points in the direction of the negative values of the distance function. In the Euclidean three-dimensional space $\mathbb{R}^3$, the signed Euclidean distance function to a closed oriented surface M naturally defines a solid by: $\{(x, y, z) \in \mathbb{R}^3 : d((x, y, z), M) \geq 0\}$.

Euclidean distance fields have numerous applications in geometric modeling [8], shape metamorphosis [5], object reconstruction from cross-sections [13], robust rendering with sphere tracing [9], generation of skeletal shape representation [32], and other areas.

Computation of the Distance Function. Let $d(\mathbf{p})$, $\mathbf{p} \in \mathbb{R}^3$ be the signed distance function to an oriented closed surface M. The function d is the viscosity solution of the Eikonal equation [31,30,25]:

$$|\nabla d| = 1, d|_M = 0 \tag{1}$$

d corresponds to the time arrival of a wave propagating from the surface boundary, with a speed of unit magnitude. Let $\mathbf{c}$ be the closest point to $\mathbf{p}$ on the surface M, the distance is then $|\mathbf{p} - \mathbf{c}|$, with a negative sign if $\mathbf{p}$ is outside M. If the surface is smooth, then $\mathbf{p} - \mathbf{c}$ is orthogonal to the surface. The signed Euclidean distance function is at least C^0, but may be not differentiable at some points.

Expressions for the distance function to most of the classic surfaces of a CSG system (sphere, cylinder, cone) are known analytically [9]. For example, the signed distance to a sphere (boundary of a ball) of radius 1 and center at the point $(0, 0, 0)$ is given by the function: $d(x, y, z) = 1 - \sqrt{x^2 + y^2 + z^2}$. The signed distance to ellipsoids can be computed by a numerical procedure [10].

In general, if the surface M is available as an oriented point-set or a mesh of triangles, it is possible to solve the Eikonal equation (Eq. 1) on a finite grid. Examples of numerical algorithms to solve that problem are: the fast marching method [25], the fast sweeping method [30,31], or the characteristics / scan conversion algorithm [14]. Algorithms, that exploit the GPU (Graphics Processing Unit), have also been designed in order to compute efficiently the Euclidean distance function [11,29]. After the signed Euclidean distance has been computed on each grid nodes, it is possible to apply spline interpolation to get an analytical expression – see for example [19] for the interpolation of volume data. These methods may suffer from numerical issues and a loss of accuracy depending on factors such as the choice of the basis, the sampling of the discrete distance field, or the quality of the input data.

Constructive Geometry with Distance Functions. In constructive geometry, complex solids are built by applying successively set-theoretic operations to primitives. When the primitives have the distance function property, we want that the resulting function for the complex solid, obtained by applying the set-theoretic operations to primitives, is again the distance function or at least its good approximation. We study in the following different implementations of the set operations: min/max and the R-functions R_0 in terms of distance approximation and smoothness.

In the following, d_1 and d_2 are two distance functions to two $(n-1)$-manifolds M_1 and M_2; practically, $n = 2$ or $n = 3$, so M_1 and M_2 are curves or surfaces, and d_1 and d_2 naturally define surfaces or solids, denoted by S_1 and S_2. The results remain valid in any dimension n.

Functions Min/Max. Sabin [24] and Ricci [18], independently proposed the use of the min/max functions to describe set-theoretic operations on on solids with implicit surfaces. Using min/max, the set-theoretic operations are given by:

$$S_1 \cup S_2 : d_1 \vee d_2 = max(d_1, d_2) \tag{2}$$

$$S_1 \cap S_2 : d_1 \wedge d_2 = min(d_1, d_2) \tag{3}$$

$$S_1 \setminus S_2 : d_1 \wedge -d_2 \tag{4}$$

The function built by applying min or max to two distance functions d_1 and d_2 is 0 on the surface defined by the corresponding set-theoretic operation applied on the solids S_1 and S_2. If the gradient is defined, its norm is equal to 1; so that both of the properties of Eq. 1 hold. However, it does not correspond exactly to the Euclidean distance function, as it can be seen in Fig. 1, with the distance to a square built as an intersection of four halfplanes. Contour lines of the distance function constructed analytically by applying the min function for the intersection and the Euclidean distance to the square boundary are shown at the left and the right respectively. Exterior contour lines are in light grey, interior contour lines in dark grey, and the square shape is in black. The exterior contour lines for the Euclidean distance function, in light grey, are different, with circular arcs centered at the vertices of the square, instead of sharp corners.

The main problem with the use of min/max in shape modeling is the smoothness of these functions. The function $(x, y) \rightarrow min(x, y)$ (and respectively max) is C^0 but not differentiable at points where $x = y$. In the geometric space, the resulting function will generally not be differentiable at any point p such that: d_1 is not differentiable, or d_2 is not differentiable, or $d_1(p) = d_2(p)$. The first two cases are inherent to the primitives, but the latter is due to the min/max functions.

These points can cause unexpected results in further operations on the object such as blending, metamorphosis, and others, and problems in engineering applications requiring non-vanishing gradients [3,4]. Figure 2 illustrates an unexpected result of the blending union between a sphere and a box, when min/max

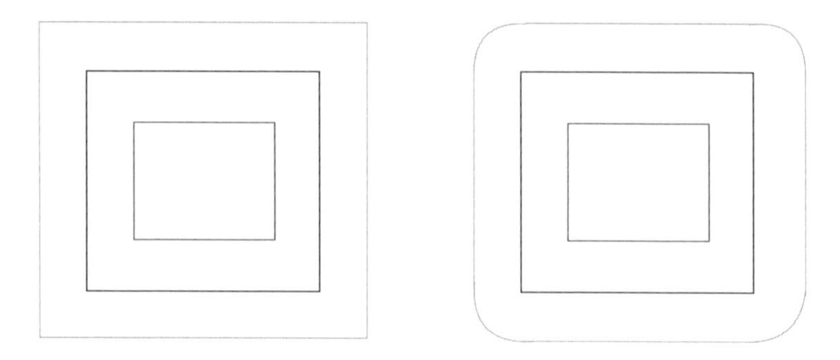

Fig. 1. Some contour lines of: the "distance" function to a unit square defined by the intersection, using min, of four infinite halfplanes (left), and the exact signed Euclidean distance function to the boundary of the square (right). See the circular arc in the exterior contour line (light grey) of the exact distance function (right), compared to the sharp corners created by using the min function (left).

are used in the modeling to implement the set-operations. The box is defined as an intersection of planes. The blending union [17] is defined by:

$$blending(d_1, d_2) = d_1 + d_2 + \sqrt{d_1^2 + d_2^2} + \frac{a_0}{1 + (\frac{d_1}{a_1})^2 + (\frac{d_2}{a_2})^2} \tag{5}$$

The unwanted edge in the material added by the blending union comes from the use of the min function to implement the intersections of the half-spaces in the cube model.

In order to remove the C^1 discontinuities of min/max, Rvachev proposed the R-functions [20,21,26], also briefly discussed in the following. Ricci [18] proposed the superelliptic approximations of min/max, which do not describe exact set-theoretic operations and suit only for blending. The elliptic approximation of min/max by Barthe et al [1] is designed for blending and the error of the distance function grows infinitely far from the boundary.

R-functions. There are various kinds of R-functions, with different order of smoothness, discussed in [20,22,26]. The most commonly used are given by:

$$S_1 \cup S_2 : d_1 \vee d_2 = d_1 + d_2 + \sqrt{d_1^2 + d_2^2} \tag{6}$$

$$S_1 \cap S_2 : d_1 \wedge d_2 = d_1 + d_2 - \sqrt{d_1^2 + d_2^2} \tag{7}$$

$$S_1 \setminus S_2 : d_1 \wedge -d_2 \tag{8}$$

The R-functions, $(x, y) \to x \wedge y$ and $(x, y) \to x \vee y$, are in C^1 over $\mathbb{R}^2 \setminus (0, 0)$. In the geometric space, the resulting function is not differentiable at all points p such that: d_1 is not differentiable, or d_2 is not differentiable, or $d_1(p) = d_2(p) = 0$.

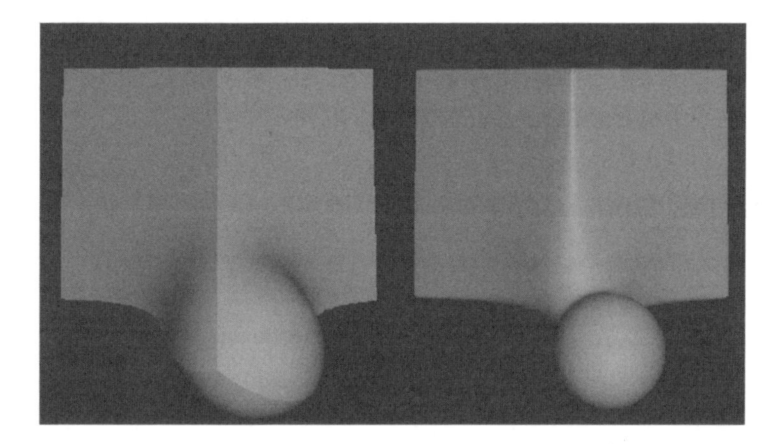

Fig. 2. Left: illustration of the C^1 discontinuity of min/max in further operations, in this case: the blending union between a sphere and a cube. Right: nice blending effect, when R-functions are used for set-theoretic operations during modeling.

The first two cases are inherent to the primitives and the latter is added by the R-functions: it corresponds to the sharp corners and sharp edges of a surface. When using R-functions to model the cube, the blending operation does not create unwanted edges as shown in Fig. 2, right.

R-functions generate however a poor approximation of the signed distance function. They suffer from a value growth's explosion when, for example, applying them to overlapping solids.

Discussion. Neither min/max nor the R-functions provide at the same time a reasonable approximation of the distance and smoothness of the resulting function. R-functions are poor approximations of the Euclidean distance, and min/max are not smooth enough for several applications. We should notice however that smoothness and Euclidean distance are contradictory properties, since the distance function is by definition not everywhere differentiable (for examle, it is not differentiable at all the points belonging to the medial axis of the solid). But we accept to loose some accuracy in favor of smoothness.

In this paper, we introduce new smooth approximations for min/max operations inspired by the works [1,12,2]. The proposed functions are C^1 continuous and keep a controllable approximation of the distance function. We call the constructed defining function of the object by the term Signed Approximate Real Distance Function (SARDF), the approximate min function can be called SARDF intersection, and the approximate max function - SARDF union.

We propose to use the proposed SARDF framework to extend the constructive hypervolume model [15] for distance-based modeling. In the latter model, both the geometry of the solid, and the shape of the definition domains for the attributes can indeed be defined in constructive ways using SARDF primitives and operations. The modified constructive hypervolume model is used to answer the question (section 5.2 of [4]) of the practical ways to compute the Euclidean distance field and then combined with the work of [15,4] to model constructive heterogeneous objects using signed distance fields.

3 SARDF Operations

R-functions have good properties of smoothness making them appreciated in solid modeling, material modeling, animation, and other areas. Unfortunately, the R-functions "destroy" quickly the distance properties of the argument functions. This effect was noticed by other researchers especially in the field of material modeling [28]. In contrary, the min/max functions keep better approximation of the distance property for the constructive shape. However, they add singularities to the constructed function in addition to the natural singularities of the true distance function.

The problem we address is to introduce new operations on functions corresponding to set-theoretic operations on solids such that these new operations have the following properties:

- they have better differential properties than min/max;
- they are a better approximation of the distance function than the R-functions and at least as good as min/max;

Some recent works proposed to modify the contour lines of the min and max functions in order to create some blending effects [12,1]. The same techniques can be used to construct some smooth versions of min/max: the sharp corners of the contour lines can be replaced by symmetric circular arcs, except the sharp corner passing through the origin, the radius of the circular arcs is either growing or bounded by two control lines, to better approximate the distance function. These smooth approximations of min/max functions are called SARDF operations, an abbreviation for Signed Approximate Real Distance Function.

We first introduce the construction of the approximate min function with a circular arc replacing a sharp corner of the contour line. Then we describe the formulation of SARDF operations as proposed in [7]. The constructions and properties of the introduced functions are similar for the intersection and union as these operations present symmetries: $intersection(x, y) = -union(-x, -y)$. We give constructions and properties only for the intersection. The difference is obtained from the intersection by the operation $f_1 \wedge (-f_2)$.

3.1 Circular Min Approximation

In this section, we describe a circular approximation of the min function for the set-theoretic operation intersection to approximate the signed real distance function. Any contour line of the min function has a sharp corner, corresponding to the union of two vertical and horizontal rays. This feature of the contour lines

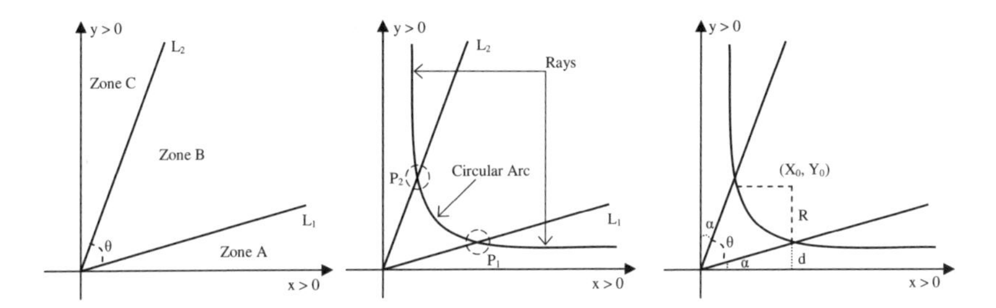

Fig. 3. Left: two straight lines with the angle θ between them break the first quadrant into three zones. Center: contour line configuration: the two rays are attached to the circular arc at the junction points P_1 and P_2. Right: unknowns of the Eq. 9 and their geometric relations.

reflects the discontinuity of the partial derivatives of the min function that occurs at any point when the two arguments are equal.

Following the general approach of [1,12], we propose to replace the sharp corner in any contour line, except the contour line passing through the origin, with a circular arc. All operations are discussed for two halfspaces $f_1 = x$, $f_1 \geq 0$ and $f_2 = y$, $f_2 \geq 0$. We consider two straight lines, symmetric with respect to the line defined by $y = x$ and with an angle θ between these lines, which act as a frontier for the circular arcs. Applying these operations to arbitrary distance functions f_1 and f_2 consists in syntactically replacing x by f_1 and y by f_2.

The Euclidean plane is divided into four quadrants; the first quadrant corresponds to $x > 0$ and $y > 0$, the second quadrant to $x < 0$ and $y > 0$, the third quadrant to $x < 0$ and $y < 0$, and finally the fourth quadrant to $x > 0$ and $y < 0$. In the second and fourth quadrants, the approximate function for min is equal exactly to min; thus we restrict the discussion to the first and third quadrants, where the sharp corners need to be smoothed.

Circular Min Approximation: Quadrant I. We discuss here the circular approximation of the function $F(x, y) = min(x, y)$ in the first quadrant, where $x > 0$ and $y > 0$. We want to replace any contour lines $F = d$ with a circular arc and two rays tangentially attached to it as shown in Fig. 3 center. The angle θ made by two straight lines L_1 and L_2 is introduced as in Fig. 3.

These two straight lines L_1 and L_2 break this first quadrant into three zones: A (below L_1), B (between L_1 and L_2) and C (above L_2), as shown in Fig. 3 (left).

The attachment points $\mathbf{P_1}$ and $\mathbf{P_2}$ of the arc and the rays are placed on the lines L_1 and L_2 correspondingly. Figure 3 (center) shows such a contour line configuration. We are interested in the contour lines $\tilde{F} = d$ of the smooth approximation $\tilde{F}$ of the min function. Given an arbitrary point $\mathbf{P} = (x, y)$, we need to calculate a function value d for it.

In zone A, $\tilde{F}$ is equal to $min(x, y)$, therefore the contour is a horizontal line going through the point $\mathbf{P}$ and defined as $\tilde{F} = y$. In zone C, $\tilde{F}$ is also equal to $min(x, y)$, so the contour is a vertical line going through the point $\mathbf{P}$ and defined as $\tilde{F} = x$.

Finally, in zone B, we want to have a circular arc passing through the point $\mathbf{P} = (x, y)$. This arc should go through the point $\mathbf{P}$ and change into the horizontal ray in zone A and into the vertical ray in zone C. Both of these rays are at the distance d from the corresponding x and y axes. Such a distance is used for the definition of the value of the function. In order to calculate this distance d, we start from the equation of the circle passing through $\mathbf{P}$:

$$(x - x_0)^2 + (y - y_0)^2 = R^2 \tag{9}$$

In this equation, x_0, y_0 and R are unknown but can be expressed in terms of the value d being searched, and α, the angle between the straight lines and the axes. Figure 3 (right) shows the unknowns and their geometric relations.

First, α is expressed using θ (a parameter left to the user, expressing the angle between L_1 and L_2): $\alpha = \frac{(\frac{\pi}{2} - \theta)}{2}$. Then, from the lower triangle in zone A (Fig. 3 right), $x_0 = d\ tan(\alpha)$. By analogy, from the upper triangle in zone C (Fig. 3 right), $y_0 = d\ tan(\alpha)$, and $R = x_0 - d$. By replacing these variables in Eq. 9, we obtain the following quadratic equation for the variable d:

$$d^2\ [cotan^2(\alpha) + 2\ cotan(\alpha) - 1] - 2\ d\ (x + y)\ cotan(\alpha) + x^2 + y^2 = 0 \tag{10}$$

The solution of Eq. 10 for the unknown d is:

$$d = \begin{cases} \frac{-b \pm \sqrt{(b^2 - 4ac)}}{2a} & \text{if } a \neq 0 \text{ and in zone B} \\ -\frac{c}{b} & \text{if } a = 0 \text{ and in zone B} \end{cases}$$

where $a = cotan^2(\alpha) + 2\ cotan(\alpha) - 1$, $b = -2\ (x + y)\ cotan(\alpha)$ and $c = x^2 + y^2$ are the coefficients of Eq. 10.

The final expression for the value of $\tilde{F}$, at $\mathbf{P}$ in the quadrant I, is summarized below:

$$\tilde{F}(\mathbf{P}) = d = \begin{cases} \frac{-b \pm (b^2 - 4ac)^{0.5}}{2a} & \text{if } a \neq 0 \text{ and } \mathbf{P} \text{ in zone B} \\ -\frac{c}{b} & \text{if } a = 0 \text{ and } \mathbf{P} \text{ in zone B} \\ y & \text{if } \mathbf{P} \text{ in zone A} \\ x & \text{if } \mathbf{P} \text{ in zone C} \end{cases}$$

where $a = cotan^2(\alpha) + 2\ cotan(\alpha) - 1$, $b = -2\ (x + y)\ cotan(\alpha)$ and $c = x^2 + y^2$, and α is an angle between L_1 and x-axis, and between L_2 and y-axis.

For the approximation $\tilde{F}$ of the min function in the third quadrant, where $x < 0$ and $y < 0$, the method is the same as for the first quadrant.

Problem of the Circular Approximation. The intersection for two given shapes, defined by the signed distance functions f_1 and f_2 is obtained by replacing x and y in the above equations by f_1 and f_2.

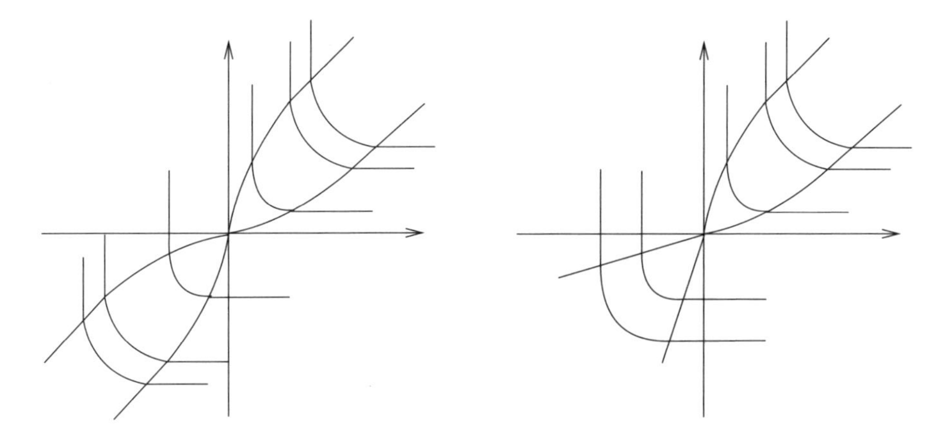

Fig. 4. Illustration of the general idea for the construction of SARDF intersection. The two possible approaches in the quadrant 3 are given left and right. Left, the growth of the radius is bounded by two straight lines; right, the growth of the radius is unbounded.

The use of the described above circular approximations for the min and max functions can provide the C^1 approximation of the resulting distance function for constructive shapes built using normalized primitives (defined by distance functions). Unfortunately, this approach has the following problem: the radius of the circular arc used to replace the sharp corners in the contour lines keeps growing with the distance from the initial surfaces. Because of this behavior of the arc radius, the error of the distance function approximation grows infinitely with the distance. We propose to prevent the radius from growing infinitely by introducing a fixed radius circular arc, and by switching to it, when some threshold for the radius is reached.

3.2 SARDF Intersection Construction

The sharp corners, in quadrant 1 and 3, are replaced as above by circular arcs with growing radius. We propose two approaches to control the growth of the radius in quadrant 3: in the first one, the radius is bounded by two parallel straight lines after a threshold radius is reached, whereas the radius is allowed to grow infinitely in the other. In quadrant 1, the growth of the radius is always bounded after some threshold. The idea behind these strategies is to mimic at best the behaviour of the distance function. The general idea is illustrated in Fig. 4 for the two different approaches.

Quadrant 1. In quadrant 1, the sharp corners of the intersection operation found in every contour lines are replaced by a circular arc with growing radius. This approach is illustrated Fig. 5. Two parabola segments symmetric in respect to the line $y = x$ are used to delimit the circular arc approximation (Zone I,B). The growth of

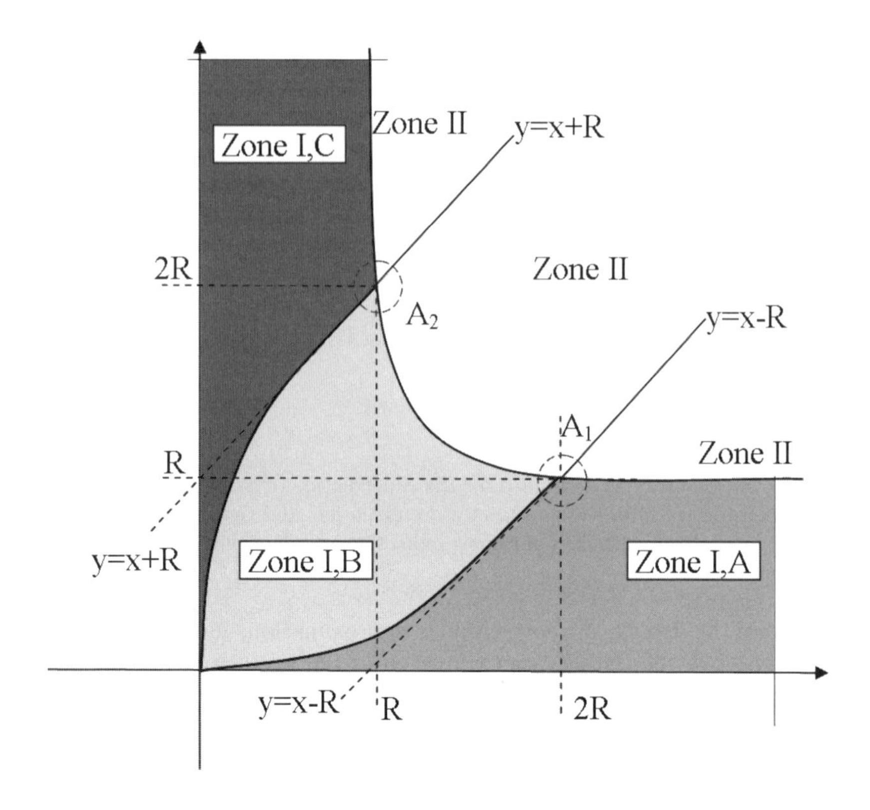

Fig. 5. The different zones of the SARDF intersection in quadrant 1

the circular arc is bounded by introducing a threshold radius R (Zone II). A bounding band is introduced by two parallel straight lines that enclose the circular arc with fixed radius. These band lines are defined by a shift of the line $y = x$ at R distance in positive and negative x directions. The two branches of the parabolas are defined to be tangent to the two parallel lines $y = x - R$ and $y = x + R$ at the connecting points $(R, 2R)$ and $(2R, R)$ and pass through the origin (0,0), it gives the expressions for these two parabolas: $y = \frac{x^2}{4R}$ and $x = \frac{y^2}{4R}$. Note that the use of parabolas to restrict the circular approximation ensures that the constructed function is C^1 on the arc of circle $A_1 A_2$.

Zone I,B. Given a point (x, y) in the first quadrant, Zone I, B, we want to calculate the iso-level value d for the SARDF intersection at this point. It belongs to a circular arc that is tangentially connected to two horizontal and vertical rays when reaching the parabola (see Fig. 5). The equation of this arc is $(x - x_0)^2 + (y - y_0)^2 = r^2$, where x_0, y_0 and r need to be expressed as functions of the searched value d. The point at the intersection of the parabola and the iso-level d of the searched function, is at a distance d from the axis $y = 0$. Because this point belongs also to the parabola, it satisfies $d = \frac{x_0^2}{4R}$. By symmetry it comes that: $d = \frac{y_0^2}{4R}$.

The coordinates of the center of the circular arc (x_0, y_0) satisfy the following equality: $x_0 = y_0 = d + r$, it follows that $r = y_0 - d = 2\sqrt{Rd} - d$. By plugging everything in the equation of the circle, using the substitution of variables $\sqrt{d} = z$ and expanding, we obtain the following algebraic equation of degree four in z:

$$z^4 - 4\sqrt{R}z^3 - 4Rz^2 + 4\sqrt{R}(x + y)z - (x^2 + y^2) = 0 \tag{11}$$

Thus in the first quadrant, in the zone I, B, the expression of the intersection is the square of one of the four roots of the algebraic equation 11. Roots of algebraic equation of degree four are known algebraically. The root of interest is found by using one of the limit conditions, for example $d(2R, R) = R$.

Zone II, inside the bounding band. Given a point (x, y) in zone II, within the bounding band (see Fig. 5), we want to compute the iso-level value d of the SARDF intersection at that point. This point belongs to a circular arc that is tangentially connected to two horizontal and vertical rays when reaching the two lines of the bounding band. The equation of this circular arc is: $(x - x_0)^2 + (y - y_0)^2 = R^2$. This time R is constant, thus only x_0 and y_0 need to be expressed as functions of d.

The coordinates (x_0, y_0) of the circular arc satisfy: $x_0 = y_0 = d + R$. After substitution into the equation of the circular arc and expanding this equation, d is one of the two solutions of the following second degree algebraic equation:

$$2d^2 + d(4R - 2x - 2y) + (x^2 + y^2 - 2R(x + y) + R^2) = 0 \tag{12}$$

The root of interest is obtained by using the limit condition: $d(2R, R) = R$.

Zone I, A and C, and II outside the bounding band. The function behaves exactly like min.

Quadrant 3. In quadrant 3, we consider two possible approaches: one is similar to the approach detailed above (Fig. 4 left), the second uses two lines symmetric in respect to the line $y = x$ and opened by an angle θ (see Fig. 4 right).

Considering a point (x, y) in the quadrant, if it is outside the angle sector made by the two lines, then the value of the function is exactly given by the min function. If it is inside, then the value of the function is computed as follows. The point belongs to the circular arc given by: $(x - x_0)^2 + (y - y_0)^2 = r^2$. x_0, y_0 and r are expressed as functions of d, the searched function value, and α the line-axis angle. It comes that d is the solution of:

$$d^2(tan^2(\alpha) + 2tan(\alpha) - 1) - 2d(x + y)tan(\alpha) + x^2 + y^2 = 0 \tag{13}$$

verifying $d(x, \frac{x}{tan\alpha}) = \frac{x}{tan\alpha}$ (for example).

Quadrants 2 and 4. In quadrants 2 and 4, the smooth min behaves exactly like min.

The final full expression for the SARDF intersection operation is given in the Appendix.

3.3 Smoothness of the SARDF Intersection

The SARDF operations are functions in $C^1(\mathbb{R}^* \times \mathbb{R}^*)$; the singularities in the resulting function are only due to the deining functions of primitives or to the need to model sharp features in the resulting solid. The discontinuity of the gradient in the origin is intentional to allow creation of sharp features.

The main steps of the proof are given for the SARDF intersection only. In quadrants 2 and 4 the function is trivially C^1. In quadrants 1 and 3 the function is symmetric with respect to the line $y = x$. Only the subset below this line needs to be studied. The function is also trivially C^1 piecewise, the discontinuity (of the function values or the derivatives values) can appear only at the boundaries between the different expressions: on the branches of the parabola, the straight lines ($y = x - R$ and $y = x + R$) or the arc boundary A_1A_2 between the growing radius and the fixed radius (quadrants 1 or 3), or on the straight lines (quadrant 3).

C^1 Continuity in Quadrant 1. The SARDF intersection is piecewise continuous. The continuity on the parabola branches, and on the lines is obvious by construction. We need to check it only on the circular arc boundary A_1A_2.

Continuity at the circular arc boundary A_1A_2. We study the continuity of the expression of the SARDF intersection at the circular arc boundary between the zones I,B, and II. Let $\mathbf{P} = (x, y) = (2R + Rcos(u), 2R + Rsin(u))$, with $u \in [\frac{5\Pi}{4}, \frac{3\Pi}{2}]$, be a point on that circular arc. The value of the function at $\mathbf{P}$ is given by its value at the point $\mathbf{A_1}$ and is $y = R$. We check that this value matches the value of the expressions in the zones I,B and II at $\mathbf{P}$, by checking that the algebraic equations 11 and 12 hold. After applying some calculus, we confirm that the relations given by Eq. 11 and 12 hold and we can conclude with the continuity at that boundary.

We give implicit definitions for the partial derivatives of the smooth function d in the quadrant 1, and then prove it is C^1 on that domain, with exception of the origin, by verifying that the partial derivatives match on the boundaries points.

Expression for the partial derivatives in zone I,B: In zone I,B, the square root of the function $\sqrt{d}$ satisfies the algebraic equation 11. Taking the partial derivative by x of 11 gives: $4z_x z^3 - 12\sqrt{R}z_x z^2 - 8Rz_x z + 4\sqrt{R}(x+y)z_x + 4\sqrt{R}z - 2x = 0$. It follows that: $z_x = \frac{2x - 4\sqrt{R}z}{4z^3 - 12\sqrt{R}z^2 - 8Rz + 4\sqrt{R}(x+y)}$. Since $z = \sqrt{d}$, it comes that $z_x = \frac{1}{2}d_x\frac{1}{\sqrt{d}}$. Combining it with the previous expression, we get a relation for the partial derivative of the function in zone I,B

$$d_x = 2\sqrt{d}\frac{2x - 4\sqrt{R}z}{4z^3 - 12\sqrt{R}z^2 - 8Rz + 4\sqrt{R}(x+y)} \tag{14}$$

with $z(x, y) = \sqrt{d(x, y)}$.

By the same procedure, we obtain an expression for the partial derivative by y in zone I,B:

$$d_y = 2\sqrt{d}\frac{2y - 4\sqrt{R}z}{4z^3 - 12\sqrt{R}z^2 - 8Rz + 4\sqrt{R}(x+y)} \tag{15}$$

with $z(x, y) = \sqrt{d(x, y)}$.

Expression for the partial derivatives in zone II, within the bounding band: In zone II, within the bounding band, the function satisfies Eq. 12. With the same method as above, we take the partial derivative by x, it gives: $4d_x d + d_x(4R - 2x - 2y) - 2d + (2x - 2R) = 0$. It follows that:

$$d_x = \frac{2d - 2(x - R)}{4d + (4R - 2x - 2y)} \tag{16}$$

Similarly an expression for the partial derivative by y can be obtained:

$$d_y = \frac{2d - 2(y - R)}{4d + (4R - 2x - 2y)} \tag{17}$$

Expression for the partial derivatives in zone I,A and zone II, below $y = x - R$ The function is exactly min in these two areas, and so the partial derivatives in the x direction is 0 and 1 in the y direction.

We verify the continuity of the partial derivatives at the different boundary curves.

Continuity of the partial derivatives at the parabolic arc. Let $\mathbf{P} = (x, y) = (u, \frac{u^2}{4R})$, with $u \in [0, 2R]$, be a point on the arc, at $\mathbf{P}$, $z = \sqrt{d} = \frac{u}{2\sqrt{R}}$; it is easy to verify that at $\mathbf{P}$ equations 14 and 15 give: $d_x = 0$ and $d_x = 1$.

Continuity of the partial derivatives at the line $y = x - R$. Let $\mathbf{P} = (x, y) = (u, u - R)$, with $u \in [2R, \infty[$ be a point on the line, at P, $z = \sqrt{d} = u - R$; it is easy to verify that at P equations 16 and 17 give: $d_x = 0$ and $d_x = 1$.

Continuity of the partial derivatives at the circular arc boundary. Let $\mathbf{P} = (x, y) = (2R + Rcos(u), 2R + Rsin(u)), u \in [\frac{5\Pi}{4}, \frac{3\Pi}{2}]$ be a point on the circle boundary between the growing radius zone and the constant radius zone. At $\mathbf{P}$, the value of the function is R. Using these informations in equation 14, we obtain after straightforward calculus $d_x = \frac{cos(u)}{cos(u)+sin(u)}$. Similarly, with equation 16: $d_x = \frac{cos(u)}{cos(u)+sin(u)}$.

Similarly for the partial derivative d_y, equation 15 gives: $d_y = \frac{sin(u)}{cos(u)+sin(u)}$. And equation 17 gives: $d_y = \frac{sin(u)}{sin(u)+cos(u)}$.

With the equality of the partial derivatives we can conclude to the C^1 continuity of the SARDF intersection on the boundary circle. The C^1 continuity still holds at A_1, with $u = \frac{3\Pi}{2}$.

***C^1* Continuity in Quadrant 3.** There are two possible approaches in the quadrant 3 to constructing the SARDF intersection. One is similar to the construction in quadrant 1, with a growing radius smoothing bounded by a threshold. The proof for the smoothness of the function is similar as in quadrant 1. We give a sketch of the proof for the second approach, which uses two lines symmetric in respect to the line $y = x$ and opened by an angle θ (see Fig. 4 right). The function is continuous by construction and piecewise C^1, except at the origin. We check only the continuity of the different partial derivatives at a boundary line.

We use equation 13 to have expressions of the partial derivatives inside the two boundary lines.

$$d_x(x,y) = \frac{2d(x,y)tan(\alpha) - 2x}{2d(x,y)(tan^2(\alpha) + 2tan(\alpha) - 1) - 2(x+y)tan(\alpha)} \tag{18}$$

$$d_y(x,y) = \frac{2d(x,y)tan(\alpha) - 2y}{2d(x,y)(tan^2(\alpha) + 2tan(\alpha) - 1) - 2(x+y)tan(\alpha)} \tag{19}$$

It is easy to check that the partial derivatives are continuous at any point on the boundary line, by evaluating d_x and d_y at a point of the line and comparing with the values of the partial derivatives from the adjacent zone (where the function behaves like min, thus has partial derivatives with values 0 along x and 1 along y). Given $\mathbf{P} : (u, \frac{u}{tan(\alpha)}), u \in \mathbb{R}^-$ a point on the lower boundary line, we have $d_x(\mathbf{P}) = 0$ and $d_y(\mathbf{P}) = 1$.

3.4 SARDF Modeling Framework

The SARDF framework is a restricted version of an FRep system where the SARDF implementations of the set-theoretic operations are used instead of the R-functions or min/max, and primitive definitions are limited to distance functions or their approximations. Of course, primitives, which are not defined by distance functions or an approximation, can be used, as well as operations, which do not conserve the distance property or a reasonable approximation, however it will result in a global defining function for the object, which does not keep the Euclidean distance property.

SARDF operations, intersection, difference, and union of two geometric objects defined by distance functions f_1 and f_2 are trivially obtained by replacing syntactically x and y in the previous SARDF functions by f_1 and f_2. Note that x and y can be seen as two real functions $(x, y, z) \rightarrow x$ and $(x, y, z) \rightarrow y$, which correspond geometrically to two orthogonal halfspaces. The smoothness of the resulting function defining the final object depends on the smoothness of the SARDF functions, which was studied above, and on the defining functions of primitives (details can be found in [6]).

4 Constructive Heterogeneous Objects Modeling with SARDF

We present in this section some examples to illustrate the use of the introduced SARDF operations and normal primitives in constructive heterogeneous object modeling. The SARDF operations are used instead of the R-functions or the min/max functions in the different constructive trees to define the geometry of the solid and the partitions where attributes are defined.

We use the term normal primitive to refer to a primitive with a defining function p, which at a given point $\mathbf{x} \in \mathbb{R}^3$, is equal to the Euclidean distance, or its approximation, from $\mathbf{x}$ to the surface $p^{-1}(0)$. A list of primitives with known expressions for the distance is given in [9].

We show in the following through two and three-dimensional examples how different expressions for the set-theoretic operations affect the material distributions and their properties.

4.1 Two-Dimensional Example

At first, we illustrate the use of SARDF for modeling a two-dimensional heterogeneous object. The geometry of the object (Fig. 6a) is defined as $f(\mathbf{X}) \geq 0$, where f is evaluated by traversing the constructive FRep tree [16] with a rectangle and disk primitives in the leaves, and the subtraction operation in the node.

This object is made of two materials and three material regions (Fig. 6b). We use the notation $m_1(\mathbf{X})$ and $m_2(\mathbf{X})$ for the scalar volume fraction component of the materials 1 and 2. For visualization purposes, the material distributions are mapped to a grey color space: a black or constant grey color is attributed to each material and the final grey color is the resulting gradient value corresponding to each material contribution, weighted by the scalar volume fraction.

Among the three material regions, there are two material features corresponding to spaces where: there is only material 1 uniformly distributed (black in Fig. 6b) and there is only material 2 uniformly distributed (constant grey area in Fig. 6b). The last material region corresponds to the functionally graded material. The geometry of each material region is defined using FRep in a constructive way, similarly to the shape's geometry, with SARDF operations in the nodes and normal primitives in the leaves. The resulting functions provide C^1 approximation of the distance to each material region. The distances to the material features are used to specify the functionally graded material.

The scalar volume fraction of each component material in the functionally graded material region is given by: $m_1(\mathbf{X}) = w_1(\mathbf{X})M_1$ and $m_2(\mathbf{X}) = w_2(\mathbf{X})M_2$, where M_1 and M_2 stand for the value of the scalar volume fraction on the boundary of the first and second material features shown in Fig. 6b.

For the weighting functions $w_1(\mathbf{X})$ and $w_2(\mathbf{X})$, we use a normalization of each inverse distance functions:

$$w_1(\mathbf{X}) = \frac{\frac{1}{d_1(\mathbf{X})}}{\frac{1}{d_1(\mathbf{X})} + \frac{1}{d_2(\mathbf{X})}} = \frac{d_2(\mathbf{X})}{d_1(\mathbf{X}) + d_2(\mathbf{X})} \tag{20}$$

$$w_2(\mathbf{X}) = \frac{\frac{1}{d_2(\mathbf{X})}}{\frac{1}{d_1(\mathbf{X})} + \frac{1}{d_2(\mathbf{X})}} = \frac{d_1(\mathbf{X})}{d_1(\mathbf{X}) + d_2(\mathbf{X})} \tag{21}$$

where $d_1(\mathbf{X})$ and $d_2(\mathbf{X})$ are the distances from point $\mathbf{X}$ to the boundary of the material features.

The distance map d_1 is illustrated in Fig. 7. Fig. 7a and Fig. 7b correspond respectively to the approximate distance map d_1 when the R-functions and the SARDF operations are used correspondingly to define the shape. The approximate distance map built using R-functions indicates that despite good smoothness properties, R-functions are not a good approximation to the distance function, making it difficult to control accurately material distributions.

The weighting functions $w_1(\mathbf{X})$ and $w_2(\mathbf{X})$ are continuous, satisfying the interpolation condition $w_i(\partial B_j) = \delta_{ij}$, where $i, j \in 1, 2$, $\delta_{i,j}$ is the Kronecker symbol[1], and ∂B_j are the boundaries of the material features. The functions $w_1(\mathbf{X})$ and $w_2(\mathbf{X})$ form a partition of unity.

The properties of $w_1(\mathbf{X})$ and $w_2(\mathbf{X})$ are illustrated in Fig. 8 with a cross section of the model through the $y-axis$ and the graph of the weighting functions $w_1(\mathbf{X}' = (x, const))$ and $w_2(\mathbf{X}')$ along the x-axis. Note that in the current example w_i and m_i, $i \in 1, 2$, have the same graph, since the values of the volume fraction on the boundaries, M_1 and M_2 have been chosen equal to 1.

There is a C^1 discontinuity at the points on the boundary of the material features (Fig. 8). This can cause the same problems as the distance function C^1 discontinuity. Fortunately, these sharp corners can be smoothened by a modification of the expressions for the coefficients (Eq. 20 and 21). Indeed the expressions used for the material feature weights correspond to a particular case of the inverse distance weighting [27]. More general expressions are: $w_1(\mathbf{X}) = \frac{d_2^k(\mathbf{X})}{d_1^k(\mathbf{X}) + d_2^k(\mathbf{X})}$ and $w_2(\mathbf{X}) = \frac{d_1^k(\mathbf{X})}{d_2^k(\mathbf{X}) + d_1^k(\mathbf{X})}$. The case $k = 1$ gives Eq. 20 and 21. The parameter k controls the smoothness of the functions on the points of the material features.

Replacing every SARDF operation by an R-function or min/max in the constructive trees for the geometry of the solid and the material regions gives different material distributions in the same cross-section (see Fig. 9, right and 9, left). Figure 9 reflects at the level of the material distribution the problems of using the R-functions (right graph), or min/max (left graph) in constructive heterogeneous modeling. Figure 9, right, shows the role played by the accuracy of the distance approximation when the distance is used to parameterize the material

[1] Equals to 1 if $i = j$ and 0 otherwise.

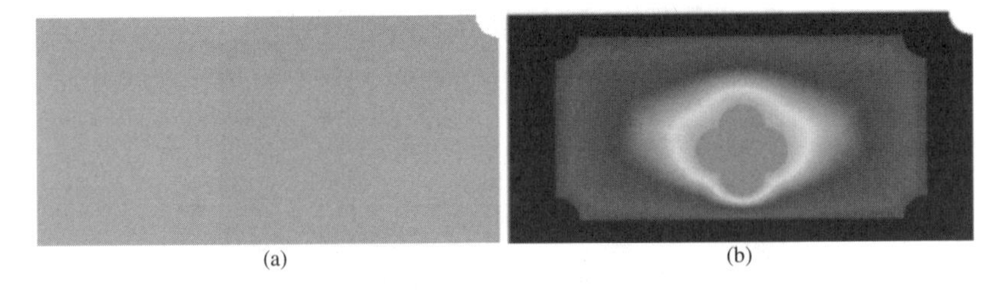

Fig. 6. Two-dimensional CAD part. (a) Geometry of the CAD part defined by an FRep model. (b) Three different material regions (outter black: material 1, inner constant grey: material 2, grey gradient: functionally graded material).

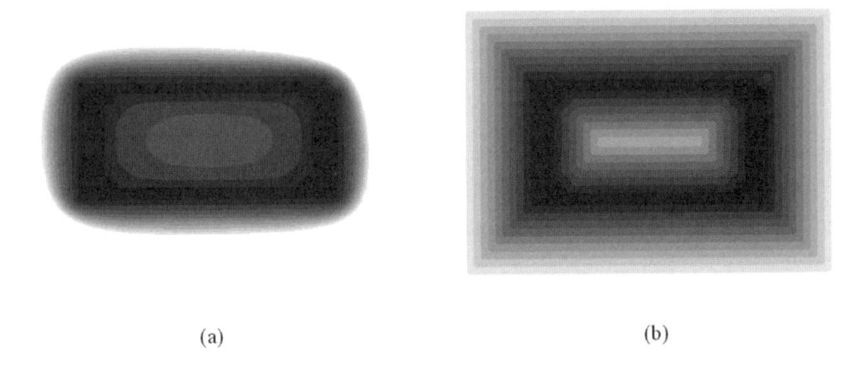

Fig. 7. Approximate distance map d_1 from point **X** to the boundary of the region where only material 1 exists. (a) Using R-functions. (b) Using SARDF operations.

distributions. The unpredictable behaviour of the distance approximation makes the task of the designer difficult. For example, we would expect that the first part of the black curve (just before the intersection with the grey curve) is linear. This behaviour of the R-functions was noticed by Shin and Dutta in [28].

Figure 9, left, illustrates the C^1 discontinuity of the min (and max) functions and its impact on the material distribution. Both distributions of material 1 (black) and 2 (grey) have two points of C^1 discontinuity (circled in Fig. 9, left). It results in problems of stress or concentrations as noticed by Biswas et al in [4].

Using SARDF for the set-theoretic operations does not introduce new points of C^1 discontinuity, and keeps a good approximation of the distance, these properties can be seen in the graph of the material distributions (Fig. 8).

In this example (as well as in the following) only two materials are in the overlapping zone. More materials can be blended and the expressions for inverse distance weighting (Eq. 20 and 21) can be extended for the cases where more than 2 materials are blended. Additional details on the inverse distance weighting used for the interpolation of materials defined over functionally defined sets can be found in [23]. More complex expressions for compositions of multiple materials,

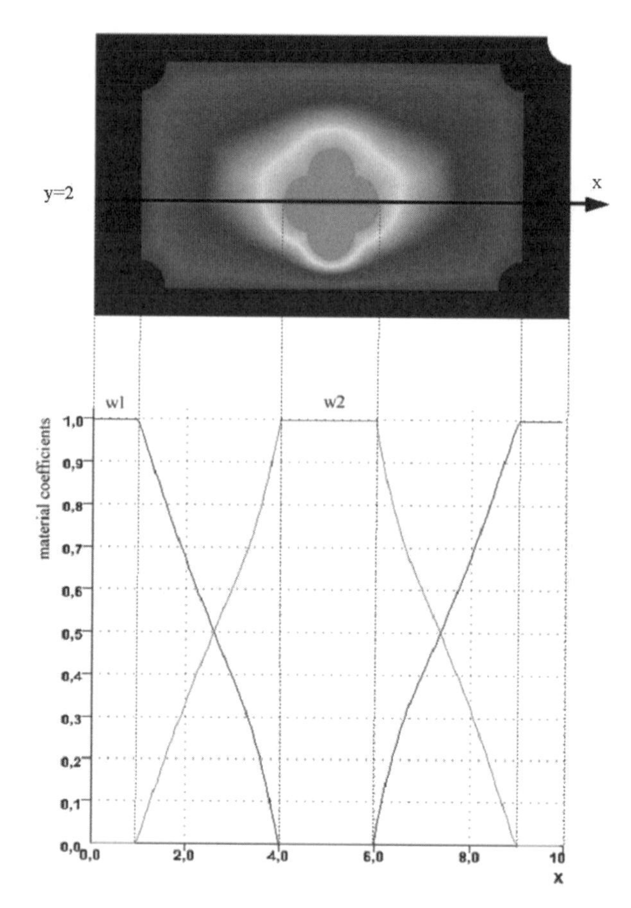

Fig. 8. A cross-section parallel to x-axis and the distribution of the materials in the cross section for the CAD part constructed with SARDF functions

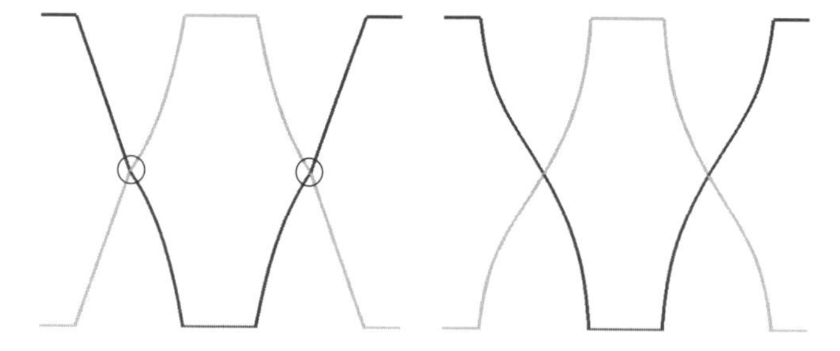

Fig. 9. Material distributions in the cross-section $y = 2$ for materials 1 and 2 using: left, min/max in the constructive trees for the geometry of the solid and the material regions. The circled points correspond to points of C^1 discontinuity of the material distributions. Right: R-functions in the constructive trees for the geometry of the solid and the material regions.

like vector valued materials, constrained and weighted interpolation of materials, can be found in [4].

4.2 Three-Dimensional CAD Part

The second example illustrates more complex three-dimensional shapes for the geometry of the object and the material features.

The overall geometry of the object is a block with two (constant) material features inside. We keep the same notation as in the previous subsection, with $m_1(\mathbf{X})$ and $m_2(\mathbf{X})$ the scalar volume fraction of the materials 1 and 2. Figure 10a shows the first material feature corresponding to the material 1 (in black); it is cut by a planar half-space for visualization purposes only. Figure 10b shows the second material feature (in red); its geometry is composed of blocks and ellipsoids, combined with SARDF unions and intersections. The right of Fig. 10b illustrates a zoom on one of the pins. Such a pin is modeled with ellipsoids as primitives and SARDF union and intersection as operations. Exactly, it is the SARDF union of four ellipsoids, which are after subtracted from a fifth ellipsoid.

To express the material behaviour in the region between the two material features (this region can be seen in Fig. 10c), we use the equations 20 and 21 for the weights for each material feature. It indicates that the closest material feature has the strongest influence. The overall distribution of the materials is

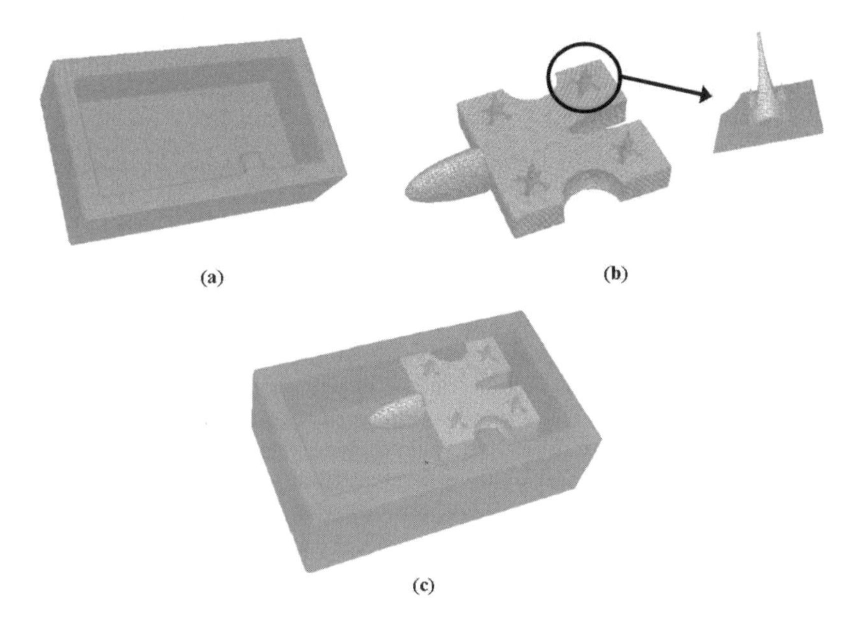

Fig. 10. Geometry of the material features of the 3D CAD part: (a) the first material feature, (b) the second material feature, with a zoom on one of the pins, on the right, (c) union of the two material features

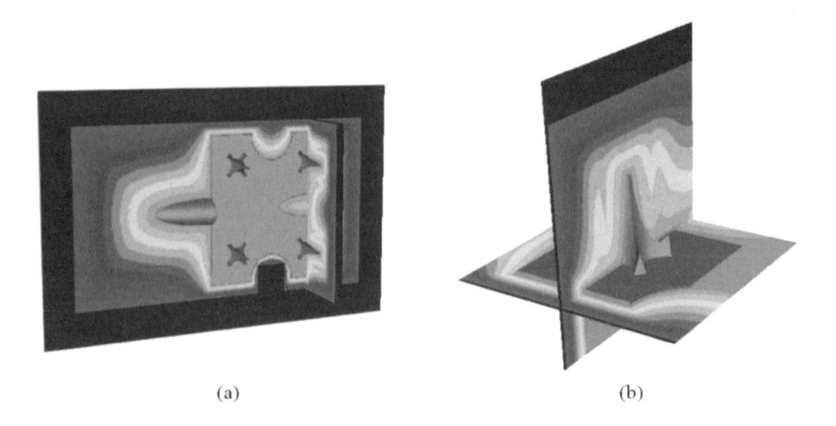

Fig. 11. Distribution of two materials. Black color corresponds to material 1, constant grey color to material 2. The grey intensity variation indicates the fraction of each material. (a) Two cross sections are made for $x = 0$ and $y = 0$ to show the material distribution. (b) A zoom is made on one of the pins with two additional cross-sections.

shown in Fig. 11a. The geometry corresponding to the second material feature is rendered, using a grey constant color, then for the visualization of the material distribution, two cross-sections are made: one for $x = 0$ and one for $z = 0$. For each of the cross-section, the material distribution is projected to a grey color.

5 Conclusion

A framework for the construction of heterogeneous volumetric objects using distance-based scalar fields has been presented. This framework is named SARDF, for Signed Approximate Real Distance Function and is a subset of the Function Representation (FRep) model [16]. Within this framework, an object is modeled in a constructive way by applying operations (SARDF operations) on primitives (defined by the Euclidean distance function or its approximation). Primitives and operations are defined functionally, with a closure property: the final object, obtained by applying operations to functions, is also described by a function.

Extra requirements on the overall defining function are imposed: its value should return an approximation of the signed Euclidean distance function, and it should be smooth. The traditional implementations of set-theoretic operations, namely the R-functions and min/max, either suffer from inaccurate Euclidean distance approximation (R-functions), or are not smooth and create discontinuities of the partial derivatives of the constructed function (min/max). Therefore, we have introduced new functions, and presented their constructions and implementations for defining set-theoretic operations that keep a reasonable approximation of the distance function and are smooth. We have proven that SARDF operations are C^1.

We have proposed to use the SARDF framework for constructive heterogeneous object modeling as an extension of the constructive hypervolume model [15]. The distance function is used here to parameterize the material distribution [4]. We have illustrated through case studies the properties and importance of distance and smoothness of SARDF in constructive heterogeneous object modeling.

Appendix: Expression for the SARDF Intersection

Let f_1 and f_2 be the distance functions defining two solids by $f_1 > 0$ and $f_2 > 0$. The intersection between the associated solids is defined by applying the SARDF intersection ($\wedge_s$) function on f_1 and f_2 as follows:

Case 1: $f_1 > 0$ and $f_2 > 0$

In the following paragraph E_1 corresponds to the boolean expression: $E_1 = f_1 < R$ or $f_2 < R$ or $(f_1 < 2R$ and $f_2 < 2R$ and $(f_1 - 2R)^2 + (f_2 - 2R)^2 > R^2)$.

- if E_1 and $f_2 > \frac{f_1^2}{4R}$ and $f_1 > \frac{f_2^2}{4R}$, then $f_1 \wedge_s f_2 = z^2$. Where z is the root of $z^4 - 4\sqrt{R}z^3 - 4Rz^2 + 4\sqrt{R}(x+y)z - (x^2+y^2) = 0$ verifying $z^2(2R, R) = R$.
- if E_1 and $f_2 \leq \frac{f_1^2}{4R}$, then $f_1 \wedge_s f_2 = f_2$.
- if E_1 and $f_1 \leq \frac{f_2^2}{4R}$, then $f_1 \wedge_s f_2 = f_1$.
- if $not(E_1)$ and $f_1 - R < f_2 < f_1 + R$, then $f_1 \wedge_s f_2 = \frac{1}{2a}(-b + \sqrt{b^2 - 4ac})$. Where $a = 2$, $b = -2f_1 - 2f_2 + 4R$ and $c = f_1^2 + f_2^2 - 2f_1R - 2f_2R + R^2$.
- if $not(E_1)$ and $f2 \leq f_1 - R$, then $f_1 \wedge_s f_2 = f_2$.
- if $not(E_1)$ and $f2 \geq f_1 + R$, then $f_1 \wedge_s f_2 = f_1$.

Case 2: $f_1 \leq 0$ and $f_2 \geq 0$

$f_1 \wedge_s f_2 = f_1$.

Case 3: $f_1 < 0$ and $f_2 < 0$

E_2 corresponds to the boolean expression: $E_2 = f_1 > -R$ or $f_2 > -R$ or $(f_1 > -2R$ and $f_2 > -2R$ and $(f_1 + R)^2 + (f_2 + R)^2 < R^2)$.

- if E_2 and $f_2 < -\frac{f_1^2}{4R}$ and $f_1 < -\frac{f_2^2}{4R}$, then $f_1 \wedge_s f_2 = z$. Where z is the root of $\frac{d^4}{16R^2} - \frac{d^3}{2R} + \frac{1}{2R}d^2(f_1 + f_2 - 2R) + (f_1^2 + f_2^2) = 0$ verifying $z(-2R, -R) = -2R$.
- if E_2 and $f_2 \geq -\frac{f_1^2}{4R}$, then $f_1 \wedge_s f_2 = f_2$.
- if E_2 and $f_1 \geq -\frac{f_2^2}{4R}$, then $f_1 \wedge_s f_2 = f_1$.
- if $not(E_2)$ and $f_1 - R < f_2 < f_1 + R$, then $f_1 \wedge_s f_2 = \frac{1}{2a}(-b + \sqrt{b^2 - 4ac})$. Where $a = 2$, $b = -2f_1 - 2f_2 + 4R$ and $c = f_1^2 + f_2^2 - 2f_1R - 2f_2R + R^2$.
- if $not(E_2)$ and $f2 \leq f_1 - R$, then $f_1 \wedge_s f_2 = f_2$.
- if $not(E_2)$ and $f2 \geq f_1 + R$, then $f_1 \wedge_s f_2 = f_1$.

Case 4: $f_1 \geq 0$ and $f_2 \leq 0$

$$f_1 \wedge_s f_2 = f_2.$$

References

1. Barthe, L., Dodgson, N.A., Sabin, M.A., Wyvill, B., Gaildrat, V.: Two-dimensional potential fields for advanced implicit modeling operators. Computer Graphics Forum 22(1), 23–33 (2003)
2. Barthe, L., Wyvill, B., De Broot, E.: Controllable binary csg operators for soft objects. International Journal of Shape Modeling 10(2), 135–154 (2004)
3. Biswas, A., Shapiro, V.: Approximate distance fields with non-vanishing gradients. Graph. Models 66(3), 133–159 (2004)
4. Biswas, A., Shapiro, V., Tsukanov, I.: Heterogeneous material modeling with distance fields. Comput. Aided Geom. Des. 21(3), 215–242 (2004)
5. Breen, D.E., Whitaker, R.T.: A level-set approach for the metamorphosis of solid models. IEEE Transactions on Visualization and Computer Graphics 7(2), 173–192 (2001)
6. Fayolle, P.-A.: Construction of volumetric object models using distance-based scalar fields. PhD Thesis, University of Aizu, Japan (2006)
7. Fayolle, P.-A., Pasko, A., Schmitt, B., Mirenkov, N.: Constructive heterogeneous object modeling using signed approximate real distance functions. Journal of Computing and Information Science in Engineering, Transactions of the ASME 6(3) (September 2006)
8. Frisken, S.F., Perry, R.N., Rockwood, A.P., Jones, T.R.: Adaptively sampled distance fields: a general representation of shape for computer graphics. In: Proceedings of the 27th annual conference on Computer graphics and interactive techniques, pp. 249–254. ACM Press/Addison-Wesley Publishing Co. (2000)
9. Hart, J.: Sphere tracing: A geometric method for the antialiased ray tracing of implicit surfaces. The Visual Computer 12(10), 527–545 (1996)
10. Hart, J.C.: Distance to an ellipsoid. In: Heckbert, P. (ed.) Graphics Gems IV, pp. 113–119. Academic Press, Boston (1994)
11. Hoff, K., Culver, T., Keyser, J., Lin, M., Manocha, D.: Fast computation of generalized voronoi diagrams using graphics hardware. In: Proceedings of ACM SIGGRAPH, pp. 277–286. ACM, New York (1999)
12. Hsu, P.-C., Lee, C.: The scale method for blending operations in functionally-based constructive geometry. Computer Graphics Forum 22(2), 143–158 (2003)
13. Jones, M., Chen, M.: A new approach to the construction of surfaces from contour data. Computer Graphics Forum 13(3), 75–84 (1994)
14. Mauch, S.: Efficient Algorithms for Solving Static Hamilton-Jacobi Equations. PhD thesis, California Institute of Technology (2003)
15. Pasko, A., Adzhiev, V., Schmitt, B., Schlick, C.: Constructive hypervolume modeling. Graphical Models 63(6), 413–442 (2001) (Special issue in volume modeling)
16. Pasko, A., Adzhiev, V., Sourin, A., Savchenko, V.: Function representation in geometric modeling: concept, implementation and applications. The Visual Computer 11(8), 429–446 (1995)
17. Pasko, A., Savchenko, V.: Blending operations for the functionally based constructive geometry. In: set-theoretic Solid Modeling: Techniques and Applications, CSG 1994 Conference Proceedings, pp. 151–161. Information Geometers (1994)

18. Ricci, A.: A constructive geometry for computer graphics. The Computer Journal 16(2), 157–E160 (1973)
19. Roessl, C., Zeilfelder, F., Nurnberger, G., Seidel, H.-P.: Spline approximation of general volumetric data. In: Proceedings of ACM Solid Modeling 2004 (2004)
20. Rvachev, V.: On the analytical description of some geometric objects, vol. 153(4), pp. 765–767 (1963)
21. Rvachev, V.: Methods of Logic Algebra in Mathematical Physics (in Russian). Naukova Dumka, Kiev (1974)
22. Rvachev, V.: Theory of R-functions and Some Applications (in Russian). Naukova Dumka, Kiev (1982)
23. Rvachev, V.L., Sheiko, T.I., Shapiro, V., Tsukanov, I.: Transfinite interpolation over implicitly defined sets. Computer Aided Geometric Design 18, 195–220 (2001)
24. Sabin, M.: The use of potential surfaces for numerical geometry. Technical Report VTO/MS/153, British Aircraft Corporation (1968)
25. Sethian, J.: Level-Set Methods and Fast Marching Methods. Cambridge University Press, Cambridge (1999)
26. Shapiro, V.: Theory of r-functions and applications: A primer. Technical report, Cornell University (November 1988)
27. Shepard, D.: A two-dimensional interpolation function for irregularly spaced data. In: Proceeding 23 National Conference, vol. 23, pp. 517–524. ACM, New York (1968)
28. Shin, K.-H., Dutta, D.: Constructive representation of heterogeneous objects. Journal of Computing and Information Science in Engineering 1, 205–217 (2001)
29. Sud, A., Otaduy, A., Manocha, D.: Difi: Fast 3d distance field computation using graphics hardware. Computer Graphics Forum 23(3), 557–566 (2004) (Proceedings of Eurographics 2004)
30. Tsai, Y.R.: Rapid and accurate computation of the distance function using grids. J. Comput. Phys. 178(1), 175–195 (2002)
31. Zhao, H.: A fast sweeping method for eikonal equations. Mathematics of Computation (2004)
32. Zhou, Y., Kaufman, A., Toga, A.: 3d skeleton and centerline generation based on an approximate minimum distance field. The Visual Computer 14(7), 303–314 (1998)

Feature-Based Material Blending for Heterogeneous Object Modeling

Kuntal Samanta and Bahattin Koc

Department of Industrial Engineering, University at Buffalo
401 Bell Hall,
Buffalo, NY 14260, USA
{ksamanta, bkoc}@buffalo.edu

Abstract. In this paper, a new feature-based material blending method is proposed to represent and design heterogeneous objects. Geometric features dictating the material variation are defined as material governing features to control material composition inside the objects. Interrelations between the material governing features and material attributes are established by constraining the geometric and material features and retained in the object model. Using these relationships, variant heterogeneous objects are developed easily by changing the geometric and material features of the heterogeneous object. Geometric methods are developed to blend not only the geometric features but also the property requirements at each of the feature. To obtain the best material variation inside the object, an optimization-based solution method based on the object's functional requirements are developed. Implementation and illustrative examples are also presented in this paper.

Keywords: Heterogeneous object design, feature-based design and modeling, material features, design optimization.

1 Introduction

In many engineering applications, the products are designed to meet one or more functional property requirements for a particular real life application. For instance, the designed part might be required to bear some load or pressure applied to it, or it might be required to withstand very high temperatures. The designed product must possess the intended functional properties to perform satisfactorily in its application. The product's properties depend on not only its geometric shape but also the properties of the material which it is made of. Therefore, the design process involves determining the object's geometric and material attributes that are most suited for the intended application. Traditionally, the design process starts with choosing the most suitable material whose properties are consistent with those desired in the product. For instance, steel is preferable to ceramic where the object must exhibit good tensile strength.

Traditionally, the underlying assumption in the design process is that the final object is made of a one single material (homogeneous material). The main reason behind this assumption is that most traditional manufacturing processes such as casting and molding are capable of producing only homogeneous products.

A. Pasko, V. Adzhiev, and P. Comninos (Eds.): HOMA, LNCS 4889, pp. 142–166, 2008.

In most applications, choosing one suitable homogeneous material satisfies all the property requirements imposed on the product. However, in many engineering applications, there can be conflicting or special functional property requirements imposed on an object such that no homogeneous objects can meet all the requirements satisfactorily. This is because homogeneous materials usually contain some desired and some undesired properties.

Use of multi material objects is a possible solution in case of conflicting or special property requirement cases. However, multi-material objects exhibit sharp interfaces between the lumps of dissimilar materials. In course of time, often stress concentration develops at these interfaces due to poor interfacial bonding strength. This eventually separate the materials creating a crack in between which ultimately leads to failure of the object [12].

Therefore, it is more desirable to eliminate the sharp interfaces and obtain a smooth variation among different materials. Objects with such kinds of material composition variation are called heterogeneous objects. These objects do not have any sharp interface of dissimilar materials and therefore exhibit a gradual change in material composition and associated properties in geometric space [26].

To satisfy the functional property requirements imposed on a heterogeneous object, material compositions must vary continuously throughout its geometric domain. This enables the object to exhibit different properties at its different regions which play a key role behind the object's performance. Therefore, besides geometry, the material variations also need to be designed so that the object can best satisfy all the functional requirements [23].

Designing heterogeneous objects is a complex task because the design process involves not only developing models for both the object geometry but also its varying material compositions. Any new method of heterogeneous object design needs to have a few basic capabilities which are indispensable. For the purposes of visualization, analysis, modifications and fabrication, the design method must allow for creation of a solid model of the object containing not only the complex geometry, but also the material composition information of the object [26].

This paper presents a novel integrated design methodology which establishes and integrates geometric and material attributes of a heterogeneous object by utilizing the principles of feature-based design. The heterogeneous object is represented as a feature-based model where both the geometry and material attributes are identified as object features. Continuous material composition variations are represented with relation to the object's geometric features as functions of parametric distances in one, two or three parametric dimensions. Given the initial geometric model of the object, the material features are determined through optimization techniques. In this research work, the focus is given to development of methodologies in design and representation of heterogeneous objects. Determination of exact overall material properties and functional requirements are beyond the scope of this research.

The rest of the paper is organized as follows: Section 2 reviews the literature. In Section 3, the developed methodology for feature-based design of heterogeneous objects is described. Section 4 explains the methodologies for material blending. Feature matching for material blending is presented in Section 5. Section 6 describes the optimization processes to establish the material features. Implementation and examples are presented in section 7. Finally, conclusions are drawn in Section 8.

2 Literature Review

In this paper, heterogeneous object modeling is defined as the process of developing and storing a heterogeneous object's geometry and material information. Modeling is an integral part of design optimization which is the process of establishing the object's geometry and material composition variation for a specific application.

In the literature, several modeling methodologies have been proposed: the r_m-object approach by Kumar and Dutta [11], its extension into constructive methods of heterogeneous object representation by Shin and Dutta [29], grading source based approach by Siu and Tan [30] and its application to fiber type reinforcement composites modeling [31]. All these methods essentially represent the material variation as functions of spatial position relative to a reference entity and material variations were assumed to be given *a priori*.

Ma *et al* [14] developed voxel based volume modeling where the object geometry is discretized into very small cubes called voxels. Constant material compositions are assigned separately to each voxel. Voxel based methods are based on discrete units and their accuracy is determined by the number of voxels used in the model. Therefore the voxel based representation may not be as accurate as a continuous representation. Moreover a voxel model after calculated cannot be used to obtain similar variant models as the original topology and surface information has been lost.

Some researches, such as design of heterogeneous flywheel [6] and injection mold cooling systems [5] have been reported to present the advantages of heterogeneous objects in specific applications. The modeling methods used in these researches appear to be rather application-specific and possible extensions and applications of these methods to generic design cases require further study.

In a heterogeneous object design problem, one of the principal tasks is to establish the material attributes of the object. In the literature, it was assumed that the variation follows a polynomial of a certain order and the job of the designer is to calculate the coefficients of the polynomial terms [15]. This method may sometimes be erroneous because the actual variation may follow a polynomial of an order higher than that of the designer's guess, or worse, the variation might not follow any polynomial at all.

B-spline volume representation [18, 16] has been used to represent free-form heterogeneous objects. Qian and Dutta [18] have presented a heterogeneous object design method that uses B-spline volumes to model the object geometry. The material variation is represented by a physic-based process called diffusion. In the paper, the material properties are specified at the object control points, assuming that the material variation would always be represented as B-spline functions of geometry. In case the properties are specified on the object geometry itself, expressing this information in terms of control points might be difficult. In [16], a new representation method to specifying attribute data across a trivariate NURBS volume has been proposed. Although, the presented method is not specifically used for heterogeneous object representation, the method could be used to represent the material composition inside a heterogeneous object.

Biswas *et al* [2] have shown that any material function can be converted to a canonical form of material variation based on Taylor series approximation. The canonical form is an approximate polynomial function of Euclidean distances.

In a prototype CAD system developed by Bhashyam *et al* [1], a designer starts with an initial r_m-object model [11] and tries to improve upon it based on his/her experience. The material variations in the model are chosen from in-built library functions which are mostly expressed as polynomials. Whenever changes are made in the model, a Finite Element Analysis is carried out to decide whether to accept the changes or not. The library functions are not derived as a part of the design process, rather they are collected from papers on manufacturing listed in the literature. Therefore, in case the chosen library functions do not work, the designer may not have any other alternatives.

Chen and Feng [3] provide an approach based on axiomatic design principles. The method creates a 3D variational geometric model of the component before it is divided into several "regions" using commercial FEA software. Material composition in each region is assumed to be constant. Optimal material constituent composition are selected using optimization techniques such as sensitivity analysis and steepest descend method. Since the material attributes are not expressed as a function of geometry, the material composition must be stored explicitly in every region. Therefore, in cases of complex objects where a large number of regions are created, memory space requirement to store material compositions for all the regions might be prohibitive.

Jackson *et al* [7] have developed a heterogeneous object modeling method with local composition control (LCC) based on a subdivision method. The barycentric Bernstein blending functions are used to define the material composition inside the object. The specification of composition is calculated as a function of the distance from the surface of a part. Liu *et al* [13] has proposed a parametric feature-based methodology for the design of solids with local composition control (LCC). Material composition features are related to the geometry of the designed which allows changing the geometry and material composition simultaneously until a satisfactory result is obtained. The Euclidean digital distance transform and Boundary Element Method are used to calculate the material composition. These methods are used to fabricate objects with three-dimensional printing (3DP) process [21]

The material variation information in the object model is also important considering the manufacturing aspect of heterogeneous objects. Recently, a number of Rapid Prototyping (RP) processes have been found capable of manufacturing these objects, such as three-dimensional printing (3DP) [21], shape deposition manufacturing (SDM) [32], multi material selective laser sintering (M^2SLS) [8], laser engineered net shaping (LENS) [4]. A new molding method proposed in [9] and [10] is also capable of fabricating heterogeneous objects with continuous material variation. Therefore, the object model also needs to contain continuous material variation.

In this paper, to model heterogeneous objects, feature-based methods are employed to retain the geometry-material relationship inside the object. To represent the continuous spatial distribution of material compositions, B-spline functions are used which can represent virtually any shape of material variation. These B-spline functions are not known *a priori* but are derived through an optimization process. In this paper, the focus is given to development of methodologies in design and representation of heterogeneous objects. Related topics such as determination of exact overall material properties and stress analysis are beyond the scope of this paper. The details are presented in the following sections.

3 Feature-Based Design of Heterogeneous Objects

In the feature-based design method, an object model is modeled using its features [27, 23]. The features are constrained by various parameters which specify their relationships with other features. In this approach, the features of an existing model can be changed to obtain a new model, called a *variant*, and the process is called *variational design*. In the variant model, all the relationships and constraints of the parent model are maintained. In this research, the material attributes are developed with relation to the geometric features for variational heterogeneous object design. These relationships are formalized and established as object-material constraints.

In feature-based modeling, a free-form object **O** is modeled as a collection of two components – a set of features (**F**) and a set of relations (**R**) as follows [23]:

$$\begin{aligned} &\mathbf{O} = (\mathbf{F}, \mathbf{R}) \\ &\mathbf{F} = \{FF_a\}_{a=0,\ldots,A} \\ &\mathbf{R} = \{R_{a_1a_2}\}_{a_1=0,\ldots,A;\,a_2=0,\ldots,A;\,a_1\neq a_2} \end{aligned} \tag{1}$$

Each feature FF_a is represented as a collection of surfaces on the boundary of the object. **R** contains the relations $R_{a_1a_2}$ between a pair of features FF_{a_1} and FF_{a_2}. These relations are considered constraints which remain valid even when the features are changed or modified.

The main driving forces in feature-based design are the object features. The objects under design consideration are required to possess some properties so that they can function properly in their respective applications. During the design process, these property requirements dictate the object's geometry and the material. Specific features of an object that dominate the design material selection are termed as *material governing features* (*GF*).

In cases where there are different property requirements at more than one feature in a single object, such that each material governing feature tries to dominate different materials, there might be a conflict of material selection. It is assumed that properties of a mixture of materials are contributed proportionally by each of the materials. Therefore, a fixed ratio composite of all the suitable materials may not satisfy the conflicting requirements. The best solution is to vary the material composition inside the object such that the most satisfactory performance is achieved. Since the material composition plays an important role behind the object's performance, the variation of material composition is defined as a *material feature*.

To include material features, the traditional feature-based model given in Equation (1) needs to be modified to include two more characteristic sets, namely, the material composition and composition variation to Equation (1). The point-set constituting a heterogeneous object made of n primary materials $M_1, M_2, \ldots, M_n$ is denoted by **O** as follows [26, 23]:

$$\begin{aligned}
&\mathbf{O} = (\mathbf{F}, \mathbf{R}, \mathbf{M}, \mathbf{C}); \\
&\mathbf{F} = \{ \{FF_a\}_{a=0,\ldots,A}, \{GF_b\}_{b=0,\ldots,B} \} \in E^3 \\
&\mathbf{R} = \{ R_{a_1 a_2} \}_{a_1=0,\ldots,A;\, a_2=0,\ldots,A;\, a_1 \neq a_2} \\
&\mathbf{M} = \left\{ \{M_k, f^{(k)}(r,s,t)\}_{k=1,\ldots,n} \right\} \qquad \{M_k\} \in M^n \\
&\mathrm{C} = \{C_b\}_{b=0,\ldots,B}
\end{aligned} \tag{2}$$

where E^3 is the three-dimensional Euclidean space and M^n is the n-dimensional material space. The material governing features (*GF*s) are identified and separated from the other form features. The governing features and their relations with other form features are still maintained in the relations set **R**. The model hierarchy is shown in Figure 1.

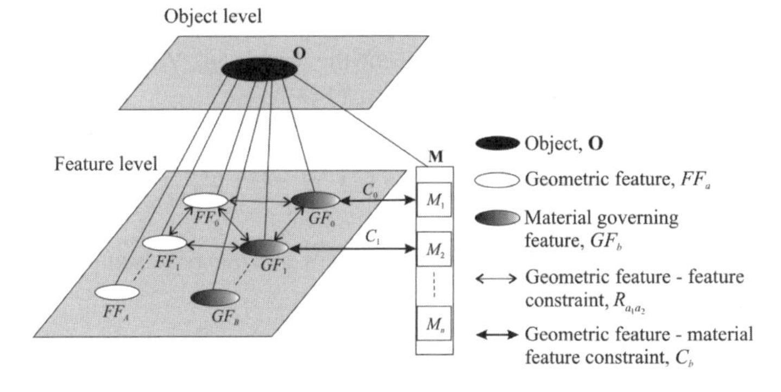

Fig. 1. Feature-based model of a heterogeneous object [23]

Form features $\{FF_a\}_{a=0,\ldots,A}$, and the material governing features $\{GF_b\}_{b=0,\ldots,B}$ are represented as B-rep models with free-form surfaces represented by B-spline functions of parameters *u* and *v*:

$$FF_a(u,v) = \sum_{i=0}^{\alpha} \sum_{j=0}^{\beta} N_{i,p}(u)\, N_{j,q}(v)\, P_{i,j,a} \tag{3}$$

As described before, the *geometric constraint set*, **R**, specify the relationships between the various form features of the object. The object **O** now has $B + 1$ number of material governing features $\{GF_b\}$, $b = 0, \ldots, B$. The object **O** is composed of n primary materials which are the elements of the *material composition vector*, **M**. Therefore, the object has n material features. Each material feature contains two pieces of information, the name of the primary material M_k and the 3-dimensional mathematical form of its variation $f^{(k)}(r,s,t)$ as a B-spline function. The material features are represented by free-form entities, such as curve, surface and volumes.

Upon evaluation, $f^{(k)}(r,s,t)$ gives the value of volume fraction of M_k as a function of the parameters r, s and t [23].

The material variation can be one-, two- or three-dimensional. For objects with one-dimensional material variation, a 2D B-spline curve is used to represent the variation. In Euclidean space E^3, one dimension — the x- coordinate gives the parametric distance from a relative feature and the other dimension — the z-coordinate gives the volume fraction of the respective material. Volume fraction $f^{(k)}$ of material M_k is given as a parametric curve as [23]:

$$f^{(k)}(t) = \sum_{j=0}^{\varepsilon} N_{j,\rho}(t)\, Q_j^{(k)} \tag{4}$$

where, $Q_j^{(k)}$ are the control points that control the volume fraction $f^{(k)}$ of material M_k. In a similar way, for objects with two-dimensional material variation, a B-spline surface of degree (ρ, θ) is used to represent the variation. Volume fraction $f^{(k)}$ of material M_k is given as a parametric surface as [23]:

$$f^{(k)}(s,t) = \sum_{j_1=0}^{\varepsilon} \sum_{j_2=0}^{\phi} N_{j_1,\rho}(s)\, N_{j_2,\theta}(t)\, Q_{j_1 \cdot j_2}^{(k)} \tag{5}$$

Similarly, three-dimensional material variation can be represented by a B-spline volume of degree (ρ,θ, φ) as [23]:

$$f^{(k)}(r,s,t) = \sum_{j_1=0}^{\varepsilon} \sum_{j_2=0}^{\phi} \sum_{j_3=0}^{\omega} N_{j_1,\rho}(r) N_{j_2,\theta}(s)\, N_{j_3,\varphi}(t)\, Q_{j_1 \cdot j_2, j_3}^{(k)} \tag{6}$$

The overall properties at any point P ∈ **O** inside a heterogeneous object are directly proportional to the volume fractions of the constituent materials. If the materials $M_1, M_2, \ldots, M_n$ have values of associated properties $\pi_{M_1}, \pi_{M_2}, \ldots, \pi_{M_n}$ respectively, the overall property at a point P is given as:

$$\Pi^P = f^{(1)}.\pi_{M_1} + f^{(2)}.\pi_{M_2} + \ldots + f^{(n)}.\pi_{M_n} \tag{7}$$

where $f^{(k)}$ is the volume fraction of material M_k. By keeping the geometry unchanged, varying the material composition can directly vary the overall properties of a heterogeneous object.

A feature-based heterogeneous object model created for the first time based on a set of design variables is called an *initial model*. Features of the initial model can easily be changed to obtain different object models, called *variants*. In Figure 2(a) shows an example design with material composition varying between two material governing features, GF_1 and GF_2 and the variation direction t is from GF_1 and GF_2. Plotted below each model are the B-spline curves that represent the material features. The first and last control points of the curve are located on GF_1 and GF_2, respectively (thereby constrained). Figure 2(b) and (c) show variant models by changing material and geometric features respectively [24].

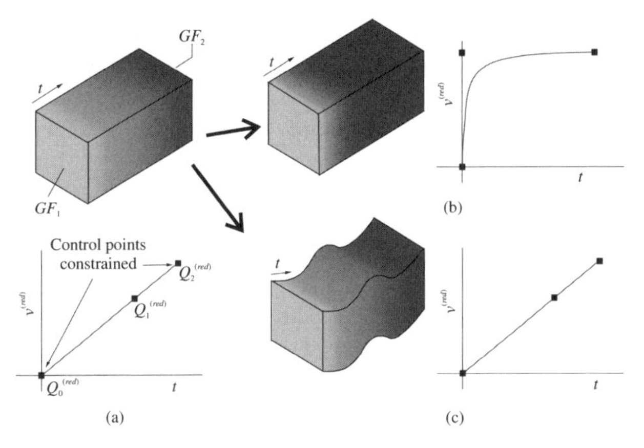

Fig. 2. 1D material feature (curves), corresponding variation in the solid model and variants, (a) initial model with 1D material features, (b) and (c) variant models [24]

In the block in Fig 3(a), material composition is varied between $GF_1 - GF_2$ and $GF_3 - GF_4$. The associated B-spline surfaces represent the red material features. Variant models from are shown in Figs. 3(b) and 3(c).

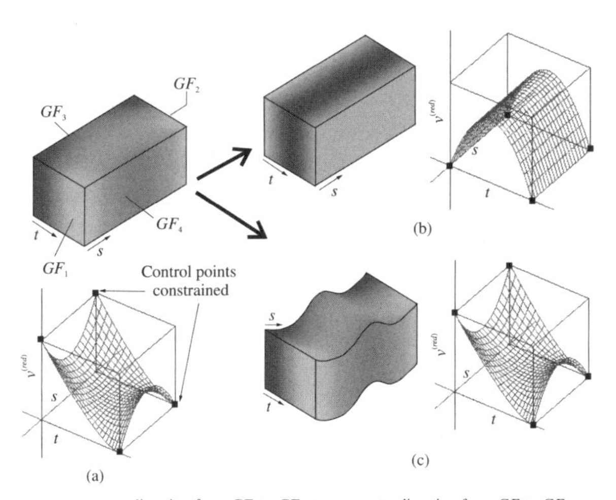

Fig. 3. 2D material features (surfaces), corresponding variations in the solid models and variants, (a) initial model, (b) and (c) variant models [24]

4 Material Blending

In this section, a mathematical model for determining the material composition variation with respect to the material governing features (*GF*s) is determined. A blending model is proposed to represent the continuous variation of property requirements (and therefore the material composition) among a set of material governing features (*GF*s). The material features are constrained to these governing features and the direction of the variation is the same as the parametric lofting direction.

Traditionally, the lofting operation [20] is used to generate a blended entity which from a set of lower dimensional entities called *generators* (curves and surfaces in 1-D and 2-D respectively). As shown in Figs. 4 and 5, the process can blend not only the geometric shape of the generators but also the property requirements at each of the generators. In the same way lofting can be used to get a smooth transition from one governing feature to another. It is assumed that each isoparametric entity in the blend direction will represent constant property requirements. Therefore, to find out and establish the material features, the loft entity must be constructed first.

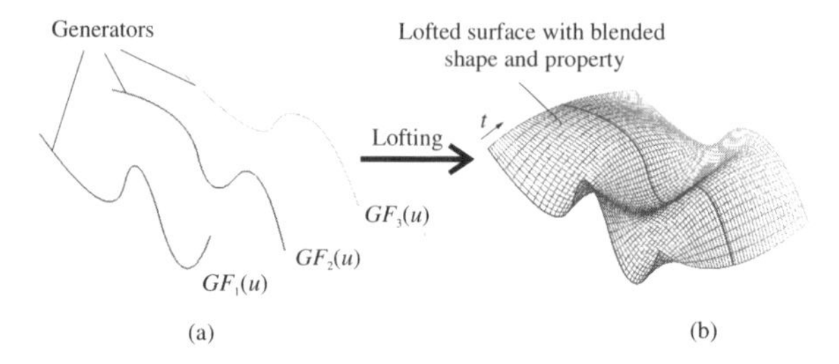

Fig. 4. (a) Generator curves with different property requirements (represented by different colors) and (b) lofted surface blends both geometric shape and property requirements [24]

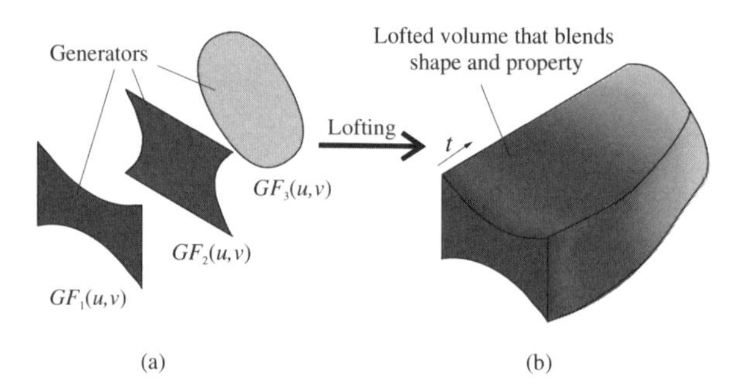

Fig. 5. (a) Generator surfaces with different property require-ments (represented by different colors) and (b) lofted volume blends both geometric shape and property requirements [24]

In this research, it is assumed that the property requirement at a point can be expressed as a function of the parametric distance from a governing feature, *GF*. In prismatic (regular) shaped objects, these functions are available.

As example, consider a pressure vessel carrying fluid at high temperature T_{in} under high pressure P_{in} and the outside of the vessel is exposed to ambient pressure P_{out} and temperature T_{out}. The vessel of length L has its inner and outer radii equal to R_{in} and R_{out}, respectively. The vessel is given as a feature model with four form features, namely inside, outside, top and bottom, as shown in Fig. 6(a). The designer identifies the inside and outside surfaces of the vessel as the material-governing features, GF_1 and GF_2, as shown in Fig. 6(a). The material variation vector t is the parametric direction from the inside surface GF_1 to the outside surface GF_2, which means, the material varies in the radial direction.

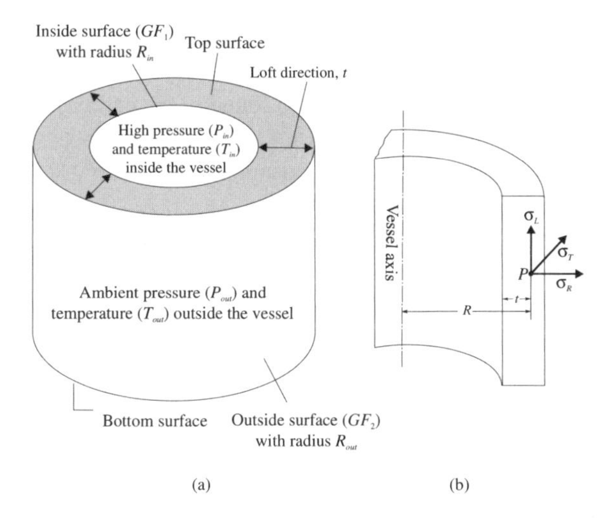

Fig. 6. (a) Feature based pressure vessel model and (b) various stress components at point P within the vessel

The pressure and temperature gradients together develop thermo-mechanical stresses inside the vessel. At a point P, which is at a radial distance R from the vessel axis and at a parametric distance t from the inner surface GF_1, there are three stress components, σ_T, σ_R and σ_L. As shown in Fig. 6(b), these components are normal to each other and they are given as [23, 24]:

$$\sigma_T = \sigma_t + \sigma_{thermal}$$
$$\sigma_t = C_1 - \frac{C_2}{t^2};\ \sigma_R = C_1 + \frac{C_2}{t^2};\ \sigma_L = +\sigma_{thermal}$$
$$\sigma_{thermal} = \alpha(t)E(t)[T(t) - T_{out}];\ \ t = \frac{R - R_{in}}{(R_{out} - R_{in})} \tag{8}$$

where C_1 and C_2 are constants. Note that the stress components are given as a function of parametric distance form the governing feature GF_1. The properties $\alpha(t)$ and $E(t)$ are the local thermal expansion coefficient and the local Young's Modulus at a point with parameter t, respectively.

So that the vessel does not fail under these developed stresses, the total yield strength σ_{Y} of the point P must be greater than the resultant von-Mises stress σ_{VM}, which is given as [23, 24]:

$$\sigma_{\mathrm{VM}} = \sqrt{\frac{(\sigma_R - \sigma_T)^2 + (\sigma_T - \sigma_L)^2 + (\sigma_L - \sigma_R)^2}{2}} \tag{9}$$

However, in cases of freeform objects, these functions are not readily available and usually cannot be derived from the geometry. In such cases, a suitable function can be assumed which adequately represents the property requirement at a point. For example, the Equation (8) can be used for cases where the pressure vessel walls are not exactly cylindrical in shape. By properly choosing the values of the constants C_1 and C_2, a function can be constructed that will more closely match the stresses.

As explained in the previous subsection (also shown in Fig. 5), a lofted volume is obtained by performing a lofting operation on the material governing features. A procedure for generating lofted volume can be found in the earlier paper [23].

As explained before, a lofted volume $L(u, v, t)$ is obtained by performing a lofting operation on $B + 1$ number of material governing features $\{GF_b\}_{b=0,\dots,B}$, with a loft direction parameter t. In [17, 23], a procedure for generating lofted surfaces is presented. To generate a volume with material variations, a new blending method based on lofting is given in this paper.

It is assumed that all governing features $\{GF_b\}_{b=0,\dots,B}$, are functions of parameter u and v of surfaces on the parametric volume $L(u,v,t)$. The features $\{GF_b\}_{b=0,\dots,B}$ have degrees (p_b, q_b) and knot sequences U_b and V_b, in u and v directions, respectively, as follows [23]:

$$GF_b(u,v) = \sum_{i=0}^{\alpha_b}\sum_{j=0}^{\beta_b} N_{i,p_b}(u)\, N_{j,q_b}(v)\, P_{i,j,b} = L(u,v,t_b^*)$$
$$[\, t_b^* = \text{constant} \,;\, 0 \le b \le B;] \tag{10}$$

where $P_{i,j,b}$ is a bidirectional control net with $(\alpha_b + 1)$ and $(\beta_b + 1)$ number of control points in u and v directions.

The lofted volume is a free-form B-spline volume with degree (p, q, r) and knot sequences U, V and T [23].

$$L(u,v,t) = \sum_{i=0}^{\alpha}\sum_{j=0}^{\beta}\sum_{k=0}^{\gamma} N_{i,p}(u)\, N_{j,q}(v)\, N_{k,r}(t)\, R_{i,j.k} \tag{11}$$

The unknown variables are the control points of the volume $R_{i,j.k}$. To set up a set of equations with the unknowns, the t_b^* values are calculated [23]:

$$t_0^* = 0;\ t_B^* = 1;$$
$$t_b^* = t_{b-1}^* + \frac{1}{(\alpha+1)(\beta+1)}\sum_{i=0}^{\alpha}\sum_{j=0}^{\beta}\frac{\left|P_{i,j,b} - P_{i,j,b-1}\right|}{\delta_{i,j}} \quad b = 1, \dots, (B-1) \tag{12}$$

where $\delta_{i,j}$ is the total chord length from $P_{i,j,0}$ to $P_{i,j,B}$.

The knot sequence in the t-direction $T = [t_0, \ldots, t_{\gamma+r+1}]$ is obtained as follows [23]:

$$t_0 = \ldots = t_r = 0; \qquad t_{\gamma+1} = \ldots = t_{\gamma+r+1} = 1$$
$$t_{j+r} = \frac{1}{r}\sum_{i=j}^{j+r-1} t_i^* \qquad j = 1, \ldots, \gamma - r \tag{13}$$

Finally the set of equations are represented in a matrix form as follows [23]:

$$\begin{bmatrix} 1 & 0 & 0 & \ldots & 0 \\ N_{0,r}(t_1^*) & N_{1,r}(t_1^*) & N_{2,r}(t_1^*) & \ldots & 0 \\ N_{0,r}(t_2^*) & N_{1,r}(t_2^*) & N_{2,r}(t_2^*) & \ldots & 0 \\ \ldots & \ldots & \ldots & \ldots & \ldots \\ 0 & 0 & 0 & \ldots & 1 \end{bmatrix} \begin{bmatrix} R_{i,j,0} \\ R_{i,j,1} \\ R_{i,j,2} \\ \ldots \\ R_{i,j,B} \end{bmatrix} = \begin{bmatrix} P_{i,j,0} \\ P_{i,j,1} \\ P_{i,j,2} \\ \ldots \\ P_{i,j,B} \end{bmatrix} \tag{14}$$

Solution of Equation (12) is the control points $R_{i,j.k}$ of the volume $L(u, v, t)$.

5 Feature Matching for Material Blending

In the previous section, a lofting based operation is used to blend material governing features. In this section, a feature matching operation between 2D (curves) material governing features is presented when material requirements change along features' normal vectors.

To be able to blend the material between two or more generator features along their normal direction, the normal vectors of the material governing features must match. While matching the normal vectors, the following conditions must be met for smooth transition:

a) The connecting normal lines must not self-intersect.
b) The length of the ruling lines must be minimum possible.

This problem can be generalized at generating ruling surface between two directices. A naïve way of constructing the ruling lines is by parametrically connecting the points on the two directrices. The rationale here is that both end points of each ruling line have the same parameter values. This does not guarantee a non-twisted ruled surface, particularly in case of directrices given as closed curves [25].

To mathematically express these two conditions, a function f can be defined that assigns a value to each ruling line as follows [25]:

$$f(\mathrm{p},\mathrm{q}) = \frac{\langle \mathrm{N}(\mathrm{p}), \mathrm{N}(\mathrm{q}) \rangle}{\left|\overline{\mathrm{p}-\mathrm{q}}\right|^2} \tag{15}$$

Without loss of generality, it can be assumed that the material governing features or curves $\mathrm{C}_1(u)$ and $\mathrm{C}_2(v)$ lie on the xy-plane and $\mathrm{C}_1(u)$ is totally contained inside $\mathrm{C}_2(v)$. The normals are calculated as follows [25]:

$$\mathrm{N}(\mathrm{p}) = \frac{\mathrm{C}_1'(u_k)}{\left|\mathrm{C}_1'(u_k)\right|} \times \mathrm{k} \text{ and } \mathrm{N}(\mathrm{q}) = \frac{\mathrm{C}_2'(v(u_k))}{\left|\mathrm{C}_2'(v(u_k))\right|} \times (-\mathrm{k}) \quad (16)$$

where **k** is the unit vector in the positive z-direction.

Now the global curve matching problem can be formulated as a continuous optimization problem where the objective is to maximize the sum of the function f over the entire parameter domain of the curve $\mathrm{C}_1(u)$, i.e. $u \in [u_{low}, u_{high}]$ [25]:

$$Maximize \int_{u_{low}}^{u_{high}} \frac{\left\langle \mathrm{N}(\mathrm{C}_1(u)), \mathrm{N}(\mathrm{C}_2(v(u))) \right\rangle}{\left|\overline{\mathrm{C}_1(u) - \mathrm{C}_2(v(u))}\right|^2} \quad (17)$$

Subject to the following constraints:

1. So that $\mathrm{C}_2(v(u))$ is a valid re-parameterization, two consecutive ruling lines $\overline{\mathrm{C}_1(u_i)\mathrm{C}_2(v(u_i))}$ and $\overline{\mathrm{C}_1(u_{i+1})\mathrm{C}_2(v(u_{i+1}))}$ should not intersect each other, i.e. $v(u_i) < v(u_{i+1})$
2. No ruling line $\overline{\mathrm{C}_1(u_i)\mathrm{C}_2(v(u_i))}$ should intersect the directrices $\mathrm{C}_1(u)$ and $\mathrm{C}_2(v)$

Note that no initial re-parameterization is specified as a constraint because the curves are not open.

The input curves $\mathrm{C}_1(u)$ and $\mathrm{C}_2(v)$ are re-parameterized into approximating polygons (or piecewise linear curves). For each polygon, the number and relative locations of the vertices are governed by the desired accuracy of approximation. In general, the higher the number of vertices, the better is the approximation. Matching is established between the vertices of the polygons.

$\mathrm{C}_1(u)$ is re-parameterized into a set of $(a + 1)$ points **P** as follows [25]:

$$\mathrm{P} = \{\mathrm{p}_i\}_{i=0,\dots,a}; \text{ where, } \mathrm{p}_i = \mathrm{C}_1(u_i); \quad u_i \in [u_{p_1}, u_{h_1 - p_1}]; \quad u_i < u_{i+1}; \quad u_0 = u_{p_1}; \quad u_a = u_{h_1 - p_1}$$

$$\mathrm{p}_0 = \mathrm{C}_1(u_0) = \mathrm{C}_1(u_{p_1}) \text{ and } \mathrm{p}_n = \mathrm{C}_1(u_a) = \mathrm{C}_1(u_{h_1 - p_1}) \quad (18)$$

Similarly, $\mathrm{C}_2(v)$ is re-parameterized into a set of $(b + 1)$ points Q as follows:

$$\mathrm{Q} = \{\mathrm{q}_j\}_{j=0,\dots,b}; \text{ where, } \mathrm{q}_i = \mathrm{C}_2(v_j); \quad v_j \in [v_{p_2}, v_{h_2 - p_2}];$$
$$v_j < v_{j+1}; \quad v_0 = v_{p_2}; \quad v_b = v_{h_2 - p_2} \quad (19)$$
$$\mathrm{q}_0 = \mathrm{C}_2(v_0) = \mathrm{C}_2(v_{p_2}) \text{ and } \mathrm{q}_b = \mathrm{C}_2(v_b) = \mathrm{C}_2(v_{h_2 - p_2})$$

Note that the parameters u_i's and v_j's are not necessarily evenly distributed in their respective domains. Moreover, a and b are not assumed to be equal. While re-parameterizing, at every point p_i and q_j, the unit normals $\mathrm{N}(\mathrm{p}_i)$ and $\mathrm{N}(\mathrm{q}_j)$ are also calculated. Now the re-parameterized version of the function f in Equation (15) becomes [25]:

$$f(\mathrm{p}_i,\mathrm{q}_{j(i)})=\frac{\langle \mathrm{N}(\mathrm{p}_i),\mathrm{N}(\mathrm{q}_{j(i)})\rangle}{\left|\overrightarrow{\mathrm{p}_i-\mathrm{q}_{j(i)}}\right|^2} \tag{20}$$

Therefore, the re-parameterized approximation of the original continuous optimization problem in Equation. (17) can be expressed as:

$$Maximize\sum_{i=0}^{a}\sum_{j=0}^{b}\frac{\langle \mathrm{N}(\mathrm{p}_i),\mathrm{N}(\mathrm{q}_j)\rangle}{\left|\overrightarrow{\mathrm{p}_i-\mathrm{q}_j}\right|^2} \tag{21}$$

Subject to the following constraints.

1. So that $j(i)<j(i+1)$ is a valid discrete re-parameterization, two consecutive ruling lines $\overline{\mathrm{p}_i\mathrm{q}_{j(i)}}$ and $\overline{\mathrm{p}_{i+1}\mathrm{q}_{j(i+1)}}$ should not intersect each other, i.e. $j(i)<j(i+1)$.
2. No ruling line $\overline{\mathrm{p}_i\mathrm{q}_{j(i)}}$ should intersect the polygons, **P** and **Q**.

An approach named *Greedy Ruled Line Construction* (*GRLC*) [25] is proposed to find a set of ruling lines, *RL* that maximizes the objective function in Equation (21). The underlying principle of the *GRLC* approach is to construct the set *RL* by adding at a time one ruling line which increases the objective function value the most. At every stage of *RL* construction, the ruling line added to *RL* is chosen from a set of candidates named *RL_candidate_list*. Each candidate in *RL_candidate_list* is called a *maximum valued ruling line* ($MVRL_i$) which satisfies both constraints 1 and 2 in Equation (21). Construction of *RL* continues until all the vertices on both polygons **P** and **Q** are connected by at least one ruling line. It can be proved that this greedy approach guarantees the global optimal solution.

A maximum valued ruling line ($MVRL_i$) represents the best match for a given vertex $\mathrm{p}_i\in\mathrm{P}$. If the ruling line $\overline{\mathrm{p}_i\mathrm{q}_j}$ satisfies both constraints in Equation. (10) and at the same time so happens that $f(\mathrm{p}_i,\mathrm{q}_j)=\max\{f(\mathrm{p}_i,\mathrm{q}_j)\}_{j=0,\dots,b}$ is true, then $\overline{\mathrm{p}_i\mathrm{q}_j}$ is designated as the $MVRL_i$. Since at every stage, a new ruling line is added to *RL*, the $MVRL_i$ may not always remain the same for the same p_i. This is the reason why the *RL_candidate_list* is emptied and reconstructed at every stage of *RL* construction. Below, the methods are described on how $MVRL_i$ is found while satisfying both the constraints. To describe these two methods in the general scenario, it is assumed that $MVRL_i$ is found while there are already some ruling lines in *RL*, none of which has p_i as an end point. In other words, *GRLC* method has already progressed to a stage when $MVRL_i$ will be one of the candidates in the *RL_candidate_list* and will be added to *RL set.*

The first constraint is met by a visibility checking method as defined below.

Definition: Visibility – A point $\mathrm{q}_j\in\mathrm{Q}$ is visible to p_i if the ruling line $\overline{\mathrm{p}_i\mathrm{q}_j}$ does not intersect any edges of either of the polygons P and Q.

While finding the $MVRL_i$ of the given vertex p_i, only those vertices in Q are considered which are visible to p_i. A function **IsVisible(**p_i, q_j) is defined which returns *true* only if q_j is visible to p_i. Let V_i be a subset of **Q** so that all vertices in V_i are visible to p_i [25]:

$$V_i = \left\{ q_j \in Q \middle| \texttt{IsVisible}(p_i, q_j) = true \right\}_{j=0,\ldots,b} \tag{22}$$

Fig. 7 shows how a ruling line $\overline{\mathbf{p}_i\mathbf{q}_j}$ enters RL. Fig. 7(a) explains Equation (22) where the vertex is connected by broken ruling lines to all the vertices in $\mathbf{V}_i$. The $MVRL_i$ is one among the broken lines, but not identified yet.

In order to meet the second constraint in Equation (21), all vertices in **P** are traversed in counterclockwise direction starting from p_{i+1} and ending at p_{i-1}. While traversing, let $p_{i'}$ and $p_{i''}$ be the first and last connected vertices encountered, i.e. $\overline{p_{i'}q_{j'}}$ and $\overline{p_{i''}q_{j''}}$ are two ruling lines already in *RL*. Then all the ruling lines $\overline{p_iq_j}; j'' \le j \le j'$ satisfy constraint 2.

A function **IsValid(**p_i, q_j) is defined which returns *true* only if $j'' \le j \le j'$. The function is so named because each $j(i), j'' \le j(i) \le j'$, qualifies to represent the discrete version of the valid re-parameterization $v_{j(i)}$. Let, R_i be a subset of **Q** containing all $q_j, j'' \le j \le j'$.

$$R_i = \left\{ q_j \in Q \middle| \texttt{IsValid}(p_i, q_j) = true \right\}_{j=0,\ldots,b} \tag{23}$$

Now $MVRL_i$ can be found from the set $V_i \cap R_i$ as shown in Fig. 7(b). If $MVRL_i = \overline{p_iq_j}$ then, by definition of $MVRL_i$, the condition $f(p_i, q_j) = \max\left(\left\{ f(p_i, q_j) \middle|_{q_j \in V_i \cap R_i} \right\}\right)$ holds.

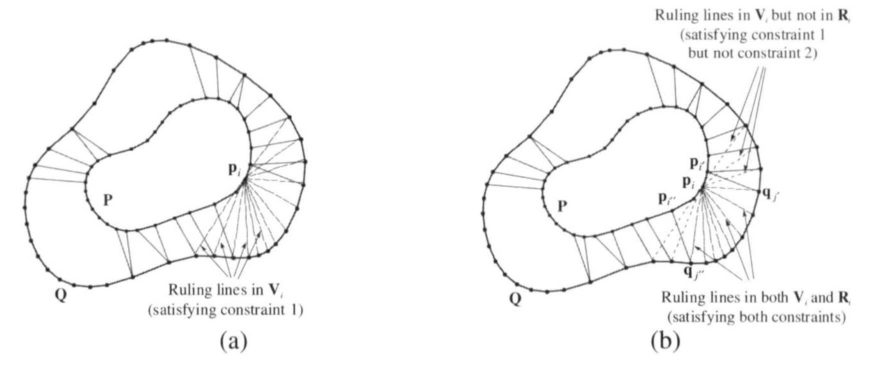

Fig. 7. (a) Visible vertices to $\mathbf{p}_i$ satisfying first constraint (b) finding ruling lines satisfying both constraints 1 and 2 [25]

Since both polygons are closed, no initial match conditions are specified. The greedy approach constructs the *RL_candidate_list*. Among all candidates in *RL_candidate_list*, the one with the maximum value is chosen and added to *RL*. The end vertices of this ruling line are marked as connected. Then the candidate set is emptied and a fresh set is reconstructed excluding all the ruling lines that are already in *RL*. Again the "best" one is chosen from the set and stored in *RL*. This is performed repeatedly until all the vertices of both polygons are connected by at least one ruling line.

It is possible that there can exist one-to-many matching of the vertices. This happens when the curvature of the curves differ significantly at the vertices. It is neither intuitive nor visually pleasing that one vertex on one directrix matches with many points on the other directrix. This means that while material blending, one vertex of the source will metamorphose into an arc on the target and vice versa. Therefore, a vertex insertion method is developed that "spreads out" the ruling lines so that all ruling lines have one-to-one correspondence. This is achieved by inserting more vertices near the vertex with degree more than one and connecting each of them with the ruling lines as shown in Fig. 8.

If $\mathtt{p}_i$ and $\mathtt{p}_{i+1}$ are two consecutive vertices on $\mathtt{P}$, both of which may have degrees more than one. Let, out of all the ruling lines connected to $\mathtt{p}_i$, the line $\overline{\mathtt{p}_i\mathtt{q}_j}$ has the highest function value. Similarly, out of all the ruling lines connected to $\mathtt{p}_{i+1}$, the line $\overline{\mathtt{p}_{i+1}\mathtt{q}_{j+k}}$ has the highest function value. Therefore, the $k - 1$ points between $\mathtt{q}_j$ and $\mathtt{q}_{j+k}$ have to be detached from their connections on **P** and be relocated because the two ruling lines $\overline{\mathtt{p}_i\mathtt{q}_j}$ and $\overline{\mathtt{p}_{i+1}\mathtt{q}_{j+k}}$ are the locally best matches. The points $\mathtt{q}_{j+1},\ldots,\mathtt{q}_{j+k-1}$ were connected to either $\mathtt{p}_i$ or $\mathtt{p}_{i+1}$ because there were no other points available in-between.

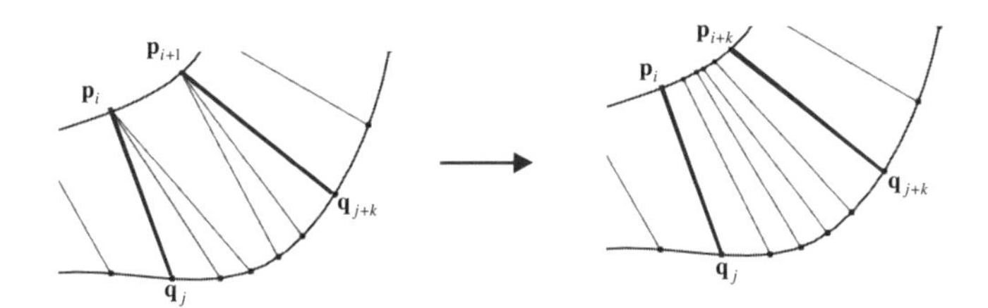

Fig. 8. The ruling lines with highest function value are identified and inserted

The insertion of points between $\mathtt{p}_i$ and $\mathtt{p}_{i+1}$ are done according to proportional parametric increments of $\mathtt{q}_j$ and $\mathtt{q}_{j+k}$. Let the parameter associated with $\mathtt{q}_{j+l}$ be $v_{j+l}, l=0,\ldots,k$. Since $k - 1$ vertices are going to be inserted between $\mathtt{p}_i$ and $\mathtt{p}_{i+1}$, index *i+1* will increase to *i+k*. Let the parameters associated with $\mathtt{p}_i$ and $\mathtt{p}_{i+k}$ be u_i and u_{i+k}, respectively. Then $k - 1$ points are sampled from $\mathbf{C}_1$ between $\mathtt{p}_i$ and $\mathtt{p}_{i+k}$ as follows.

$$\mathrm{p}_{i+l} = \mathrm{C}_1(u_{i+l}) \qquad \textbf{where,} \quad u_{i+l} = u_i + (u_{i+k} - u_i)\frac{v_{j+l} - v_j}{v_{j+k} - v_j}; l = 1,\ldots,k-1 \tag{24}$$

Using the proportional parameter increment approach of inserting points is only a discrete way of re-parameterization only. The vertex insertion is done in two stages. The first stage involves marching along C_1 and inserting points using the equation 10. Later, at the second stage points are inserted on C_2 using the same procedure discussed above.

After the material governing features are matched, the next step is to find the material composition function or material feature between the material governing features.

6 Determining Material Features

After establishing the object model and the object-material constraints, the material features need to be established. The task of establishing the material features is considered to be an optimization problem because only the optimal material feature will ensure that all the requirements are met and the objective achieved. The design methodology is depicted as a flowchart in Fig. 9.

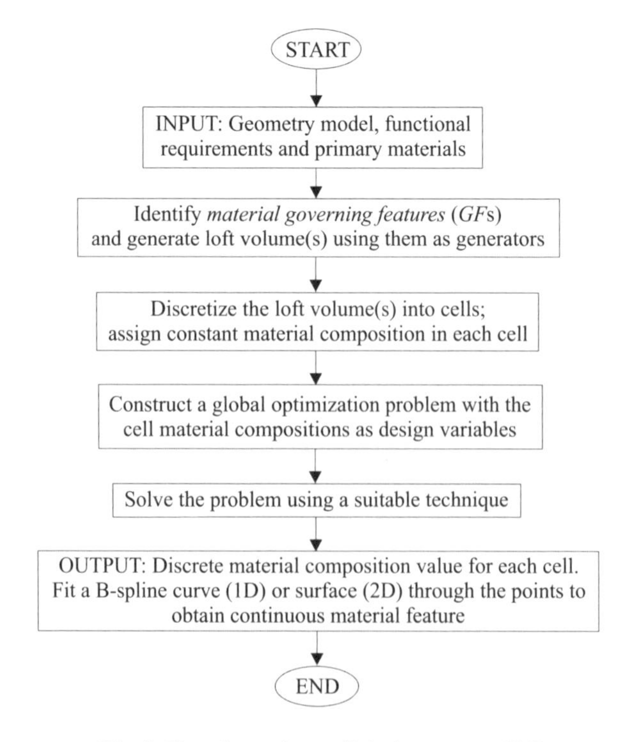

Fig. 9. Flowchart of overall design process [24]

The actual design variables are the control points of the material variation function for each material. However, finding out all the control points simultaneously is too computationally expensive to be considered as a part of an interactive design system. Therefore, the lofted volume(s) are discretized into a set of disjoint cells along the lofting directions. All points in a cell are assumed to exhibit a constant property requirement. Fig. 10 shows an example of cell formation along the lofting direction t for the pressure vessel model mentioned in section 4. Details of cell generation can be found in our earlier work [23, 24].

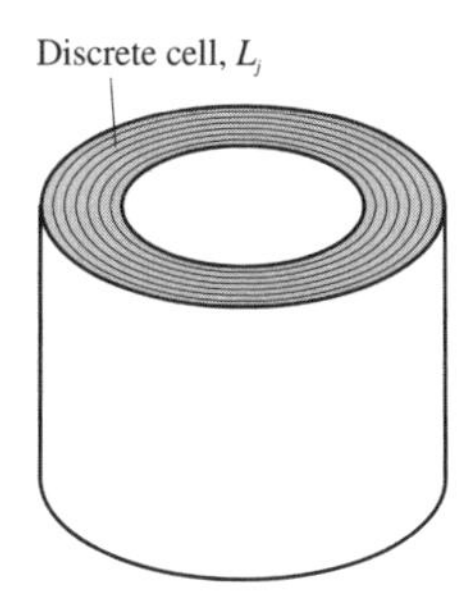

Fig. 10. Lofting and cell formation of pressure vessel model

The material composition in a cell should be such that the resulting material property can meet the requirement at that cell. Therefore, the design variables are denoted as $\mathrm{DV} = [\mathrm{V}_j]$, where, V_j is the material fraction vector for the j-th cell L_j and is given as

$$\mathrm{V}_j = \left[v_j^{(1)}, v_j^{(2)}, \ldots, v_j^{(n)}\right]^T_{j=0,\ldots,\varepsilon} \tag{25}$$

and

$$\mathbf{V}_{j1,j2} = \left[v_{j1,j2}^{(1)}, v_{j1,j2}^{(2)}, \ldots, v_{j1,j2}^{(n)}\right]^T_{j1=0,\ldots,\varepsilon;\, j2=0,\ldots,\phi;} \tag{26}$$

for 1-D and 2-D, respectively. After the cells are formed, the design problem is formulated as an optimization problem as follows [23]:

Min (Max): Objective function $f = (\mathbf{DV})$
Subject to:
(i) All material volume fractions must add to unity.

$$\sum_{k=0}^{n} v_j^{(k)} = 1 \qquad \forall j;$$

(ii) Inequality constraints:

$$G_p(\mathbf{DV}) \le 0 \qquad p = 1, \cdots, g$$

(iii) Equality constraints:

$$H_q(\mathbf{DV}) = 0 \qquad q = 1, \cdots, h$$

There are g numbers of inequality design constraints $G_1, \cdots, G_g$ and h numbers of equality design constraints $H_1, \cdots, H_h$. Examples of design constraints include upper limit of weight of object, minimum failure stress etc.

The optimization problem can be solved using a suitable solving algorithm or a commercial solver. In case of two materials M_1 and M_2, *incremental search* algorithm can be implemented to solve the problem [23].

After the optimization problem is solved, the optimum values of $[V_j]$ for each cell are known in the 1-D case. To represent the material variation as a continuous function, B-spline curves are fitted through each set of values of $v_j^{(k)}$. There will be one curve for each material M_k. A curve fitting algorithm, adapted from [17], is given in our earlier paper [23]. Similarly, for the 2-D case, the optimum values of $[\mathbf{V}_{j_1,j_2}]$ will be obtained after solving the optimization problem. To represent the material function as a continuous function, a B-spline surface is fitted through each set of values of $v_{j_1,j_2}^{(k)}$. There will be one surface for each primary material. Algorithm for surface fitting can be found in [17, 23].

The resulting object model obtained at the end of the process is called an *initial model*. Unlike the initial model which has the optimized features for a specific set of design constraints, the variants will not necessarily be the most suitable models for their respective sets of design conditions. Therefore, it might be required to construct and solve a new optimization problem to establish the material features of a variant.

However, since the variant model already has material attributes, the re-optimization will take less time than it did for the initial model. This is because the optimization process for the initial model starts form scratch where the input model had no material attribute at all. In case of re-optimization of the variant model, the existing material feature will represent an upper bound (in case of minimization) or a lower bound (in case of maximization). Therefore, the variant optimization process will be faster. This property of feature based modeling and design is very useful in case a large number of similar shaped models need to be created at an interactive rate.

7 Implementation and Examples

The proposed design methodology is implemented on a PC using Microsoft Visual C++. OpenGL library functions have been used for displaying the model along with their material variations.

Example part I is the simplified heterogeneous pressure vessel mentioned in Section 4. An optimization problem is modeled where the objective is to minimize heat flow H from inside to outside of the vessel and the constraints are to withstand the stresses in each cell. Two materials M_1 and M_2 are chosen as primary constituents. M_1 has low heat conductivity and low mechanical strength. Material M_2 has high mechanical strength but has high heat conductivity.

The incremental search technique is used for solving the problem. Smooth B-spline curves are fitted through the points which represent the variation function $v^{(1)}$ for material M_1 as shown in Fig. 11. The first and last control points of the B-spline curves are constrained to be on material governing features GF_1 and GF_2, respectively. Fig. 12 shows the optimum heterogeneous model.

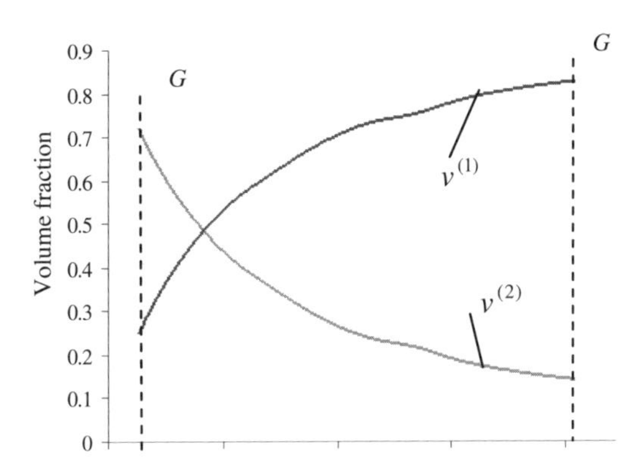

Fig. 11. Material variation curves for Example part I. [24]

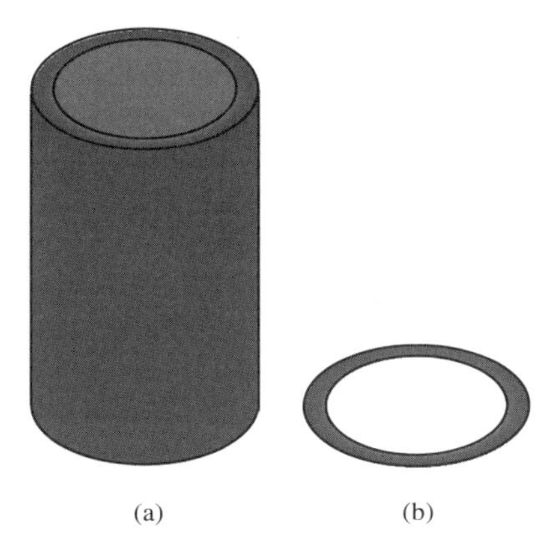

Fig. 12. Example part I: optimized heterogeneous pressure vessel, (a) initial model and (b) cross section of initial model [24]

After the optimal heterogeneous model is designed, variant models are created by modifying the features as shown in Figs. 13(a) and 13(b). In Fig. 13(a) the variant model is cylindrical shaped where the wall thickness and the height have been changed keeping the material feature unaltered. In Fig. 13(b), a free-form model is obtained by repositioning some of the geometric feature control points from Fig. 12, while maintaining the material variation profile. As the geometry control points are repositioned to change the shape, the material feature control points automatically reposition themselves to maintain the feature relationships.

Example part II is a mold with a freeform shaped cavity, as shown in Fig. 14(a). The molten metal is poured at a high temperature T_H and high pressure P_H. The outside surface of the cube is exposed to coolant which is at low temperature T_L and pressure P_L. The design requirement for the part is that the mold must dissipate the heat quickly to allow for rapid cooling of the molded part while withstanding the thermo-mechanical stresses developed due to the pressure and temperature gradients.

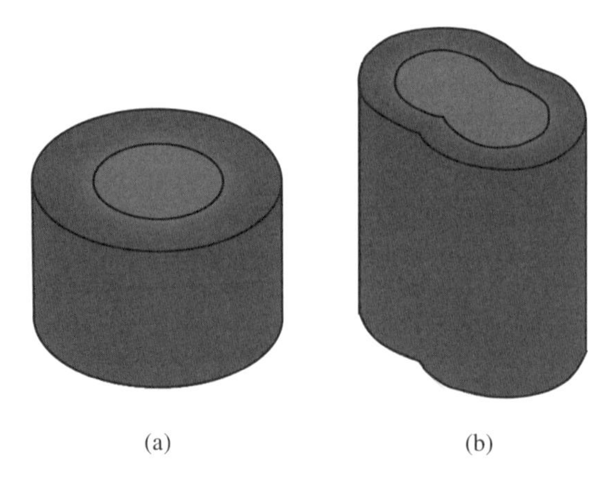

Fig. 13. Variant models of Example part I, (a) cylindrical and (b) free-form [24]

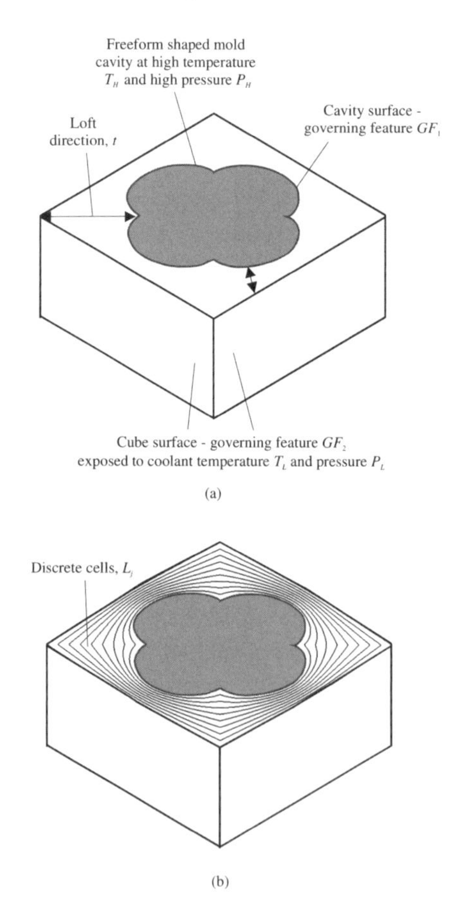

Fig. 14. (a) Example part II: mold with a freeform cavity and (b) blending and cell formation [24]

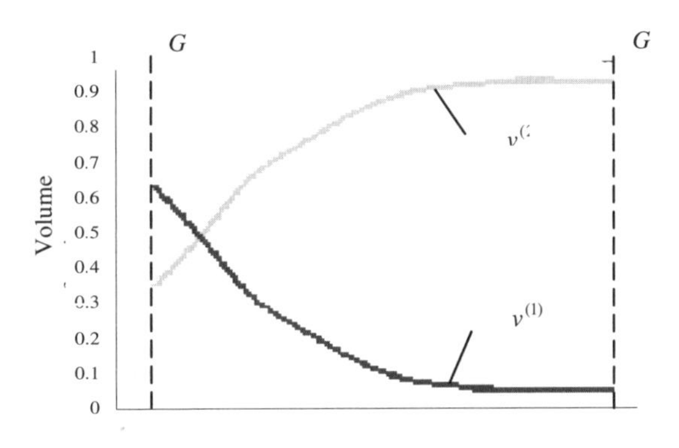

Fig. 15. Smooth B-spline curves representing material variation for Example part II [24]

Two candidate materials, M_1 and M_2 are chosen. M_1 has a high mechanical strength but low heat conducting properties whereas heat conductivity of M_2 is higher. An optimum material variation needs to be calculated to achieve the design requirements.

The material governing features are identified as the cavity surface (GF_1) and the outside cube surface (GF_2) and the lofting direction t is from GF_1 outwards to GF_2 as shown in Fig. 14(a). All the stress components are the same as Equations (8) and (9). A sectional view of the lofting and cell generation is shown in Fig. 14(b). An optimization problem is modeled with the design variables as explained before.

After solving the problem, smooth B-spline curves are fitted to represent the material variations which are shown in Fig. 15. The output design with material variations is shown in Fig. 16.

The solid model of example part III is shown in Fig. 17(a). Property requirement at governing features GF_1 and GF_2 is different from the property requirement at governing features GF_3 and GF_4. Therefore, this is a case of 2-D material feature design. Two loft volumes are generated in the parametric directions s and t, respectively, and the cells are shown in Fig. 17 (b).

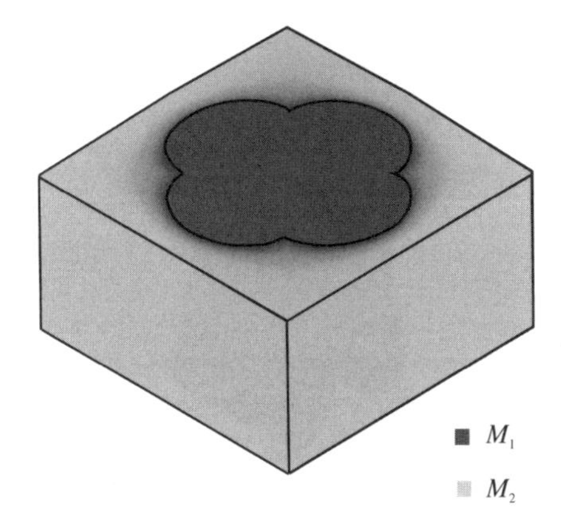

Fig. 16. Example part II: heterogeneous mold [24]

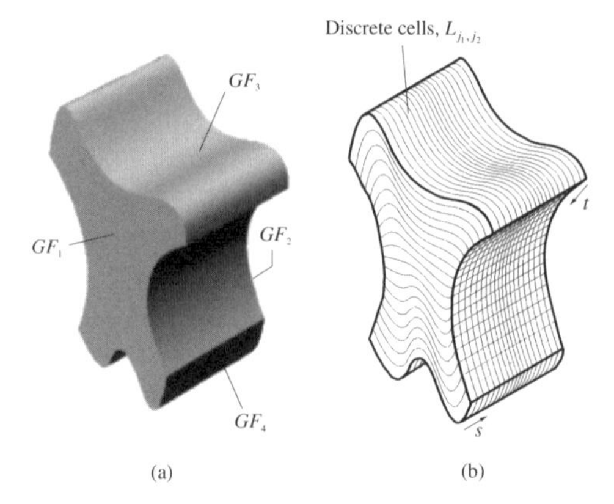

Fig. 17. (a) Example part III: Freeform solid object and (b) lofting and cell formation [24]

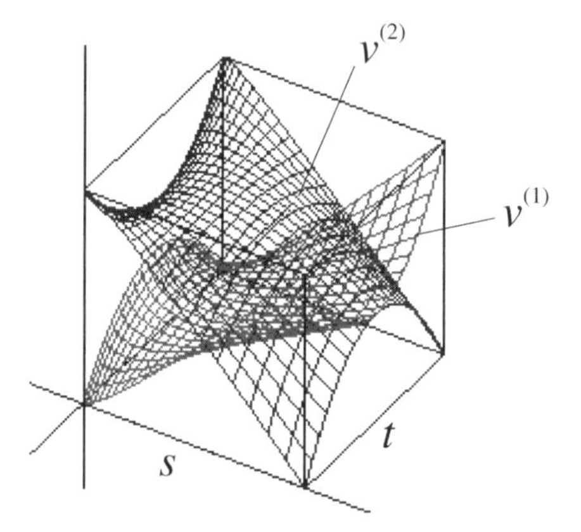

Fig. 18. Smooth B-spline surfaces representing two dimensional material features for Example part III [24]

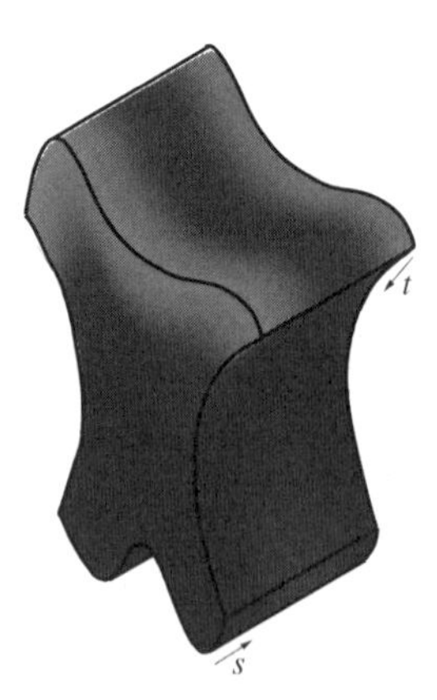

Fig. 19. Example part III: heterogeneous freeform solid object model [24]

Two different materials M_1 and M_2 are selected as primary materials. M_1 satisfies one set of property requirement but doesn't satisfy the other whereas the reverse is the case for M_2. The design variables are same as in Equation (26). Solution of the optimization process gives the optimum values of the design variables. Two fitted B-spline surfaces that represent material features are shown in Fig. 18. The solid model with continuous material variation is shown in Fig. 19.

8 Conclusions

In this paper, feature-based design methodologies have been developed to the design freeform heterogeneous objects. Freeform (sculptured) object features has been used to model and represent heterogeneous objects and material features. A new method is developed to generate matching lines between two material governing features (directrices. Each connecting line represents a match between two points on the directrices. A new method named Greedy Ruling Line Construction (GRLC) is developed to match the directices such that their material features (properties) are blended each other along their matched normal direction. Given the initial object geometry, property requirements and candidate materials, a suitable optimization problem is formulated and solved to construct the material features. Under the assumption that the property requirement is given as a function of parametric distance from a material governing feature, this methodology will generate valid feature based objects where all the features relations are retained. Variant models are created easily by changing material or geometric features.

References

1. Bhashyam, S., Shin, K.H., Dutta, D.: An integrated CAD system for design of heterogeneous objects. Rapid Prototyping Journal 6(2), 119–135 (2000)
2. Biswas, A., Shapiro, V., Tsukanov, I.: Heterogeneous material modeling with distance fields. Computer Aided Geometric Design (in Press)
3. Chen, K., Feng, X.: Computer-aided design method for the components made of heterogeneous materials. Computer-Aided Design 35(5), 453–466 (2003)
4. Griffith, M.L., Harwell, L.D., Romero, J.T., Schlienger, E., Atwood, C.L., Smugeresky, J.: Multi-material processing by LENSTM. In: Proceedings of Solid Free-form Fabrication, Austin (August 1997)
5. Huang, J., Fadel, G.M.: Bi-objective optimization design of heterogeneous injection mold cooling systems. ASME Transactions 123(2), 226–239 (2001)
6. Huang, J., Fadel, G.M.: Heterogeneous flywheel modeling and optimization. Materials & Design 21(2), 111–125 (2000)
7. Jackson, T.R., Liu, H., Patrikalakis, N.M., Sachs, E.M., Cima, M.J.: Modeling and designing functionally graded material components for fabrication with local composition control. Materials and Design 20(2/3), 63–75 (1999)
8. Jepson, L., Beaman, J., Bourell, D.L., Wood, K.L.: SLS processing of functionally gradient materials. In: Proceedings of Solid Free-form Fabrication, Austin (August 1997)
9. Kelkar, A., Koc, B., Nagi, R.: Geometric algorithms for rapidly re-configurable mold manufacturing free-from objects. Computer-Aided Design (in press)
10. Kelkar, A., Koc, B., Nagi, R.: Rapidly re-configurable mold manufacturing. In: Proceedings of ASME Design and Manufacturing Conference, Chicago (September 2003)

11. Kumar, V., Dutta, D.: An approach to modeling and representation of heterogeneous objects. ASME Journal of Mechanical Design 120(4), 659–667 (1998)
12. Lee, Y.-D., Erdogan, F.: Interface cracking of FGM coatings under steady state heat flow. Engineering Fracture Mechanics 59(3), 361–380 (1998)
13. Liu, H., Maekawa, T., Patrikalakis, N.M., Sachs, E.M., Cho, W.: Methods for feature-based design of heterogeneous solids. Computer-Aided Design (in press)
14. Ma, D., Lin, F., Chua, C.: Rapid prototyping applications in medicine. Part 1: NURBS-based volume modeling. International Journal of Advanced Manufacturing Technology 18(2), 103–117 (2001)
15. Markworth, A.J., Saunders, J.: A model of structure optimization for a functionally graded material. Materials Letters 22, 103–107 (1995)
16. Martin, W., Cohen, H.: Representation and extraction of volumetric attributes using trivariate splines: a mathematical framework. In: Proceedings of the 6th ACM Symposium on Solid Modeling and Applications, pp. 234–240 (2001)
17. Piegl, L., Tiller, W.: The NURBS Book, 2nd edn. Springer, New York (1997)
18. Qian, X., Duta, D.: Physics based B-spline heterogeneous object modeling. In: Proceedings of ASME DETC and Computers and Information in Engineering Conference, Pittsburgh (September 2001)
19. Qian, X., Dutta, D.: Design of heterogeneous turbine blade. Computer-Aided Design 35(3), 319–329 (2002)
20. Rhinoceros 2.0, Robert McNeel & Associates, Seattle., `http://www.rhino3d.com`
21. Sachs, E.M., Cima, M.J., Williams, P., Brancazio, D., Cornie, J.: Three dimensional printing: rapid tooling and prototypes directly from a CAD model. ASME Journal of Engineering for Industry 14(4), 481–488 (1992)
22. Samanta, K., Koc, B.: Feature-based design and NURBS modeling for heterogeneous objects. In: Proceedings of 12th Annual Industrial Engineering Research Conference, Portland (May 2003)
23. Samanta, K., Koc, B.: Feature-based design and material blending for free-form heterogeneous object modeling. Computer-Aided Design 37(3), 287–305 (2005)
24. Samanta, K., Koc, B.: Heterogeneous object design with material feature blending. Computer-Aided Design and Applications 1(1-4), 429–439 (2004)
25. Samanta, K., Koc, B.: Optimum Matching of Geometric Features for Material Metamorphosis in Heterogeneous Object Modeling. Computer-Aided Design and Applications 4(1-4) (2007)
26. Samanta, K.: Feature-Based Design and Freeform Modeling for Heterogeneous Objects, M.S. Thesis, Industrial Engineering, University at Buffalo (2003)
27. Shah, J.J., Mantyla, M.: Parametric and feature-based CAD/CAM. John Wiley and Sons, New York (1995)
28. Shigley, J.E., Mitchell, L.D.: Mechanical Engineering Design, 4th edn. McGraw Hill, New York (1977)
29. Shin, K.H., Dutta, D.: Constructive representation of heterogeneous objects. Journal of Computing and Information Science in Engineering 1(3), 205–217 (2001)
30. Siu, Y.K., Tan, S.: 'Source based' heterogeneous solid modeling. Computer-Aided Design 34(1), 41–55 (2002)
31. Siu, Y.K., Tan, S.T.: Modeling the material grading and structures of heterogeneous objects for layered manufacturing. Computer-Aided Design 34(10), 705–716 (2002)
32. Weiss, L.E., Merz, R., Prinz, F.B., Neplotnik, G., Padmanabhan, P., Scultz, L., Ramaswami, K.: Shape Deposition Manufacturing of heterogeneous structures. Journal of Manufacturing Systems 16(4), 239–248 (1997)

Constructive Hypervolume Modeling Using Extended Space Mappings

Benjamin Schmitt[1], Alexander Pasko[2], and Christophe Schlick[3]

[1] Digital Media Professionals, Hosei Research Institute Tokyo, Japan
hfschmitt@gmail.com
[2] Bournemouth University, United Kingdom
apasko@bournemouth.ac.uk
[3] LaBRI, Laboratoire Bordelais de Recherches Informatiques Talence, France
schlick@labri.u-bordeaux.fr

Abstract. In this paper, we propose a framework for modeling and deforming heterogeneous volumetric objects defined as point sets with attributes. We propose to use constructive hypervolume objects, where the function representation (FRep) is used as the basic model for both object geometry and attributes represented independently using real-valued scalar functions of point coordinates. While FRep directly defines object geometry, for an attribute it specifies space partitions used to define the attribute functions.

In this model, an extended space mapping is applied in a geometric space extended by a functional coordinate to transform volumetric objects. We describe in this chapter three different applications of this mapping. First we propose an approach to constructive modeling of 3D solids defined by real-valued functions using B-spline volumes as primitives. A 4D uniform rational cubic B-spline volume is employed to define a 3D solid, which can serve as object geometry or as a space partition for defining volumetric attributes.

Second application of the extended space mapping is deformation of an existing object. We propose to define a new node in the FRep tree based on shape-driven deformations. These deformations can be controlled by additional shapes (points, curves, surfaces, or solids) and can be applied to object geometry and attributes during any modeling step.

The last application is a new operation for a locally controlled metamorphosis between two functionally defined shapes. To implement this operation, a set of non-overlapping space partitions is introduced, where the metamorphosis occurs locally. The sequence of local metamorphosis processes is controlled by a specific time schedule. The definitions of the partitions, of the time schedule, and finally of the local metamorphosis operation, are described and illustrated by examples.

1 Introduction

Volume modeling has become an important research topic. In contrast to homogeneous volumes with uniform distribution of properties, a heterogeneous volumetric object has a number of attributes assigned at each point and varying in 3D

A. Pasko, V. Adzhiev, and P. Comninos (Eds.): HOMA, LNCS 4889, pp. 167–192, 2008.

space. It is not necessary for an attribute to be continuous. Heterogeneous volumetric objects can be modelled as 3D point sets with non-uniform distribution of attributes of an arbitrary nature (photometric, material, physical, statistical, etc.). Heterogeneous objects are considered in such different areas as CAD/CAM and rapid prototyping of objects with multiple materials and varying material distribution, representing results of physical simulations, geological and medical modeling, volume modeling and rendering.

Different models based on the boundary representation [24,23], volumetric [10,34,30], and function-based constructive techniques [35] have been proposed for heterogeneous objects. As it was shown in the mentioned publications, real-valued functions serve well for modeling both geometry and attributes. The work presented in this paper is based on the function representation, FRep for short [37], and the generalised constructive hypervolume model [35,36] (see other chapters), which supports uniform constructive modeling of point set geometry and attributes using real-valued functions of several variables.

Tradionnally, only the constructive approach is used in the creation of FRep models and contructive hypervolume models. Building a constructive tree to generate an object is usually a time-consuming and sometimes difficult process. Usually, a complex shape contains both regular parts that can be easily decomposed into a set of primitives, and some other parts more difficult to decompose. An alternative to the constructive approach is the free-form design. With this general method, parts that can be hardly decomposed into elementary elements are modelled directly, using a sculpting approach. Various sculpting metaphors have been proposed, often referred as *clay sclupting* in the litterature [15]. In those modeling environements, artists do not manipulate polygons but rather add or remove material at different scales until the desired shape is obtained. A different aspect of free-form modeling consits in deformations, where starting from an initial object, it is deformed by various operations, such as twisting, bending, tapering among severals. Another way to create new shapes and animations can be achieved by metamorphosis. Given two initial functionnally defined shapes, the metamorphosis operation transforms the first shape into another according to a dynamic time variable.

Those approaches are based on the same mathematical framework called *extended space mapping*. In the next section, we first recall its definition and show how it naturally includes sculpting, modeling by deformations, and metamorphosis. Later sections respectively illustrate the extended space mapping with a primitive based on the trivariate B-spline, then a set of shape driven deformations, and finally local metamorphosis. In each section, we propose a short survey of exising works, a primitive or operation definition, and finally application examples.

2 Introduction to Extended Space Mapping

The concept of extended space mapping introduced in [43] is used as the underlying framework for modeling complex heterogeneous objects. A space mapping

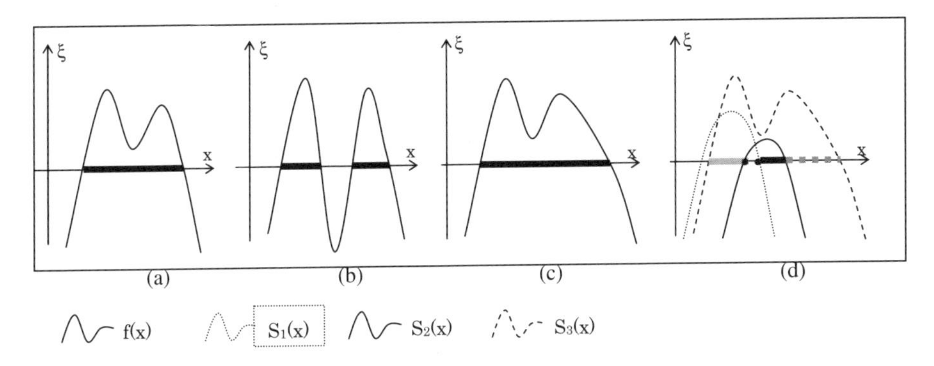

Fig. 1. Extended space mapping. (a) 1D point set (bold line) is defined by a curve $\xi = f(x)$, and the projection of its positive part onto the x-axis. (b) Modification using function mapping (the function f is modified).(c) Modification using space mapping (the x coordinate is scaled non-linearly) (d) Heterogeneous object with attribute functions added to the model. A different attribute corresponds to each function S_i.

establishes a one-to-one correspondence between points of a given space and, if applied to one point set in the space, it changes this set to a different one. A mapping can be defined by the functional dependence between the new and old coordinates of points. A formal definition of the extended space mapping can be found in [43]. It generalises both space mapping and function mapping (i.e., manipulations done on the defining function values) by considering an extended space with geometric coordinates and the additional functional coordinate.

Let us discuss the example given in Fig. 1. The intention is to model a 1D point set with attributes, i.e., a segment along the x-axis. Let f be a real-valued function of one variable such as $\xi = f(x)$. The inequality $f(x) \geq 0$ defines a segment. At the same time, a curve is defined in the (x, ξ) plane. This plane is called an *extended space*, because it has a geometric coordinate and an additional function coordinate. Fig. 1*a* shows such a curve and the corresponding 1D segment at some modeling step. The function $f(x)$ is drawn in black, and its corresponding point set in as a bold line. Different transformations can then be achieved, and are called under the general term *extended space mapping.* For instance, if a translation is applied to the curve, the extended space is mapped onto itself, and thus defines an extended space mapping. As said, extended space mapping combines two main families of mappings, namely the *function mapping* and the *space mapping.* The first one gathers transformations that occur at the function level (see Fig. 1*b*), and the second type stands for transformations at the coordinate level (see Fig. 1*c*).

In a similar way, extended space mappings can be used for attribute functions, as Fig. 1d shows. The function $S_i(x)$ defined in the extended space corresponds to an attribute A_i. In this example, the function S_3 is similar to the function used to define the point set, and the functions for the other attributes are arbitrarily defined.

As point sets and the space partitions for attributes are defined in a similar way, it naturally appears that they can be modelled in a uniform manner. The sculpting paradigm we propose can thus be used for both of them. Then, in the remaining part of this chapter, we use the term *object* to identify without any difference either the geometry of the point set or its spacepartitions with attributes.

In the next section, we propose to model an object using a B-spline function using only the function mapping. Then, in the following sections, the space mapping will be used to define deformations of functionally defined objects.

3 Volume Sculpting Using Trivariate B-spline Primitive

3.1 Previous Works

The term virtual sculpting was first introduced in [33], and several works have been conducted in this direction. An object is sculpted while adding or removing some material at the desired location. A very good survey on the existing sculpting techniques can be found in the PhD thesis [15]. Several existing tools are dedicated to a surface manipulation only, i.e., polygonal [33,5] and parametric [18,21].

In [20,51], an environment in which a 2D painting scheme is proposed to add or to carve an existing object, and in [1,16] a 3D interaction is proposed using a haptic device to sculpt the object. Those methods rely on discretely sampled function, resulting into a 3D grid of uniform voxels, called *voxmap*, containing a function value at each node of the grid. The major advantage is the constant time evaluation procedure, i.e., a trilinear interpolation, used to visualise the object after a modification. Another approach of the sculpting metaphor using real-valued functions has been proposed by Elber et al. [41] and by Schmitt et al. [46]. In those works, the object is represented as a zero set of a trivariate B-spline [41] or trivariate Bézier function [32,46].

A valuable approach would be to combine constructive modeling and volume sculpting, because both approaches contain useful features. To combine these approaches, one has to define a primitive that can be both sculpted and included in a constructive tree [47]. To add a new primitive to a FRep tree, one has to verify the continuity of the defining function (at least C^0 continuity everywhere in the space). The characteristic functions used in [20,51,1,16] are C^0 continuous as the representation is piecewise linear. In the three first works, the grid resolution is fixed making the sculpting process difficult. In the latter one, [16], the sculpting process can be achieved at any level of resolution, and results in a powerful tool.

In [41], the sculpting area depends on the space where the parametric function is defined, i.e., it is restricted within the boundary of the parameter space of the trivariate B-spline function. Therefore, such definition makes it difficult to use the sculpted object in another context, and especially to use it as a constructive primitive. We propose hereafter a similar approach for trivariate Bézier functions as in [32,46] and extend it to B-spline functions as in [45,47] to satisfy the continuity requirement induced by the FRep model and then propose to apply it to define both geometry and attributes in the constructive hypervolume model.

3.2 Trivariate B-spline Primitive

To create FRep objects, we use a sculpting scheme close to the one proposed in [41,45], where uniform cubic trivariate B-spline functions are used. The basic definition of the B-spline is similar, but as it will be shown in the next subsections, additional properties are required in order to be able to provide a new primitive for a FRep constructive tree.

Primitive Definition. Let the defining function f be a trivariate cubic uniform B-spline function defined in a parametric space by a set of $l \times m \times n$ scalar coefficients λ_{ijk}, called control coefficients or control points:

$$f(u,v,w) = \sum_{i=0}^{l}\sum_{j=0}^{m}\sum_{k=0}^{n} N_i(u)N_j(v)N_k(w)\lambda_{ijk} \tag{1}$$

where $N(t)$ are the cubic B-spline basis functions [14] (or blending functions) defined over a uniform knot vector, and u, v and w belong to the parametric space $[0, 1]$.

The control points λ_{ijk} are regularly placed in the space to insure that along each axis the following equalities are verified (using the property of the cubic B-spline basis functions) :

$$\begin{cases} f^x(u,v,w) = u \\ f^y(u,v,w) = v \\ f^z(u,v,w) = w \end{cases} \tag{2}$$

The tensor product used for the B-spline definition will be applied only to the ξ coordinate of each control point. The resulting 3D point set will be defined as :

$$G = \{(x,y,z)/f^{\xi}(x,y,z) \geq 0\} \tag{3}$$

One can recall the example given in Fig. 1, and consider the curve S in the extended space as a traditional B-spline curve. Then, to model a 1D object, one has to use a B-spline curve defined by a set of 2D control points. By analogy, to model a 3D object, the use of 4D control points is required. While the first three coordinates are used to locate a control point in the space (i.e., the usual xyz coordinates), the fourth coordinate ξ contains the scalar coefficient.

Functional Clipping Definition. To insure that the B-spline primitive remains negative and decreasing outside the parametric space, we use the functional clipping defined in [46]. We can force the trivariate B-spline function, or more generally any functions, to become negative outside a certain domain, i.e., the parametric space in our case. This space can be considered as a unit cube with the use of some simple scaling operations. Then, the B-spline function has to be negative outside this cube.

The functional clipping is defined as follows. Let & be the intersection operation defined with the R-function as:

$$f(x) \,\&\, g(x) = f(x) + g(x) - \sqrt{f^2(x) + g^2(x)} \tag{4}$$

and let ω be the defining function of the unit strip of one variable:

$$\omega(t) = t(1 - t) \tag{5}$$

The defining function of the unit cube can be expressed as follows:

$$\Omega(u, v, w) = \omega(u) \,\&\, \omega(v) \,\&\, \omega(w), \tag{6}$$

The inequality $\Omega(u, v, w) \geq 0$ defines a closed subset, which corresponds to the parametric space.

The functional clipping can now be defined as the intersection of this subset with the trivariate B-spline function :

$$F_{clip}(u, v, w) = F(u, v, w) \,\&\, \Omega(u, v, w), \tag{7}$$

The application of the functional clipping insures that outside the parameter space, i.e., the unit cube, the trivariate B-spline function will remain negative. Furthermore, the functional clipping provides a distance property of the trivariate B-spline function outside the domain, but does not change the solid primitive.

3.3 Geometry and Attributes Modeling and Visualisation

We have implemented an interactive modeller on the base of the proposed primitive. Interactive rates are obtained as the B-spline functions are evaluated in constant time, due to the local support of the basis functions.

To visualise the object, we polygonise the iso-surface defined as $F(x, y, z) = 0$, where F is a function evaluated using the constructive tree. Many different algorithms have been proposed for this task. We choose the polygonalisation algorithm based on hyperbolic arcs proposed in [38]. As in the classical Marching Cube algorithm [28], an exhaustive enumeration of the 3D grid cells is applied, but instead of using a look-up table to generate the polygons belonging to a given cell, this algorithm uses the trilinear interpolation inside the cell combined with the bilinear interpolation on the cell faces, and resolves topological ambiguities using hyperbolic arcs. The strength of the algorithm is that the polygonal model it generates is topologically correct without unexpected holes. In our implementation, with the use of this algorithm, the polygonal surface is updated in real-time, which leads to an interactive modeling tool.

Fig. 2 shows a snapshot of the modeling tool we developed, in grey scale. In this environment, to model a shape using a trivariate B-spline function, one can use a 3D cursor to select a position in the set of control points. To help

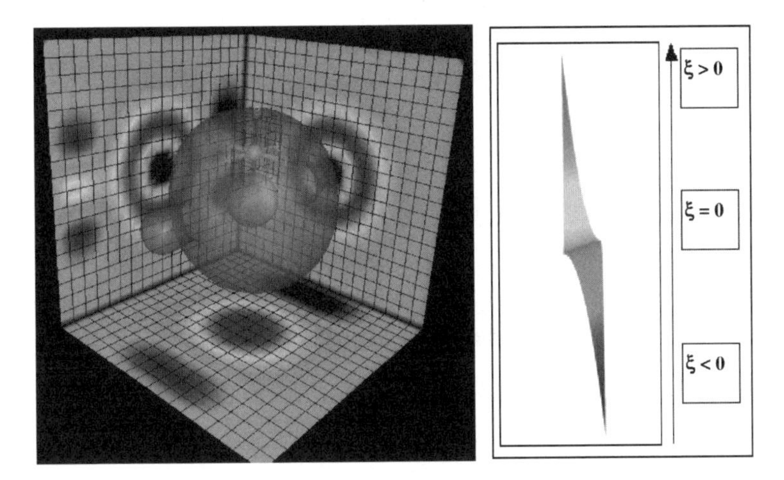

Fig. 2. Snapshot of the user interface. (Left) The iso-surface is polygonised in real-time while modifying the scalar coefficients of the trivariate cubic B-spline function using the 3D cursor. To help the navigation, each face of the bounding cube is coloured according to a colour scale (Right) and to the function values, depending on the location of the cursor.

Fig. 3. Complex heterogeneous object modelled using only trivariate B-spline primitives. (Left) Complete model of a light-bulb; (Right) Zoom on the filament of the light-bulb.

the artist while modeling, we chose to map the parameter space on a cube with a front face culling. Each visible side of the cube is coloured according the function value of the B-spline. A "heat" colour scheme is employed, where the blue gradient ("cold") corresponds to negative function values, and the red gradient ("hot") corresponds to positive values. In Fig. 2, different grey scales

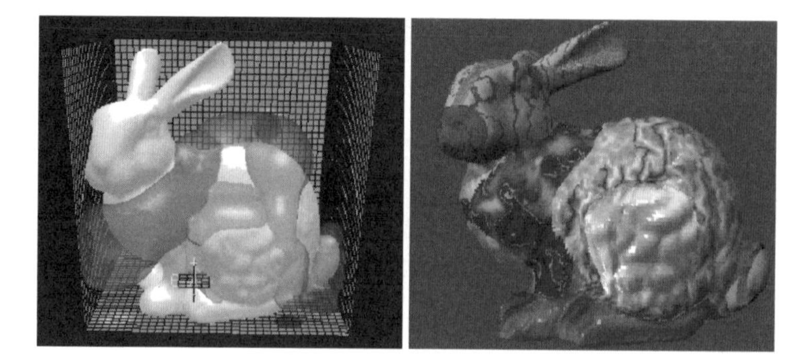

Fig. 4. Interactive modeling applied to texture a BRep object. (Left) Several B-spline primitives are used to define the space partitions. (Right) The corresponding textured BRep object rendered using a surface ray-tracing engine.

are used to represent those colors; a light grey color for positive values and a darker grey color for negative values. Green colours (half toned greys in the figure) mean that the function values are around zero. The aim of this colour scheme is not to provide the exact function value, but rather an approximation to help the user to navigate in the set of control points.

Using this modeling tool, and thus the proposed primitive, one can model 3D objects to be used as both object geometry and space partitions for attributes modeling, as proposed in [45]. An example is given in Fig. 3, using the opacity as an attribute. The object is a light-bulb where the geometry is defined using only B-spline primitives. Twisting operations where applied to the mouthpiece and to the filament, which are two different B-spline primitives. The space partition for the attributes is simple, as the constructive geometric tree coincides with the attributes one.

In the case of a geometric model other than FRep, the only difference in the modeling process is that the geometric model should be imported first, as Fig. 4 shows. In this example, a standard B-rep object (the polygonal "Stanford Bunny") has been loaded, and different space partitions were modelled using the B-spline modeller. In this example, several cubic B-spline primitives were used to create space partitions for photometric attributes (colours and other shading parameters). While modeling, simple colours are used as in Fig. 4(*left*) to show which partition a vertex belongs to. One can then export the object to POV-Ray [39] or other formats for rendering. For each vertex defined in the B-rep model, a tree traversing procedure is executed, and depending on the partition the vertex belongs to, a texture index is defined.

After the modeling process is finished, the object with FRep geometry can be saved as a HyperFun script to be used later by other components of the HyperFun software environment [40]. HyperFun is a high level language supporting exchange of FRep and hypervolume models. The proposed trivariate B-spline primitive has been included in the FRep library of HyperFun. A HyperFun script can also be used to save space partitions for attributes.

4 Deformations in the Constructive Hypervolume Model

4.1 Previous Works

One of the first deformation scheme, called *warping*, was proposed in [33]. Given a polygonal surface, a vertex is selected, and moved towards the outside of the object. Neighbour vertices of the mesh are then displaced according to a distribution function depending on their distance to the displaced vertex. Another technique, called FFD [49,11] which stands for *Free-Form Deformation*, embeds the object to be deformed into a rectangular volume defined using a control point lattice. Then, while moving the control points, the embedding volume and the embedded object are deformed. Several extensions have followed, namely EFFD for *Extended Free-From Deformation* [12], where the embedding volume is replaced by some more complex one, or the RFFD (*Rational Free-Form Deformation*), where another degree of freedom is provided while adding a weight factor to the control points of the lattice. One of the last extensions of the FFD model is presented in [29], where a subdivision volume is used to embed the object.

Several other deformation techniques exist. For instance, instead of embedding the object into a volume, an axis can be defined in the object. Then, as one deforms the axis, the object deformation follows. The axis can be a polyline [33,25], a Bézier curve [9] or even a Bézier surface [31]. Other approaches to deformation include the "simple constrained deformation" using ellipsoids *Scodef* [7,8,3], and its extension with generalised metaballs [22], the "implicit free-form deformation" *IFFD* [13], and the "Wires" model [50].

Control of 2D image deformations using feature shapes (points, segments) was described in [4]. Similar approaches were proposed independently for controlling 3D deformations [44,42] and 3D metamorphosis of homogeneous volumes [26].

4.2 Deformations in the FRep Model

Most of the techniques for deformations presented above can not be directly used in constructive modeling. Indeed, those techniques are applicable only to a polygonal surface, as they are defined using a forward mapping. Even if some deformation tools deform the whole space using a function from $\Re^3 \rightarrow \Re^3$, such as in [13,3], they are applied directly to the vertices of a polygonal surface.

In [46], the constructive approach and the volume sculpting approach were combined using the unifying FRep model. In this section, we describe a new generalized deformation node for the FRep tree data structure. The goal is to provide the possibility for the user to model an object without the traditional separation between the constructive approach, the sculpting process, and the deformation steps. Usually, the modeling scheme is as follows: first one models an object in some way, and then uses deformation tools to obtain the desired shape. As most of the existing tools for deformations are using forward mapping, once the object is deformed, one can hardly return to the constructive modeling step, and combine the existing object with another one.

In order to be able to switch from the constructive modeling step to the deformation step whenever it is needed, one solution is to define the deformation using inverse mapping and to use it as an operation node in the FRep tree. To calculate the inverse mapping from the forward mapping of the previous works is a very difficult operation, and is even sometimes impossible. A preferable solution is to define a new deformation node from the scratch. Of course, several similarities can be found with the previous works, and we used general ideas for defining this new node. But one has to keep in mind that the definition we propose is based on inverse mapping, and thus, even if the visual result of the deformation is close to existing deformations, the underlying idea is different.

Non-linear deformation nodes already exist in the FRep tree. One can easily twist, taper, or stretch an object along an axis. These deformations are based on the work presented in [2], but the set of available deformations is quite limited. A general framework for deformations in the FRep model has been proposed in [43], where the extended space mapping was defined. In [44], deformations were defined using point-controlled space mapping, but it had a too global character due to the interpolation with radial-basis functions. Our goal is to provide localized and intuitively controlled deformations with a general uniform definition.

4.3 Shape Driven Deformations

Simple Deformations Using Space Mapping. In this subsection, we present the underlying idea of deformations using space mappings. Let us first consider a simple translation. Let f be a defining function of some 2D geometric object (Fig. 5*a*), and T a translation vector defined as (dx, dy). Then, the inverse mapping for this transformation is defined as:

$$T : f(x, y) \rightarrow f(x - dx, y - dy) \tag{8}$$

This operation is globally applied to the entire object as dx, dy are constants (Fig. 5*b*). Now, let us define the deformation by non-linear space mapping. Consider the same 2D object and the displacement of a point A towards a point A'. The variables (dx, dy) become functions of point coordinates. To define a local deformation centred at the point A, we need to satisfy the two following requirements:

- functions (dx, dy) have maximum values at A'.
- (dx, dy) drop uniformly to zero when (x, y) is far from A'.

To satisfy these requirements, dx and dy have to become two bell-shaped functions. We propose to use the following functions for $dx(x, y)$ and $dy(x, y)$:

$$\begin{cases} if \gamma \leq 1 & \\ & dx(x,y) = e^{-\gamma^2} \times (x_{A'} - x_A) \\ & dy(x,y) = e^{-\gamma^2} \times (y_{A'} - y_A) \\ else & \\ & dx = 0 \\ & dy = 0 \end{cases} \tag{9}$$

where:

$$\gamma^2 = \frac{(x - x_{A'})^2 + (y - y_{A'})^2}{r^2} \tag{10}$$

As it can be observed, the displacement is maximum when the considered point is placed at A' (Fig. 5(c)). We explicitly set the displacement to zero when γ is greater than one. This insures a localised deformation that depends on the size of the curve "bell". Note that the point A is arbitrarily selected, and thus, a space mapping can be defined for any given point in the space. The function $\gamma(x)$ is similar to the Euclidean distance function.

The parameter r is a real value given by the user, and it defines the area of influence of the space mapping, (and thus the shape of the "bell"). Figure 6 shows the deformations corresponding to different values of r in the 3D case. As it can be seen, different levels of deformation, local or global, can be obtained. The base shape is an ellipsoid. Each row in the figure represents a different value of the r parameter. The first column represents the effect of the displacement of a single control point inside the object, and the second column represents the displacement of the control point outside the object.

Let us illustrate an application of such a mapping to a constructive hypervolume model. In Fig. 7(*left*), the geometry is defined as a single semi-transparent sphere. The additional space partition for the attributes is defined as a union of four smaller spheres. The attributes of these four spheres of the space partition are constructed as successive fully opaque red and green layers. Then, two space mappings are defined in order to deform the object (shown by the arrows). Located along the vertical axis and inside the geometric sphere, two control points of the non-linear space mapping are moved vertically towards the outside of the object. In the case of the upper point, the defined space mapping is applied both to the geometry and to the space partition for the attributes. As one can see, the two upper internal spheres are also deformed, and the red and green colour strips follow the deformation. We choose deliberately to apply the deformation only to the two upper spheres. No space mapping was applied to the two lower spheres, and they remain unchanged, whereas the bottom of the geometric sphere has been deformed.

The equation 9 can be reformulated as follows:

$$\begin{cases} X' = X - \Delta(X) \\ \Delta(X) = f \circ \gamma(X) \times (A' - A) \end{cases} \tag{11}$$

where X is an input point of the Euclidean space, X' is its image after applying the inverse mapping, A and A' are the source and the target points of the deformation.

This definition looks exactly the same as the basic formulation of blobby objects proposed in [6], where f is a potential function, and γ a distance function. One can change either the potential function f or the distance function γ to obtain different behaviors in the deformation. A list of available choices and results can be found in [48].

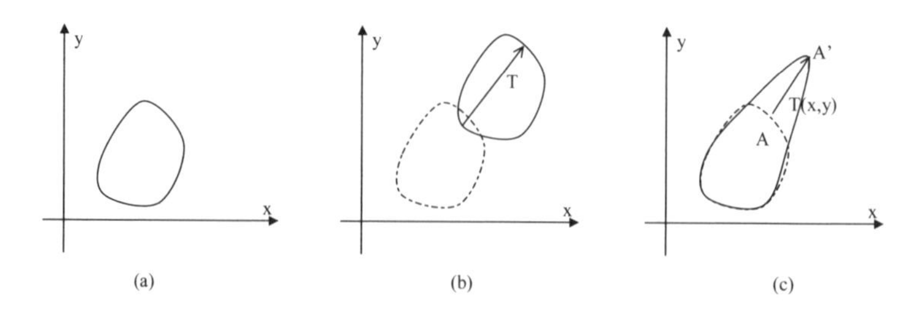

Fig. 5. Simple example of space mapping. (a) The iso-contour of a 2D object is defined in the xy plane. (b) Deformation of the object using a linear space mapping, i.e., a constant translation vector T. (c) Deformation of the object using a non linear space mapping.

Framework for Shape Driven Deformations. Given a source point A and a target point A', let us define two special areas corresponding to an area of influence and an area of projection. The area of influence is defined by a real-valued function $\tilde{Z}$, and takes value in the interval $[0, 1]$. The area of projection is defined explicitly, and can consist in a line segment, a plane or other objects. Let us denote by H the projection of a given point X on this area. Using the same notation as in the previous section, we propose to define a general deformation as follows:

$$\begin{cases} X' = X - \Delta(X) \\ \Delta(X) = f(\gamma(X))\tilde{Z}(X)(H - A) \end{cases} \tag{12}$$

The use and the influence of each term of this equation are explained in the following sub-sections. The next sub-section illustrates the use of different functions for the area of influence $\tilde{Z}$, and then different areas of projection are considered. We suppose that the potential function and the distance function are already defined.

Area of Influence. In this subsection, our interest is turned towards the function $\tilde{Z}$ defining the area of influence. It can consist in any shapes, such as a block, a cone, or any other FRep object. Let Z be the defining function of this FRep object. Z is a real-valued function, at least C^0 continuous, that takes value in $\Re$. To insure that the resulting real value is in the interval $[0, 1]$, we propose to use the following mapping function:

$$\tilde{Z}(X) = \frac{1}{2}\left(1 + \frac{Z(X)}{\sqrt{(\vartheta + Z^2(X))}}\right) \tag{13}$$

The variable ϑ is an important feature in the behaviour of the deformation. In some sense, it can be compared to the hardness factor as it controls blending between the deformation area of influence and the initial object. Figure 8 shows a deformed ellipsoid, where Z is defined as a FRep tree, composed of an intersection of a block and a torus. Therefore, the deformation that can be achieved

Fig. 6. Influence of the parameter r in the definition of the space mapping. An ellipsoid is deformed using space mapping with one control point displaced towards the inside of the object (left column) and outside (right column) with three different values of the parameter r.

Fig. 7. Deformation of a constructive hypervolume. The geometry of the object is defined as a semi-transparent yellow sphere. Inside, a space partition is defined as a union of four smaller spheres. Non-linear space mappings are defined while moving two points (dark dots and arrows). In the upper part, the space partition follows the deformation; in the bottom it is independent.

looks like a torus. To emphasise the influence of the parameter ϑ, three different deformations are shown, with different values for ϑ. As one can see, when ϑ is large, the resulting deformation is close to the area of influence with less blend, and as ϑ gets smaller, the deformation is smoother and blends more with the initial object. One important feature is that one can change the topology of the initial object using the proposed deformation method. None of the existing methods based on forward mapping can handle this problem.

Despite the topology change, the result is also interesting if one considers the attributes of the object, i.e., the texture in this case. In the torus example and for a given ϑ, the result is similar to a blending union of the initial object and a torus. However one can notice that the texture is stretched along the deformation.

Extensions to arbitrary deformation directions are straightforward. Considering the displacement vector AA', one can apply a set of two inverse rotations and a translation to obtain the desired orientation and location. Furthermore, as the area of influence is defined as a FRep object, all the set of available operations and primitives can be combined to define it. These possibilities extend considerably the set of possible deformations.

Area of Projection. The area of projection increases considerably the set of available deformations. We consider in the following two areas of projection,

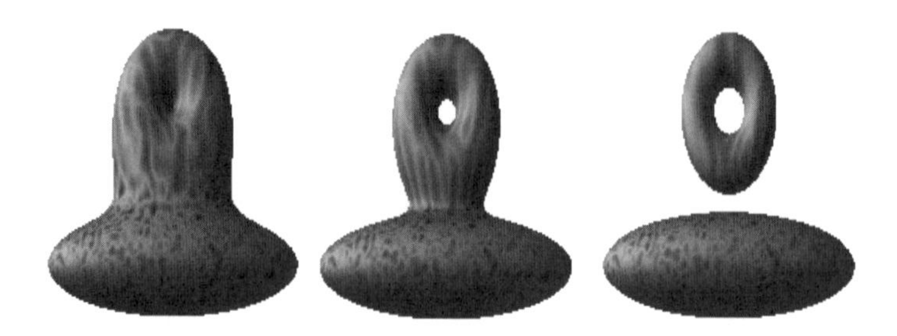

Fig. 8. Changing the area of influence and ϑ. The Z function is defined as a FRep tree, composed of an intersection of a block and a torus. From left to right, the ϑ parameter in the $\tilde{Z}$ function is defined as $\vartheta < 1$, $\vartheta = 1$ and $\vartheta > 1$.

i.e., a plane and a line segment. Furthermore, the proposed examples also consider projections of different natures, respectively a perspective and a parallel projection.

Given a source point A and a target point A', one can define a cone of height AA', and a plane orthogonal to AA' and containing A'. The cone is the area of influence of the deformation, and corresponds to the Z function. For each given point X, one can calculate its projection H onto the plane. In this example, we choose a projection similar to a perspective one, where the vanishing point is A, and H is defined as the intersection of the line AX and the plane. The distance function is then calculated depending on H (and not A' as previously). The use of the area of influence takes now its full meaning. Indeed, if Z is not included in the definition of the inverse mapping, an infinite deformation occurs. Figure 9a shows the result of such deformation. The object to be deformed is an ellipsoid. The source point A coincides with its centre, and the displacement of the target point A' is along the vertical axis, z. The area for the projection is defined as the xy plane, translated along the z axis of AA'. The result is an infinite deformation. It naturally comes that to cut the unwanted part of the deformation, one has to define the area of influence Z. Different results can be obtained depending on the choice of Z. Figure 9b illustrates the previous explanations, as we choose a cone with axis AA' for the area of influence, and in Fig. 9c, we choose to replace the cone by a block.

Instead of considering a plane as the projection area, one can also consider a line segment. In the example shown in Fig. 10, the initial configuration for the source and the target points is identical to the previous example of Fig. 9. The difference comes from the projection area and from the area of influence. The projection area is defined as a line segment, and we choose to use a parallel projection to map every point X on it. The area of influence is defined as a convolution triangle. In Fig. 10a, one can see the initial state, where the object to be deformed is an ellipsoid, the source and the target points are positioned, as well as the line for the projection, and the triangle for the area of influence. Figure 10b shows the projection of three different points of the space, X_1, X_2 and X_3. The three new ellipsoids in the figure correspond to the distance function (as usual, the internal part of the ellipsoid

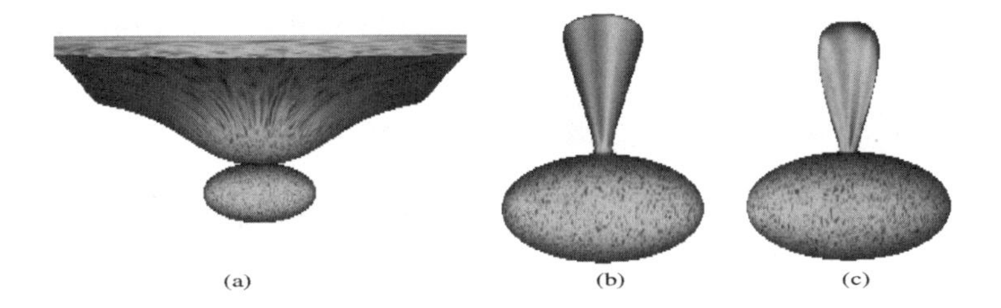

Fig. 9. Shape driven deformation of an ellipsoid. Effect of applying the area of influence. The area of projection is a plane. (a) No area of influence is defined, resulting is an infinite deformation. (b,c) The areas of influence are defined respectively using a cone function and a block function.

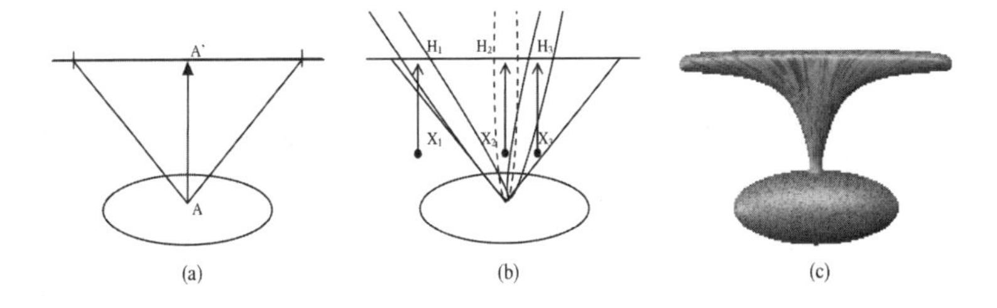

Fig. 10. Shape driven deformation of an ellipsoid. Influence of the projection. The area of projection is straight line. (a) Initial configuration. (b) Parallel projection of some points. (c) Resulting deformation.

represents distance to centre below or equal to 1). The point X_1 is mapped onto the projection line, parallel to AA'. Its image is H_1, but as X_1 is located outside the area of influence, the function $\tilde{Z}(X_1)$ is equal to zero and X_1 will be mapped onto itself. The second point X_2 is mapped onto the line, and its image H_2 corresponds to the centre of the ellipsoidal distance function γ. As $\gamma(X_2)$ is lower than 1, and as X_2 lies inside the area of influence, the point X_2 is mapped to some other location (close to A to be more precise). The third point X_3 emphasises the importance of the choice of the projection. As we mentioned previously, in this example, the projection is a parallel one. The point X_3 lies inside the area of influence. Its image generates an ellipsoidal distance field, but as one can see, the value $\gamma(X_3)$ is greater than one. Thus, the corresponding potential value $f(\gamma(x_3))$ is equal to zero, and the point X_3 will be finally mapped onto itself. Figure 10c shows the resulting deformed object.

4.4 Examples

The shape driven deformations can be easily applied to traditional implicit objects. Nevertheless, the most intriguing results are achieved while using the constructive hypervolume model. Indeed, if the proposed deformations are applied

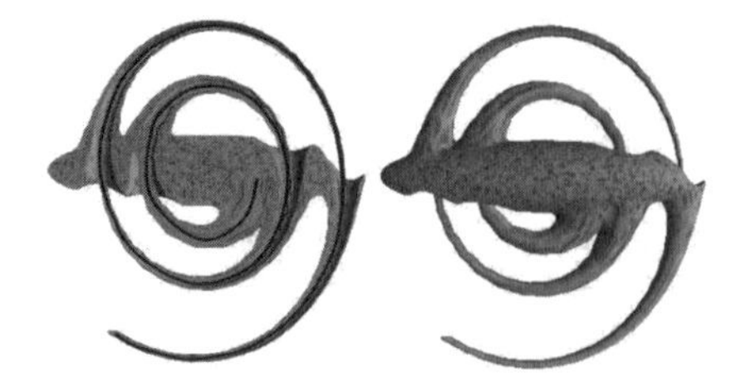

Fig. 11. Self intersecting deformations. An ellipsoid is deformed using a spiral curve. Several intersections between the original shape and the deformation occur. The left picture shows the inside of the deformed object.

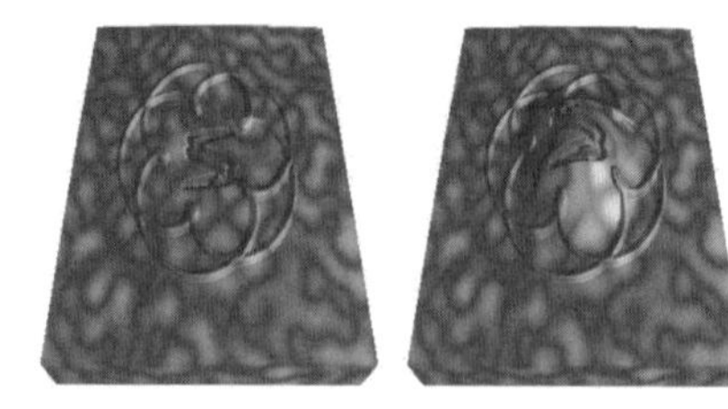

Fig. 12. Deformations as carving tool. (Left) A block is carved according to a path defined a long a curve. (Right) An additional global deformation is then added, and the carving follows it.

Fig. 13. Space mapping as the new node of the FRep tree. Several deformations are applied to an ellipsoid, which then is used in a FRep tree (intersection operation with a cylinder and another ellipsoid).

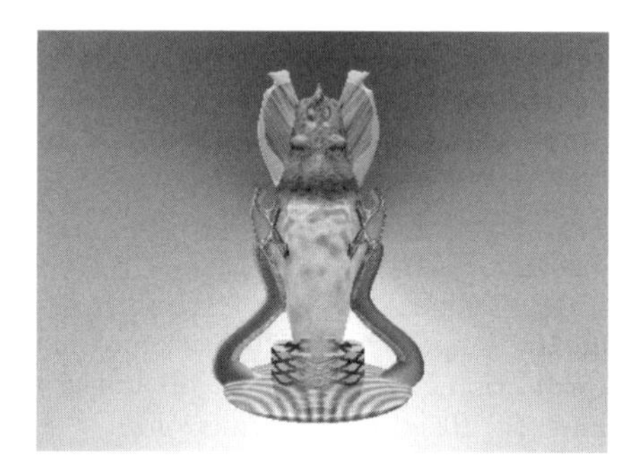

Fig. 14. Example of a shape-driven deformation. Constructive modeling and deformation steps were performed in arbitrary order while designing the shape and photometric attributes of this object.

only to the geometry of the object, similar visual results can be obtained with traditional set-theoretic and blending operations. The justification of the proposed work takes its full meaning when both attributes and geometry are considered. If the deformation node is included in the geometric and attribute trees, the deformation of geometry is followed by the corresponding deformation of attributes. The complex examples that have been given show that the texture intuitively follows the deformation of the objects geometry.

The proposed deformation framework can be used in various directions. One interesting method is to use a curve as a path for the deformation. Source points and target points are obtained while sampling regularly the curve. The resulting deformation is expressed in terms of FRep, i.e., successive small deformations correspond to the nodes of a FRep tree. One important feature of such tree structure is that the proposed deformation supports self-intersection. Consider for instance an ellipsoid deformed along a spiral, as shown in Fig. 11. The spiral intersects the ellipsoid several times. Figure 11*a* shows a vertical cut of the ellipsoid, and Fig. 11*b* shows the whole object.This deformation can be thought as a dual effect, where first, the object is locally deformed along the spiral while pulling objectfs point inside the area of influence, and in a as a second effect, in the vicinity outside the area of influence, objectfs points are locally repelled, resulting in a cavity around the spiral, inside the object. The resulting shape is similar to a shell.

Figure 12 is given to show that the deformation scheme we propose can be also used to carve object, and even if the most of the examples show major deformation of an object, subtle details can be also defined. At the left part, an object has been carved (using two parametric Lissajous curves). At the right, an additional deformation has been applied. As one can see, the carved details also follow the deformation. Figure 13 provides another example, where an ellipsoid is deformed using four deformations, and is then combined with a cylinder and another ellipsoid using the set-theoretic intersection.

The last example is a vase (Fig. 14). To model this object, we combined the constructive approach with the sculpting and deforming steps. First, the body was created using a B-spline object, and then its top was deformed. The next step was to combine it with an ellipsoid. Once both parts were combined, other deformations are applied along two curves, such as they get close to the body. The final step was to create the handles of the vase. If the deformation step was the last one as it is usually the case, to find the correct location for the handles may be difficult, but as we built the tree node by node, regardless of the nature of the operation, this task was easy.

5 Local Metamorphosis of Functionally Defined Shapes

5.1 Previous Work

Metamorphosis between two functionally defined shapes is considered in this section. A metamorphosis is a smooth transformation of an initial shape to a final shape. This animation technique is popular for polygonal objects. Nevertheless, for functionally defined shapes, few techniques have been proposed, especially for the locally controlled metamorphosis.

One of the first metamorphosis operations was proposed by Wyvill [52], where initial and final shapes were defined as soft objects. Surface in-betweening is based on weights assigned to the initial shapes. The force property of each key point is weighted. The weighting of the source keys is gradually changed from one to zero as the in-betweening progresses. The weighting of the force property of the destination keys changes from zero to one. This results in the interpolation between defining functions of source and destination soft objects.

In the advanced modeling technique proposed in [53], still based on skeletal implicit surfaces, metamorphosis is well defined by the use of a warping operator [19]. This operator allows for the fine control of the metamorphosis in some cases. In this context, initial and final shapes are defined by a constructive approach and are represented using a tree structure called a BlobTree. To define a metamorphosis with the proposed warping operators, one needs to define a correspondence between subtrees of the initial shape with some other subtrees of the destination shape. A subtree is a subset of the initial BlobTree, and can include several primitives (leaves of the tree) based on a distance function and a potential function, and several operations (nodes of the tree; set-theoretic, algebraic transformations, etc.). However, the warping operator is practically applicable only to blobby objects. It can be hardly applied to other functionally defined shapes. In this section, we intend to fill this gap and try to provide a new operation that enables the definition of local metamorphosis for more general functionally defined shapes. As a framework for defining the shapes, we use the function representation. Given two FRep objects with the corresponding defining functions f_1 and f_2, a metamorphosis can be defined as a linear interpolation between these two real valued functions, where the time t is the parameter used for the interpolation:

$$F = (1 - t) \times f_1 + t \times f_2 \tag{14}$$

The metamorphosis operator in eq. 14 is defined in the extended space. Indeed, this formula is equivalent to a linear interpolation of two function values that is in other words a special function mapping defined in the extended space.

It assumes that the shapes defined by f_1 and f_2 overlap in the space; otherwise in the resulting animation, one shape disappears and another appears somewhere else, without actual shape transformation. This definition of the metamorphosis is global as the interpolation is defined for all points in space. Using this definition, one can not define metamorphosis locally.

However, one may want to have a finer control of the metamorphosis process. Hereafter, we define a new operation for metamorphosis that allows one to

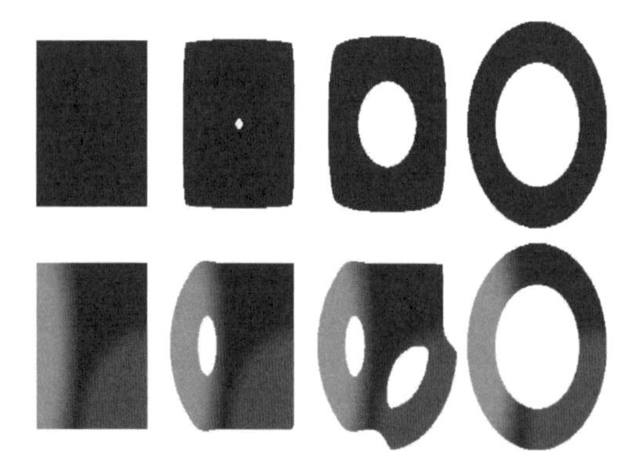

Fig. 15. Metamorphosis of a block to a ring. (top) Global metamorphosis using the linear interpolation. (bottom) Local metamorphosis using two user defined partitions shown in different gray colors.

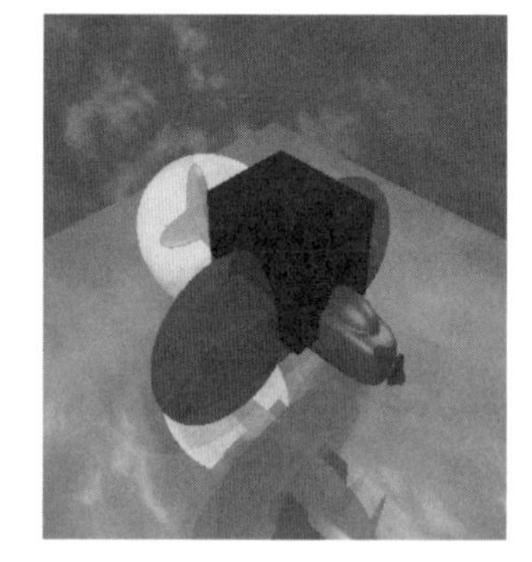

Fig. 16. Metamorphosis partitions used for the animation shown in Fig. 17

control locally its behavior. To define this operation, we need first to define a set of space partitions, where each local metamorphosis occurs and then a corresponding time schedule. With the use of these two items, we finally define the new metamorphosis operation.

5.2 Local Metamorphosis

This section is organized as follows. First, we propose a means to specify partitions of the Euclidean space where the initial and final shapes are defined. Such partitions are called in the following *metamorphosis partitions*, and are denoted Λ_i, where the lower script i indicates the partition number.

Second, we specify a time schedule. We use in the following the global and local dynamic variables which can be mapped to time in order to generate animation. To control the time interval when a metamorphosis occurs inside a partition and the behavior of the metamorphosis in each partition, we assign a local dynamic variable to each partition. Finally, in the last part, we define the local metamorphosis operation, using the metamorphosis partitions, the global dynamic variable, and the local dynamic variables.

Models of Partitions. Metamorphosis partitions are defined using any real valued functions that fit to the FRep model. To define a metamorphosis partition, two different functions are needed. The first function defines the FRep model and takes values in $\Re$. The second function is defined depending on the first function, and takes values in the interval $[0, 1]$.

Let P_i be the defining function of the ith partition Λ_i in the Euclidean space. The function P_i defines two subsets of this space, one where P_i is greater or equal to zero and another where it is negative.

Let F_{P_i} be a real valued function defined upon the defining function P_i. It is constructed to meet the following requirements. Given a user defined threshold value Γ_i (greater or equal to zero), the behavior of F_{P_i} is as follows:

$$F_{P_i} = \begin{cases} 0 & if\ P_i(X) < 0 \\ \Omega\left(\frac{P_i(x)}{\Gamma_i}\right) & if\ 0 \leq P_i(x) < \Gamma_i \\ 1 & if\ P_i(X) \geq \Gamma_i \end{cases} \tag{15}$$

with the function Ω being one of the functions below:

$$\Omega(X) = \begin{cases} X \\ 3X^2 - 2X^3 \\ 6X^5 - 15X^4 + 10X^3 \end{cases} \tag{16}$$

Loosely speaking, F_{P_i} is a simple step function that maps P_i to the interval $[0, 1]$, and Ω allows one to control the continuity of this mapping. Other definitions for such a function can be found in the literature, see, for instance, [27].

In the general case, a set of metamorphosis partitions is defined. From the user side, he/she models a set of partitions P_i using FRep and associates each partition with a threshold value Γ_i. After he/she has defined N partitions, an additional metamorphosis partition, called the remaining partition Λ_r, must be defined in order to cover the entire Euclidean space. The remaining partition is defined as the set-theoretic difference between the Euclidean space and the union of all the partitions defined by the P_i functions. The mapping function associated with this remaining function is expressed as:

$$F_{P_R}(X) = 1 - \sum F_{P_i}(X) \tag{17}$$

Only for this special partition, the function F_{P_R} may take values that are not within the interval $[0, 1]$, but, as it will be shown later in this document, this property does not influence the desired result. Once a set of metamorphosis partitions has been defined, one needs to define a time schedule to indicate when a metamorphosis starts and stops evolving inside a partition.

Time Schedule. The whole metamorphosis is controlled by the time parameter, a global dynamic variable t, defined for simplicity in the interval $[0, 1]$. We define for each partition Λ_i a local dynamic variable t_i. Let us assume that N partitions have been defined. Then, for each partition Λ_i, each local dynamic variable t_i takes value in the interval $[0, 1]$, which corresponds to the interval $[T_i, T_{i+1}]$ of the global dynamic variable t, where T_i and T_{i+1} are user defined boundary interval values, $T_0 = 0$ and $T_N = 1$. These boundaries correspond to the beginning and ending time values, when the metamorphosis starts and finishes inside the given partition Λ_i.

For instance, let us assume that two partitions have been defined, plus the third one, i.e., the remaining partition. At the beginning, the time t_0 corresponding to the first partition changes linearly from zero to one; then when t_0 is equal to one, t_1 starts to grow until it reaches one; finally, once t_0 and t_1 are equal to one, the dynamic variable t_r corresponding to the remaining partition starts to increase, until it reaches one. When t_r is equal to one and the metamorphosis is finished.

In the previous two subsections, we explained how to define space partitions and the corresponding time schedule. In the next subsection, we define the local metamorphosis operation for two given shapes.

Local Metamorphosis Definition. The local metamorphosis operation is defined by a mapping function that maps the global dynamic variable t of Eq. 14 to another global dynamic parameter t'. Given a set of N metamorphosis partitions Λ_i, a time schedule with local dynamic variables t_i, the remaining partition Λ_R with its own local dynamic variable t_R, and a global dynamic variable t, the mapping of t can be described as follows:

$$t'(X) = \sum_{i=0}^{N-1} t_i F_{P_i}(x) + t_R F_{P_R}(X) \tag{18}$$

Then, the mapped global dynamic variable t' is simply used instead of t in the linear interpolation for the global metamorphosis operation (see Eq. 14):

$$F = (1 - t') \times f_1 + t' \times f_2 \tag{19}$$

Equation 19 defines the mapping of the global dynamic variable depending on the metamorphosis partitions. It implies a special requirement to the metamorphosis partitions. They should be non-overlapping. This requirement is not very strong, because partitions are modeled for metamorphosis. To model overlapping partitions is conceptually difficult to understand, as the aim of the partition is to localize metamorphosis, i.e., to specify when a part of the shape starts and stops to evolve. If this part belongs to two or more metamorphosis partitions, the result that one expects is not clear.

Thus, using Eq. 18, a set of non-overlapping metamorphosis partitions modeled with FRep, and a time schedule, one can model the metamorphosis of two functionally defined objects with local control. The next subsection presents several examples. As a last remark concerning the definition of the local metamorphosis, and especially of the time schedule, one may wish to change the linear growth of the dynamic variables (both local and global) to some more complex behavior. As the Ω function provides certain continuity, existing animation techniques can be easily included to this definition. Indeed, in the literature, non-linear time control has been widely discussed, and can be based on a 1D mapping function, using a spline function [17]. Such a fine control of the time schedule can be included as an additional mapping of the dynamic variables as long as the resulting mapped variable remains in its dedicated interval.

5.3 Results and Examples

In this section, we give several examples of local metamorphosis. The first example of the metamorphosis between a 2D block and a 2D ring is shown in Fig. 15. Two partitions are defined, an ellipse and a disk. This figure shows some frames of three different animations. The top row shows a simple linear interpolation between the initial block shape and the final 2D ring (Eq. 14). The bottom row shows the resulting frames while using the local metamorphosis operation defined in this section. The assigned greycolor corresponds to the given metamorphosis partitions. The threshold values for both partitions are different. The value corresponding to the partition defined by an ellipse is close to one, and for the other partition it is close to zero. As one can see, in the first partition, the metamorphosis is smoother than in the second one. According to the time schedule, first, the metamorphosis occurs in the area defined by an ellipse, then in the area corresponding to the second partition, and finally the remaining space is metamorphosed.

The given definition for local metamorphosis provides the possibility to apply metamorphosis anywhere in space and completely independent of the constructive trees of the objects under consideration. This feature is very important for

Fig. 17. Metamorphosis of a tank to a plane. (top) Global metamorphosis. (bottom) Local metamorphosis using the partitions shown in Fig. 16.

the creation of a complex metamorphosis. Figure 17 shows the metamorphosis between FRep models of a tank and a plane. The upper set of frames shows the result of the global metamorphosis, while the lower set of frames illustrates the usage of the local metamorphosis. Figure 16 shows the metamorphosis partitions. The metamorphosis occurs first in the partition defined by a block, then in the darkest spheres, then in the brightest sphere and finally in the remaining space. As one can see, we choose to morph some parts of the object that do not correspond to a precise subtree of the initial constructive trees.

So far, we have defined the local metamorphosis operation and considered only geometry. But as a matter of fact, nothing prevents one from using this operation in the constructive hypervolume model [36]. Indeed, in this model, attributes, such as color, temperature or any other abstract point-wise attribute, are also functionally defined. The definition of local metamorphosis for attributes is straightforward, as one need to simply replace the defining function of the geometric shapes with the real valued function of an attribute. The only restriction on the metamorphosis is that the initial and destination attributes should be of the same nature (a color is changed to another color, for instance).

6 Conclusion

We presented different modeling approaches for constructive hypervolume models based on extended space mappings. It allows us to combine two different modeling paradigms, namely interactive sculpting and constructive modeling, with metamorphosis operations.

First, we defined a new primitive based on the trivariate B-spline function, and then included it to the set of primitives of the FRep constructive tree. An object modeled using this new primitive is defined by a single real-valued function and can be used in further modeling stages in the FRep model. Such an object can also be used to defined space paritions in constructive hypervolume models.

Then, a new set of operations for deforming constructive hypervolumes is proposed relying on the definitions of areas of influence and of a target areas. A large number of new deformations can be obtained with this method, including the possibility to easily change the topology of the initial shape.

Finally, a new metamorphosis operation was described that permits local control of the metamorphosis behavior. To define this operation, a set of non-overlapping metamorphosis partitions and a corresponding time schedule are introduced.

All the described techniques can be easily applied to traditional implicit objects. Nevertheless, the most intriguing results are achieved while using the constructive hypervolume model, where both geometry and attributes undergo simultaneous transformations.

In our future research, we would like to include the SARDF operations (see another volume chapter) and distance based primitives to check if the distance field provides better control and localization of operations.

References

1. Avila, R., Sobierajski, L.: A haptic interaction method for volume visualization. In: Yagel, R., Nielson, G. (eds.) IEEE Visualization 1996, pp. 197–204. IEEE Computer Society Press, Los Alamitos (1996)
2. Barr, A.H.: Global and local deformations of solid primitives. Proceedings of SIGGRAPH 1984, Computer Graphics 18(3), 21–30 (1984)
3. Bechmann, D.: Space deformation models survey. Computer and Graphics 18(4), 571–586 (1994)
4. Beier, T., Neely, S.: Feature-based image metamorphosis. In: SIGGRAPH 1992 Proceedings, pp. 35–42. ACM Press, New York (1992)
5. Bill, J.R., Lodha, S.: Sculpting polygonal models using virtual tools. In: Graphics Interface 1995, pp. 272–278. Morgan Kaufmann Publihsers, San Francisco (1995)
6. Blinn, J.: A generalization of algebraic surface drawing. ACM Transactions on Graphics 1(3), 235–256 (1982)
7. Borrel, P., Bechmann, D.: Deformations of n-dimensional objects. Internat. J. Comput. Geom. App. 1(4), 427–453 (1991)
8. Borrel, P., Rappoport, A.: Simple constrained deformations for geometric modeling and interactive design. ACM Transactions on Graphics 13(2), 137–155 (1994)
9. Chang, Y.K., Rockwood, A.P.: A generalised de casteljau approach to 3d free-form deformation. In: SIGGRAPH 1994 proceedings, pp. 257–260 (1994)
10. Chen, M., Tucker, J.: Constructive Volume Geometry. Computer Graphics Forum 19(4), 281–293 (2000)
11. Coquillart, S.: A sculpting tool for 3d geometric modelling. In: Computer Graphics (SIGGRAPH 1988 Proceedings), vol. 24, pp. 205–212 (1988)
12. Coquillart, S.: Extended free-form deformation: A sculpting tool for 3d geometric modelling. Computer Graphics (SIGGRAPH 1990 Proceedings) 24(4), 187–193 (1990)
13. Crespin, B.: Modélisation et déformation de forme libre à base de surfaces splines équipotentielles. PhD thesis, Bordeaux University I (1998)
14. Farin, G.: Curves and Surfaces for Computer Aided Geometric Design: A Practical Guide, 2nd edn. Academic Press, London (1990)
15. Ferley, E.: Sculpture virtuelle. PhD Thesis, iMAGIS-GRAVIR laboratory, Grenoble, France (2002)
16. Ferley, E., Cani, M.-P., Gascuel, J.-D.: Practical volumetric sculpting. Visual Computer 16(8), 469–480 (2000)
17. Foly, J.D., Van Dam, A., Feiner, S.K., Hugues, J.F.: Computer Graphics: Principles and Practices, 2nd edn. Addison Wesly, London (1995)
18. Forsey, R.D., Bartels, R.H.: Hierarchical bspline refinement. Computer Graphics 22(4), 205–211 (1988)
19. Galin, E., Leclercq, A., Akkouche, S.: Morphing the blobtree. Computer Graphic Forum 19(4), 257–270 (2000)
20. Galyean, T., Hughes, J.: Sculpting: an interactive volumetric modelling technique. SIGGRAPH 1991, Computer Graphics Proceedings 25(4), 267–274 (1991)
21. Gonzales-Ochoa, C., Peters, J.: Localized-hierarchy surface spline (less). In: ACM Symposium on Interactive 3D Graphics, April 1999, pp. 7–16 (1999) ISBN 1-584-13-0821
22. Jin, X., Li, Y.F., Peng, Q.: General constrained deformations based on generalized metaballs. In: Proceedings of Pacific Graphics 1998, pp. 115–124 (1998)

23. Kumar, V., Burns, D., Dutta, D., Hoffmann, C.: A framework for object modeling. Computer-Aided Design 31(9), 541–546 (1999)
24. Kumar, V., Dutta, D.: An approach to modeling multi-material objects. In: Fourth Symposium on Solid Modeling and Applications, ACM SIGGRAPH, pp. 336–345 (1997)
25. Lazarus, F., Coquillart, S., Jancène, P.: Axial deformations: an intuitive deformation technique. Computer Aided Design 26(8), 607–613 (1994)
26. Lerios, A., Garfinkle, C.D., Levoy, M.: Feature-based volume metamorphosis. In: SIGGRAPH 1995, Computer Graphics Proceedings, pp. 449–456 (1995)
27. Li, Q.: Blend implicit shapes using smooth unit step functions. In: WSCG short Communication papers proceedings, pp. 297–304 (2004) ISBN 80-903100-6-0
28. Lorensen, W.E., Cline, H.E.: Marching cubes: A high resolution 3d surface construction algorithm. Computer Graphics, Siggraph 21(4), 163–196 (1987)
29. MacCracken, R., Joy, K.I.: Free-form deformation with lattices of arbitrary topology. In: SIGGRAPH 1996 Proceedings, pp. 181–188 (1996)
30. Martin, W., Cohen, E.: Representation and extraction of volumetric attributes using trivariate splines: a mathematical framework. In: Anderson, D., Lee, K. (eds.) Sixth ACM Symposium on Solid Modeling and Applications, pp. 234–240. ACM Press, New York (2001)
31. Mikita, M.: 3d free-form deformation: Basic and extended algorithms. In: Purgathofer, W. (ed.) 12th Spring Conference on Computer Graphics, Comenius University, Bratisalava, pp. 183–191 (1996)
32. Miura, K., Pasko, A., Savchenko, V.: Parametric patches and volumes in the functional representation of geometric solids, Set-theoretic Solid Modeling: Techniques and Applications. In: Proceedings CSG 1996, Winchester, UK, April 17-19, 1996, pp. 217–231. Information Geometers, UK (1996)
33. Parent, R.: A system for sculpting 3d data. Computer Graphics 11(8), 138–147 (1977)
34. Park, S.M., Crawford, R., Beaman, J.: Volumetric multi-texturing for functionally gradient material representation. In: Anderson, D., Lee, K. (eds.) Sixth ACM Symposium on Solid Modeling and Applications, pp. 216–224. ACM Press, New York (2001)
35. Pasko, A., Adzhiev, V., Schmitt, B.: Constructive hypervolume modelling. Technical Report TR-NCCA-2001-01, National Centre for Computer Animation, Bournemouth University, UK, p. 34 (2001), ISBN 1-85899-123-4, URL: `http://wwwcis.k.hosei.ac.jp/~F-rep/BTR001.pdf`
36. Pasko, A., Adzhiev, V., Schmitt, B., Schlick, C.: Constructive hypervolume modelling. Graphical Models 63, 413–442 (2002) (Special issue on volume modeling)
37. Pasko, A., Adzhiev, V., Sourin, A., Savchenko, V.: Function representation in geometric modelling: concept, implementation and applications. The Visual Computer 11(8), 429–446 (1995)
38. Pasko, A., Pilyugin, V.V., Pokrovskiy, V.V.: Geometric modelling in the analysis of trivariate functions. Computers and Graphics 12(3/4), 457–465 (1988)
39. PovRay. The Persistance of Vision. `http://www.povray.org/`
40. HyperFun Project. Language and Software for FRep Modelling, `http://www.hyperfun.org`
41. Raviv, A., Elber, G.: Three dimensional freeform sculpting via zero sets of scalar trivariate functions. Technical Report CIS9903 (1999)
42. Ruprecht, D., Mueller, H.: Spatial free form deformation with scattered data: interpolation methods. Computers and Graphics 19(1), 63–71 (1995)

43. Savchenko, V., Pasko, A.: Transformation of functionnaly defined shapes by exented space mapping. The Visual Computer 14(5/6), 257–270 (1998)
44. Savchenko, V., Pasko, A., Kunii, T., Savchenko, A.: Feature based sculpting of functionally defined 3d geometric objects. In: Chua, T.S., Pung, H.K., Kunii, T.L. (eds.) Multimedia Modeling. Towards Information Superhighway, pp. 341–348. World Scientific, Singapore (1995)
45. Schmitt, B., Kazakov, M., Pasko, A., Savchenko, V.: Volume sculpting with 4D spline volumes. In: CISST 2000, September 2000, vol. 2, pp. 475–483 (2000)
46. Schmitt, B., Pasko, A., Savchenko, V.: Extended space mapping with Bézier patches and volumes. In: Hughes, J., Schlick, C. (eds.) Implicit Surfaces 1999, Eurographics/ACM SIGGRAPH Workshop, September 1999, pp. 25–31 (1999)
47. Schmitt, B., Pasko, A., Schlick, C.: Constructive modelling of FRep solids using spline volumes. In: Anderson, D., Lee, K. (eds.) Sixth ACM Symposium on Solid Modeling and Applications, pp. 321–322. ACM Press, New York (2001)
48. Schmitt, B., Pasko, A., Schlick, C.: Shape driven deformations of functionally defined heterogeneous volumetric objects. Graphite2003, Publication of ACM SIGGRAPH 12, 127–134 (2003)
49. Sederberg, T.W., Parry, S.R.: Free-form deformations of solid geometric models. Computer Graphics (SIGGRAPH 1986 proceedings) 20(4), 151–160 (1986)
50. Singh, K., Fiume, E.: Wires: A geometric deformation technique. SIGGRAPH 1998, 405–414 (1998)
51. Wang, S., Kaufman, A.: Volume sculpting. In: Symposium on Interactive 3D Graphics, pp. 151–156. ACM Press, New York (1995)
52. Wyvill, B.: A computer animation tutorial. In: Rogers, D.F., Earnshaw, R.A. (eds.) computer graphics techniques: theory and practice, pp. 235–282. Springer, New York (1990)
53. Wyvill, B., Galin, E., Guy, A.: The BlobTree. warping, blending and Boolean operations in an implicit surface modelling system. Computer Graphics Forum 18(2), 149–158 (1999)

Optimization of Continuous Heterogeneous Models

Jiaqin Chen and Vadim Shapiro

Spatial Automation Laboratory*
University of Wisconsin - Madison
jiaqinchen@wisc.edu, vshapiro@engr.wisc.edu

Abstract. A heterogeneous model consists of a solid model and a number of spatially distributed material attributes. Much progress has been made in developing methods for construction, design, and editing of such models. We consider the problem of optimization of a heterogeneous model, and show that its representation by a continuous function defined over a constructively represented domain naturally leads to simple and effective optimization procedures. Using minimum compliance optimization problem as an example, we show that the design sensitivities are directly obtainable in terms of material and geometric parameters, which can be used in any standard gradient-based optimization procedures. The proposed approach allows both local control of the material properties and global control of geometric variations, and can be used with many existing techniques for material modeling. Numerical experiments are given to demonstrate these representational advantages.

1 Introduction

1.1 Motivation

The term *heterogeneous model* refers to a general computer representation of a (typically solid) geometric domain with one or more spatially varying attributes. It is common to view such a model as a tuple [18,1]:

$$\langle \Omega, \mathbf{F} \rangle,$$

where $\Omega \subset \mathbb{E}^3$ is a solid model, and $\mathbf{F}$ is a collection of attribute material functions $F_i : \Omega \rightarrow \mathbb{R}^m$ which may include scalar- and tensor-valued properties, such as density, volume fractions, modulus of elasticity, conductivity, and so on.

Over the last twenty years, much of the research in geometric modeling focused on construction, design, and editing of such models. Early approaches recognized that the material attribute modeling problem is an instance of a boundary value problem and developed material representation schemes based on finite element meshing and other spatial discretizations [25,20,27]. But advances in design and manufacturing of functionally graded materials and related

* Complete address: 1513 University Avenue, University of Wisconsin, Madison, WI 53706, USA.

A. Pasko, V. Adzhiev, and P. Comninos (Eds.): HOMA, LNCS 4889, pp. 193–213, 2008.

technologies also led to new modeling requirements. Discrete changes in material properties implied that both the geometric domain Ω and material properties F_i are modeled, composed, and edited in a piecewise continuous fashion. Furthermore, material properties are usually defined by and associated with material *features* and their geometric parameters that must be explicitly available in any geometric representation of Ω. Typical engineering features may include partial or complete boundaries, regions, and datums (references). Thus, dependence on artificial spatial discretizations becomes both awkward and inefficient. Many interpolation and composition approaches for constructing and editing such feature-based material models have been proposed in the literature (see [17] for a recent and comprehensive survey of heterogeneous modeling methods and techniques).

If the representation of a geometric domain $\Omega(\mathbf{b})$ is parameterized by a set of parameters $\mathbf{b} = \{b_i\}$, it may be convenient that the attribute model $F(\Omega(\mathbf{b}))$ should inherit this parameterization. In interactive modeling situations, or when the material attribute is completely determined by the geometric features, any changes to a parameter b_i are then reflected not only in the geometric model Ω but also in the accompanying material model F. On the other hand, there are at least two practical situations where this supremacy of geometric model over the material model is undesirable:

- A number of shape design and optimization methods determine the shape Ω based on material properties F in some larger fictitious design domain $D \supset \Omega$. Popular examples in this category include homogenization and SIMP methods for topology optimization.
- Typically, a product performance measure $J(\Omega, \mathbf{F})$ is function of both geometry Ω *and* material attributes F_i, and it is important to be able to modify them independently and/or simultaneously until an optimal heterogeneous model is found.

In both of these situations, it is more reasonable to assume that the geometric domain $\Omega(\mathbf{b})$ and material attributes $F_i(\mathbf{c})$ are independently represented and parameterized, so that neither relies on or restricts the modeling space of the other. In this paper, we make this assumption and study the general problem of optimizing such heterogeneous models.

1.2 Approach and Outline

It should be intuitively clear that material modeling approaches based on the spatial discretizations are not appropriate because they limit allowable shape changes and explicitly tie material representation to that of the shape model. It is less clear whether feature-based approaches may be adopted for our purposes. To simplify the exposition, we will assume specific but common representations for the geometry Ω, material functions F, and the optimization problem. In particular, we will assume that geometric domain Ω is represented implicitly as the positive hyper-halfspace $\Phi \geq 0$. We will represent the material distribution F using a linear combination of B-spline basis functions over a reference domain

D which contains the geometry Ω. The B-spline representation for the material field allows continuous material variations and local control of material properties. As we will discuss in Section 2, the material field does not need to conform to the underlying geometry, reducing significantly the remodeling cost caused by geometric changes. The implicit representation for the geometry allows us to combine the material and geometry in a single formulation using the characteristic function, therefore supporting simultaneous optimization of material properties and geometry variations.

In Section 3, for the sake of concreteness, we completely formulate and solve one of the most common shape and material optimization problems using our assumed representation: minimization of compliance. We then show that both material and shape sensitivities are readily obtained and computed from the assumed material model representation. Our prototype implementation and numerical experiments demonstrating effectiveness of the approach are discussed in Section 4. We demonstrate the application of the described optimization procedure to the SIMP material model which is commonly used in the area of topology optimization, and demonstrate a non-trivial extension of the method using simultaneous material and shape variations.

As we explain in the concluding Section 5, the proposed approach is not restricted to implicit representations of domain Ω. In fact, it applies with minimal modifications to *most* geometric representations that are constructive in the sense that they rely on a finite set of primitives Ω_i. We also discuss briefly how our approach may be combined with other feature-based material modeling approaches.

2 Continuous Material Field over Implicitly Defined Domain

2.1 Continuous Representation of the Material Field

As we discussed above, we separate the material representation from the geometric representation, i.e. the material representation does not need to conform to the actual geometry Ω. The material field $F(x)$, x is the spatial coordinate, is represented as a linear combination of basis functions $\{\chi_i(x), i = 1, \ldots, N\}$ from some complete space:

$$F(x) = \sum_{i=1}^{N} c_i \chi_i(x). \tag{1}$$

Choices of the basis functions $\{\chi_i(x)\}$ may include polynomials, trigonometric, B-splines, radial basis functions, etc. The appropriate choice of basis functions allows us to obtain desired properties. We choose linear B-spline basis functions due to their well-understood local control properties [12]. The B-spline basis functions are distributed over a uniform grid subdividing the reference domain D which contains the actual geometry Ω. The coefficients $\{c_i\}$ uniquely determine the associated material field in Expression (1). The basis function representation parameterizes the continuous material field in terms of the coefficients $\{c_i\}$, effectively transforming the material optimization problem to the problem of determining the optimal values for parameters $\{c_i\}$.

2.2 Implicit Representation for the Geometry

Implicit representations of shapes have a long tradition in geometric modeling and computer graphics, as described in several recent books [9,47]. All such representations define a shape $\Omega \subseteq D$ implicitly in terms of non-negative values of some function $\Phi(x)$ of the spatial variable x as $\Omega = \{x \in D \,|\, \Phi(x) \geq 0\}$, where D is some predefined reference domain that contains all possible shapes Ω of interest. The boundary $\partial\Omega$ of the shape Ω is the zero level set of the function $\partial\Omega = \{x \in D \,|\, \Phi(x) = 0\}$. This definition is consistent with the notion of level set function in [36,37,2,49,48,3], but also includes many other representations used in geometric modeling. Many techniques and transformations for constructing such representations are described in [9], including Ricci's function [29], theory of R-functions [30,31,38,41], and convolution methods. More recent notable methods include exact and approximate distance fields [14,8], blending of implicit primitives like blobs, spheres, quadrics, and local quadratics that have been fit to the points [22,19,24], radial basis functions with both global [46] and compact support [34,16], and multi-variate B-splines to represent scalar fields whose zero-sets represent the boundary of sculpted geometry [28,35]. Implicit representations may be constructed from both Constructive Solid Geometry and Boundary Representations of geometric objects [38,39,40].

We adopt the implicit representation for geometric domain Ω parametrized by geometrical parameters $\mathbf{b} = \{b_j, j = 1, \ldots, M\}$. Familiar examples of implicitly defined parametric shapes include conic sections and quadric surfaces, super-ellipses and super-quadrics, tori, as well as local and global transformations of these simple shapes [9]. The corresponding functions Φ for these primitive shapes are well known. The geometric parameters (radii, focal distances, angles, positions, etc.) of these implicit representations serve as the design variables that evolve during the search for optimal shape. Parametric implicit representations for more complex shapes can be built from primitive shapes using a variety of blending, convolution, and set-theoretic techniques [9,43,42].

2.3 Material Fields over Implicitly Represented Geometry

We consider a material optimization problem where both the material properties $F(x)$ and the geometric domain Ω are subject to change. The optimization problem has two sets of variables, one is the set of B-spline coefficients $\{c_i\}$ representing the material distribution, the other one is the geometric parameters $\{b_j\}$ defining the geometric domain Ω. If we use the usual (Heaviside) characteristic function

$$H(\Phi) = \begin{cases} 1, \text{ if } \Phi(x) \geq 0 \\ 0, \text{ if } \Phi(x) < 0 \end{cases}, \tag{2}$$

as an indicator of whether a given point belongs to Ω or not, we have

$$\Omega = \{x \,|\, x \in D, H(\Phi) = 1\}. \tag{3}$$

Then the actual material distribution of interest can be obtained as $\{F(x)H(\Phi), x \in D\}$. Notice that by separating the material representation from the geometry representation, we can write the actual material field as the product of two independent functions: one is the material properties defined by B-spline basis function coefficients $\{c_i\}$ on a fixed reference domain D, the other is an indicator function of the implicit representation Φ for geometry Ω defined by geometric parameters $\{b_j\}$. As we shall see below, this decouples the material sensitivity and the shape sensitivity in the optimization process, and allows us to perform material optimization over varying geometric domains.

3 Optimization Problem

3.1 Formulation

For demonstration purposes, we focus on a compliance (strain energy) minimization problem in linear elasticity with volume constraint that has been studied by many others and is well understood[6]. We seek an optimal shape Ω and material properties such that the compliance of the structure is minimized. Suppose we use the material density $\rho(x)$ as the design variable, and the stiffness tensor E_{ijkl} is assumed to be a known function of the density, the optimization problem can be formulated as:

$$\begin{aligned} \min_{\Omega,\rho(x)} \quad & J_0(u) = \iint_\Omega \frac{1}{2} E_{ijkl}(\rho(x))\epsilon_{ij}(u)\epsilon_{kl}(u)d\Omega \\ s.t. \quad & a(u,v) = l(v), \ \forall v \in U \\ & u|_{\Gamma_1} = u_0 \\ & \iint_\Omega \rho(x)d\Omega = V_0 \\ & 0 \le \rho(x) \le 1, \end{aligned} \tag{4}$$

where $J_0(u)$ is the total strain energy, u is the displacement field, $\epsilon_{ij}(u) = \frac{1}{2}\left(\frac{\partial u_i}{\partial x_j} + \frac{\partial u_j}{\partial x_i}\right)$ is the elastic strain, v is the virtual displacement and U is the space of all admissible displacements. The boundary $\Gamma = \Gamma_1 \cup \Gamma_2$ consists of two parts, with Dirichlet boundary condition $u = u_0$ specified on Γ_1 and boundary traction p specified on Γ_2, f is the body force. The physics of the problem is governed by the equilibrium equation $a(u,v) = l(v)$, where $a(u,v) = \iint_\Omega E_{ijkl}(\rho)\epsilon_{ij}(u)\epsilon_{kl}(v)d\Omega$, and $l(v) = \iint_\Omega fvd\Omega + \int_{\Gamma_2} pvd\Gamma$. In addition, $\iint_\Omega \rho(x)d\Omega = V_0$ is the volume (weight) constraint, and the bound constraint $0 \le \rho(x) \le 1$ reflects the fact that the material density has to be between 0 and 1.

We represent the material density $\rho(x)$ as a combination of linear B-spline basis functions $\rho(x) = \sum_{i=1}^{N} c_i \chi_i(x)$, so Problem (4) can be written as the following:

$$
\begin{aligned}
\min_{\Omega, c_i} \;& J_0(u) = \iint_{\Omega} \frac{1}{2} E_{ijkl}(\rho) \epsilon_{ij}(u) \epsilon_{kl}(u) d\Omega \\
s.t. \;& a(u, v) = l(v), \; \forall v \in U \\
& u|_{\Gamma_1} = u_0 \\
& \iint_{\Omega} \sum_{i=1}^{N} c_i \chi_i(x) d\Omega = V_0 \\
& 0 \le c_i \le 1, i = 1, \ldots, N
\end{aligned}
\tag{5}
$$

The bound constraint for density $\rho(x)$ is automatically satisfied by setting bounds on the linear B-spline's coefficients.

The geometry Ω is to be determined in Problem (5). With the implicit representation Φ (with parameters $\{b_j\}$) for the domain Ω, we can utilize the characteristic function $H(\Phi)$ to transform the integrals in Problem (5) to integrals over the reference domain D. Therefore, Problem (5) can be reformulated as:

$$
\begin{aligned}
\min_{b_j, c_i} \;& J_0(u) = \iint_{D} \frac{1}{2} E_{ijkl}(\rho) \epsilon_{ij}(u) \epsilon_{kl}(u) H(\Phi) d\Omega \\
s.t. \;& a(u, v, \Phi) = l(v, \Phi), \; \forall v \in U \\
& u|_{\Gamma_1} = u_0 \\
& \iint_{D} \left(\sum_{i=1}^{N} c_i \chi_i(x) \right) H(\Phi) d\Omega = V_0 \\
& 0 \le c_i \le 1, i = 1, \ldots, N,
\end{aligned}
\tag{6}
$$

where $a(u, v, \Phi) = \iint_D E_{ijkl}(\rho) \epsilon_{ij}(u) \epsilon_{kl}(v) H(\Phi) d\Omega$, $l(v, \Phi) = \iint_D f v H(\Phi) d\Omega + \int_{\Gamma_2} p v d\Gamma$. Notice that in this formulation, all integrations are now on domain D. Problem (6) is a fully parametrized optimization problem in terms of geometric parameters $\{b_j\}$ and B-spline coefficients $\{c_i\}$. The explicit parametrization allows easy sensitivity analysis, as shown in Section 3.3.

3.2 Algorithm

Problem (6) is an explicitly parameterized optimization problem. In principle, many optimization methods may be used to solve the problem. Since the constraints in Problem (6) address different design concerns, we choose to treat them differently in the optimization procedure.

The equilibrium equation and boundary conditions are determined by the underlying linear elasticity problem, which is typically solved by some external structural analysis method. In the optimization process, we use a meshfree analysis technique (see Section 4.1) to solve the elasticity problem at each step so that the equilibrium equation and boundary conditions are automatically satisfied. The solution field also provides the evaluations of the objective function and its sensitivity at each step.

While an equality volume constraint is usually difficult to enforce during the optimization process, we use the augmented Lagrangian multiplier method, which is well understood and is widely used (for example, see [23]). By imposing the volume constraint as a penalty term in the objective function, we obtain the following augmented Lagrangian subproblem:

$$\begin{aligned}
\min_{b_j, c_i} \; J(u) = & \iint_D \frac{1}{2} E_{ijkl}(\rho)\epsilon_{ij}(u)\epsilon_{kl}(u)H(\Phi)d\Omega \qquad (7)\\
& + \lambda\left(\iint_D \sum_{i=1}^{N} c_i\chi_i(x)H(\Phi)d\Omega - V_0\right) + \frac{1}{2\gamma}\left(\iint_D \sum_{i=1}^{N} c_i\chi_i(x)H(\Phi)d\Omega - V_0\right)^2 \\
s.t. \; & a(u, v, \Phi) = l(v, \Phi), \; \forall v \in U \\
& u|_{\Gamma_1} = u_0 \\
& 0 \le c_i \le 1, i = 1, \ldots, N,
\end{aligned}$$

where λ is the Lagrangian multiplier and γ is a pre-defined parameter (typically a very small number). At each iteration, we fix λ and solve the subproblem (7) for $\{c_i\}$ and $\{b_j\}$, then we update λ and check for termination criteria. If the termination criteria are not satisfied, we go to the next iteration.

To solve subproblem (7), we still need to consider the bound constraints $0 \le c_i \le 1$. The number of these constraints is very large in our problem. For example, if we represent the material field on a 50×50 grids in two dimension, then we have 2500 B-spline coefficients, therefore 5000 constraints! It is very challenging for most optimization algorithms to handle such a large number of constraints. In our implementation, we choose to modify the stiffness tensor as the following:

$$E_{ijkl}(\rho(x)) = \begin{cases} E_{ijkl}(\rho = 1) & \text{if } \rho(x) > 1 \\ E_{ijkl}(\rho) & \text{if } 1 \ge \rho(x) \ge 0 \\ E_{ijkl}(\rho = 0) & \text{if } \rho(x) < 0 \end{cases}, \qquad (8)$$

and handle these constraints as a post process. If the update of some coefficient results in the violation of the corresponding bound constraint, we set it to be the corresponding upper or lower bound. This is physically intuitive, since reaching the upper bound implies adding as much material as possible (therefore we set $c_i = 1$); similarly, reaching the lower bound suggests removing the material (and therefore setting $c_i = 0$). This strategy is similar to the gradient-projection method (see [23]) from the optimization point of view. The main difference is that every time we hit the bound along the search direction, we restart searching from the hitting point instead of bending the search direction.

The augmented Lagrangian subproblem (7) is solved by the conjugate gradient method. Conjugate gradient method is one of the most useful techniques for solving large scale linear systems of equations and can also be adapted to solve nonlinear optimization problems. It is very appealing because in each iteration only the evaluations of the objective function and its gradient are required, no matrix operations are performed and only a few vectors need to be stored [23]. The method is well suited for the formulated large scale optimization problem, and the gradient information can be computed as shown in Section 3.3. The Polak-Ribière conjugate gradient method is adopted in our implementation. The details and convergence studies of Polak-Ribière conjugate gradient method can be found in many standard textbooks, for example, see [23].

The main algorithm consists of the following steps:

Step 1: Initialize the B-spline coefficients c_i and geometrical parameters b_j, choose λ and γ.

Step 2: Use conjugate gradient method to solve Problem (7)

(2.1) Solve the equilibrium equation with boundary conditions.

(2.2) Calculate gradient ∇J and use it as the initial search direction.

(2.3) Construct a series of search directions until the solution is found. The termination criteria is defined as $\left|\frac{\Delta J}{J}\right| \leq \epsilon$, where ϵ is a predefined small positive number.

(2.4) Reset the values for c_i to be the corresponding bounds if they are violated.

Step 3: Update Lagrangian multiplier

$$\lambda = \lambda + \frac{1}{\gamma}\left(\iint_D \sum_{i=1}^{N} c_i \chi_i(x) H(\Phi) d\Omega - V_0\right)$$

Step 4: Check termination condition. If not satisfied, go to *Step* 2. The termination criteria is defined as $\left|\frac{\Delta\lambda}{\lambda}\right| \leq \delta$, where δ is a predefined small positive number.

3.3 Sensitivity Analysis

We now present the sensitivity analysis for the augmented Lagrangian subproblem (7). Since our design variables are material parameters $\{c_i\}$ and geometric parameters $\{b_j\}$, we seek the gradient $\nabla J = \left[\frac{dJ}{dc_1}, \ldots, \frac{dJ}{dc_N}, \frac{dJ}{db_1}, \ldots, \frac{dJ}{db_M}\right]$. We assume that the body force f and the boundary traction p are independent of the design.

Since $\{c_i\}$ and $\{b_j\}$ are two independent sets of variables, we can separate $\{\frac{dJ}{dc_i}\}$ and $\{\frac{dJ}{db_j}\}$ during differentiation. In Problem (7), the displacement field u depends on the variables c_i and b_j as well and it is not obvious how to obtain the derivatives $\frac{du}{dc_i}$ and $\frac{du}{db_j}$. However, we can use the adjoint method [15], where these derivatives are not computed explicitly. In addition, the compliance minimization problem for linear structures is self-adjoint, and the derivative of the compliance J_0 with respect to c_i can be obtained as [6]:

$$\frac{dJ_0(u)}{dc_i} = \iint_D -\frac{1}{2}\frac{dE_{ijkl}(\rho)}{dc_i}\epsilon_{ij}(u)\epsilon_{kl}(u)H(\Phi)d\Omega. \tag{9}$$

Since $\frac{dE_{ijkl}(\rho)}{dc_i} = \frac{dE_{ijkl}(\rho)}{d\rho} \cdot \frac{d\rho}{dc_i} = \frac{dE_{ijkl}(\rho)}{d\rho} \cdot \chi_i(x)$, we have

$$\frac{dJ_0(u)}{dc_i} = \iint_D -\frac{1}{2}\frac{dE_{ijkl}(\rho)}{d\rho}\chi_i(x)\epsilon_{ij}(u)\epsilon_{kl}(u)H(\Phi)d\Omega. \tag{10}$$

Using the result from [11], the sensitivity $\{\frac{dJ_0}{db_j}\}$ can be obtained as:

$$\frac{dJ_0(u)}{db_j} = \int_{\partial\Omega_j}\left[fu + div(pun) - \frac{1}{2}E_{ijkl}(\rho)\epsilon_{ij}(u)\epsilon_{kl}(u)\right]\frac{1}{|\nabla\Phi|}\frac{d\Phi}{db_j}d\Gamma, \tag{11}$$

where Ω_j is the portion of the zero set of Φ corresponding to parameter b_j, i.e., the moving boundary of Ω with respect to parameter b_j.

Since

$$\frac{d}{dc_i}\left(\iint_D \sum_{i=1}^{N} c_i\chi_i(x)H(\Phi)d\Omega - V_0\right) = \iint_D \chi_i(x)H(\Phi)d\Omega \tag{12}$$

and

$$\frac{d}{db_j}\left(\iint_D \sum_{i=1}^{N} c_i\chi_i(x)H(\Phi)d\Omega - V_0\right) = \int_{\partial\Omega_j}\sum_{i=1}^{N} c_i\chi_i(x)\frac{1}{|\nabla\Phi|}\frac{d\Phi}{db_j}d\Gamma, \tag{13}$$

it is easy to see that

$$\frac{dJ(u)}{dc_i} = \iint_D \left[-\frac{1}{2}\frac{dE_{ijkl}(\rho)}{d\rho}\epsilon_{ij}(u)\epsilon_{kl}(u) + \lambda + \frac{1}{\gamma}\left(\iint_D \sum_{i=1}^{N} c_i\chi_i(x)H(\Phi)d\Omega - V_0\right)\right]\chi_i(x)H(\Phi)d\Omega \tag{14}$$

and

$$\begin{aligned}\frac{dJ(u)}{db_j} = \int_{\partial\Omega_j}\Bigg[&fu + div(pun) - \frac{1}{2}E_{ijkl}(\rho)\epsilon_{ij}(u)\epsilon_{kl}(u)\\ &+\lambda\sum_{i=1}^{N} c_i\chi_i(x) + \frac{1}{\gamma}\sum_{i=1}^{N} c_i\chi_i(x)\left(\iint_D \sum_{i=1}^{N} c_i\chi_i(x)H(\Phi)d\Omega - V_0\right)\Bigg]\frac{1}{|\nabla\Phi|}\frac{d\Phi}{db_j}d\Gamma.\end{aligned} \tag{15}$$

Once the displacement field u is known, the computation of Expression (14) and (15) involves only differentiation, boundary and volume integration. Therefore, the computation of the gradient ∇J is straightforward and can be implemented in any systems which support these operations.

4 Experimental Results

In this section we briefly discuss our numerical implementation of the above material optimization procedures and show numerical experiments to illustrate the effectiveness of the proposed method for the problem of topology optimization.

In all examples, we adopt the SIMP (Solid Isotropic Material with Penalty) stiffness model for the material (tensor) properties $E_{ijkl}(\rho)$ as a function of material density. The SIMP method has been widely accepted in the area of topology optimization due to its conceptual simplicity and computational efficiency [4,6,5,44]. The SIMP model uses the power-law

$$E_{ijkl}(\rho(x)) = E^0_{ijkl}\rho(x)^\alpha, \quad 0 \le \rho(x) \le 1, \tag{16}$$

where E^0_{ijkl} is the stiffness tensor of the base material. The derivative $\frac{dE_{ijkl}(\rho)}{d\rho}$ in Expression (14) can be easily obtained as $\frac{dE_{ijkl}(\rho)}{d\rho} = \alpha E^0_{ijkl}\rho(x)^{\alpha-1}$. The power $\alpha > 1$ has the effect of penalizing intermediate densities. The necessary conditions of α for the material to be physically realizable has been studied in [5], which we will not discuss in this paper. By making the intermediate densities less economic in the power-law model, the SIMP penalizes the intermediate densities and drives the structure to a near 0-1 design during the optimization process. Note however that the formulation of the sensitivity does not depend on SIMP and can be used with other models of material properties.

4.1 Meshfree Implementation

The algorithm described in Section 3.2 can be implemented in many meshfree environments that support stress/strain analysis, allow some programmability for parametric functions, and provide tools for differentiation, and boundary and volume integration. Here we briefly describe how the proposed approach is implemented using the RFM method (R-function method) [45,41] and used earlier to solve the problems with shape deformations and moving boundary conditions [43].

Since we represent the boundary of a geometric domain by the zero level set of an evolving scalar function, it is natural to use an engineering analysis method that can work with the same geometric representation. The RFM method, a meshfree method with approximate distances, is well suited for the task. This method is based on the idea that a physical field can be represented by a generalized Taylor series expansion by powers of an approximate distance field to the boundary [31,32,33]. Once such distance fields are constructed, they can be used to construct solutions to boundary value problems that satisfy the prescribed boundary conditions exactly on all points where the distance field vanishes. The remainder term in the Taylor series contains degrees of freedom necessary to approximate differential equation(s), and it also assures completeness of the solution [33]. The method is essentially meshfree, though a background mesh may be used for integration and visualization purposes. A complete programming environment supporting construction, differentiation, and integration of all required functions at run time is described in [45].

In the context of the structural analysis problem solved in this paper, we represent components of the displacement vector $\mathbf{u} = (u_1, u_2)$ as products of two functions $u_i = \omega_i \Psi_i$, $i = 1, 2$, where ω_i are distance functions to the fixed portions of the boundary of the domain Ω, and functions $\Psi_i = \sum_{j=1}^{k} a_j^{\Psi_i} \xi_j$ are linear combinations of basis functions used to approximate the solution of the differential equation. Basis functions $\{\xi_j\}_{j=1}^{k}$ can be chosen from B-splines, polynomials, radial basis functions or even finite elements. Generally, these basis functions can be defined on a grid that does not conform to the geometric domain and are not related to the basis functions used to construct the material field ρ. In this paper we approximate components of the displacement vector using a uniform cartesian grid of bilinear B-splines. Numerical values of the coefficients $a_j^{\Psi_i}$ are determined by a standard technique that requires minimization of an energy functional [50]. As a result, we obtain a system of linear algebraic equations whose solution gives numerical values of the coefficients $a_j^{\Psi_i}$. Assembly of the matrix and vector of this system of equations requires differentiation of the approximate distance fields ω_i and basis functions with respect to spatial coordinates and integration over the non-meshed geometric domain and its boundary represented by a level set function. Use of B-splines as basis functions results in matrices that possess block-diagonal sparse structure. Algebraic systems with such matrices can be efficiently solved by a conjugate gradient algorithm [26]. Once numerical values of the coefficients $a_j^{\Psi_i}$ are computed, they are substituted into the expressions for components of the displacement vector $\mathbf{u}$.

4.2 SIMP Examples

The first example is a short cantilever beam, the second example is a Messerschmitt-Boelkow-Blohm (MBB) beam. Both examples are benchmark problems which have been widely used in the literature [13]. The third example is a 3-hole bracket design problem. Compared to the cantilever beam and MBB beam which are defined on a rectangular domain, the bracket has a more complicated design domain. In contrast to the spatially discretized representations that require complex meshes for complicate geometries, the continuous material representation over the implicitly represented geometry does not demand any additional effort as the complexity of the geometry increases. The fourth example is also a cantilever beam, but it is defined on a varying geometric domain with a moving circular hole. Both the material distribution and the position of the hole need to be determined. The fifth example is a stepped cantilever beam where the material density and the heights of the steps need to be determined. In all examples, uniform rectangular grids are imposed on the background for supporting the basis functions representing the material field (Section 2.1) and the basis functions representing the displacement field (Section 4.1). This grid is also used for numerical integration. All examples are plane stress problems with material properties as follows: Young's elasticity modulus for base material $E^0 = 1$, and Poisson's ratio $\nu = 1/3$. The power α in the SIMP model is chosen to be 3. The body force is assumed to be zero.

Example 1: Cantilever beam

Figure 1 shows a classic short cantilever beam design problem defined on a rectangular design domain D of length $L = 0.1$, height $H = 0.05$. The thickness of the plate is $t = 0.0025$. A distributed force $p = 200$ is applied in an interval of 0.005 around the middle point of the right edge of D and the left edge of D is fixed. The volume constraint (area of the structure) is set to be one half of the design domain. We use uniformly distributed material $\rho = 1$ (i.e. $c_i = 1, i = 1, \ldots, N$) over the design domain as the initial design.

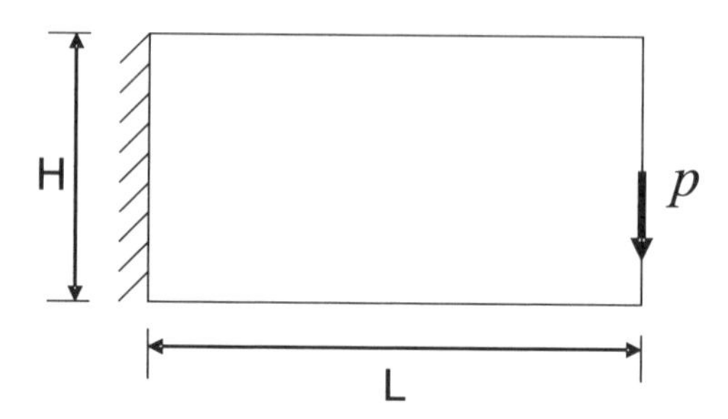

Fig. 1. Problem definition of a cantilever beam

Figure 2 shows the optimal material distribution of the cantilever beam from different grid sizes. The final designs are near 0-1 designs due to the penalization of the SIMP material model. In contrast to results from finite element based methods [6,13], the structures obtained are free of checkerboard patterns due to the continuous material representation. We notice that material distributions on finer grids (therefore, with more degrees of freedom) tend to generate finer structures with more complex topology and smoother boundaries. This phenomenon is often referred to as mesh-dependence in the literature [6,13]. But visually, this mesh dependence appears to be less prominent than that observed in similar computations with from classical finite element based methods. Table 1 lists values of the objective function and the area of the optimal structures in Figure 2.

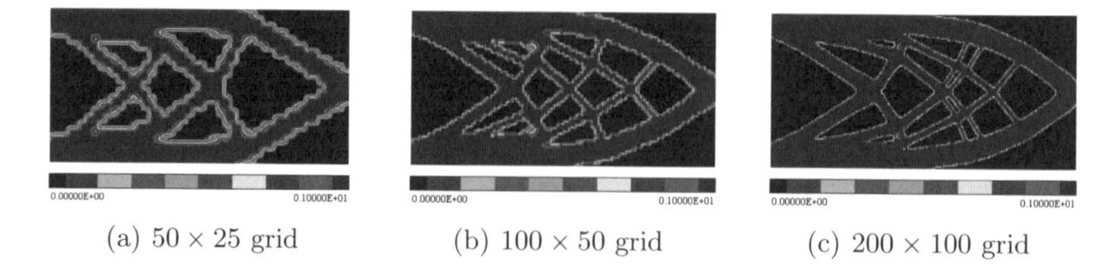

(a) 50×25 grid (b) 100×50 grid (c) 200×100 grid

Fig. 2. The optimal structures of the cantilever beam from different grid sizes

Table 1. The objective (total strain energy) and the area of the optimal cantilever beam

grid size	area	total strain energy
50×25	2.50013e-3	1.35270e+4
100×50	2.50115e-3	1.28726e+4
200×100	2.49983e-3	1.22711e+4

We see that with finer grids, the structure has a slightly better performance (lower value of the objective function), as expected.

Example 2: MBB beam

Figure 3 shows a MBB beam design problem on D of length $L = 6$, height $H = 1$. The thickness of the plate is $t = 0.1$. A distributed force $p = 100$ is applied in an interval of 0.12 around the middle point of the top edge of D. The volume constraint is half the area of the design domain. Uniformly distributed material $\rho = 1$ over the design domain is used as the initial design.

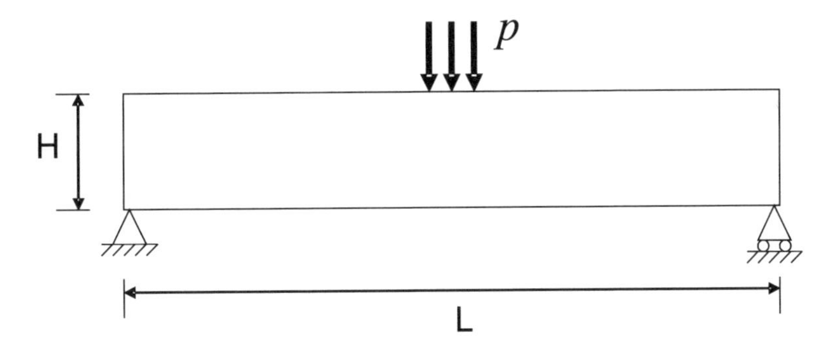

Fig. 3. Problem definition of a MBB beam

Figure 4 shows the optimal structures of the MBB beam from different grid sizes. As in the first example, the optimal structures are free of checkerboard patterns and almost 0-1 designs. The mesh dependence is even less noticeable. Table 2 lists values of the objective function and the area of the optimal structures in Figure 4. Again we see that the structure generated from fine grids has slightly better performance.

Example 3: 3-hole bracket

Figure 5 shows the design domain of a 3-hole bracket. The rectangle is of length $L = 0.1$, height $H = 0.1$. The radius of the three holes is $r = 0.01$ and the holes are fixed with $d = 0.02$. The thickness of the plate is $t = 0.0025$. A distributed

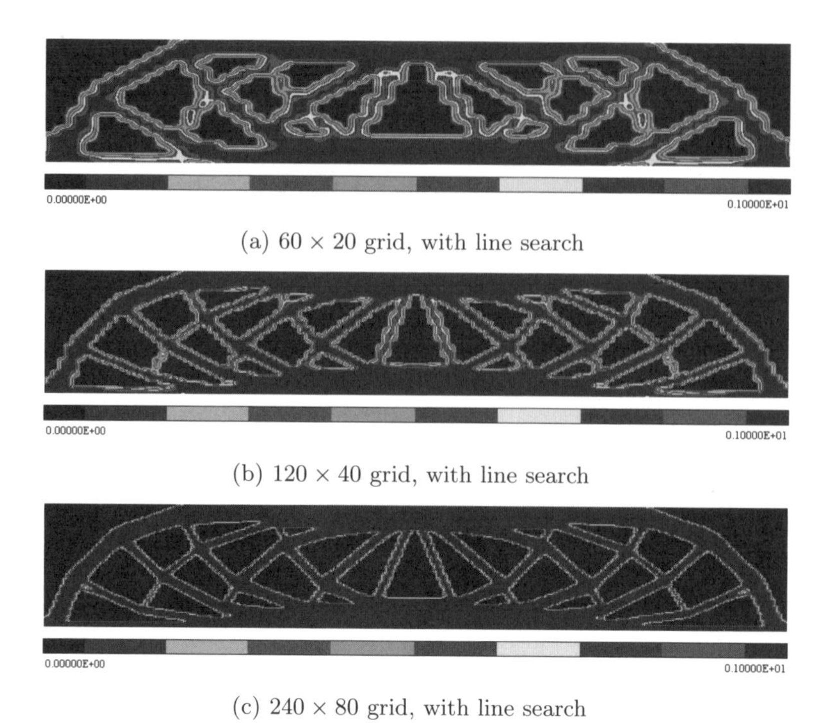

(a) 60×20 grid, with line search

(b) 120×40 grid, with line search

(c) 240×80 grid, with line search

Fig. 4. The optimal stuctures of the MBB beam from different grid sizes

Table 2. The objective and the area of the MBB beam

grid size	area	total strain energy
60×20	3.00386	6.35452e+4
120×40	3.00278	6.27174e+4
240×80	2.99862	6.05220e+4

force $p = 200$ is applied along the bottom half circle of the right hole, the left two holes are fixed, as shown in Figure 5. The volume constraint is $V = 0.003$. We use uniformly distributed material $\rho = 1$ as the initial design.

Figure 6 shows the optimal structures of the 3-hole bracket from different grid sizes. Though the design domain is more complicated, it is treated in our method in the same way as previous examples: B-spline basis functions are posed on background rectangular grids to approximate the material field and the meshfree analysis satisfies the boundary conditions automatically. To "protect" the three holes, a tolerance zone is put around the three holes and the coefficients of B-splines basis functions that have support intersecting with this tolerance zone are fixed during the optimization process. Table 3 lists values of the objective function and the area of the optimal structures in Figure 6.

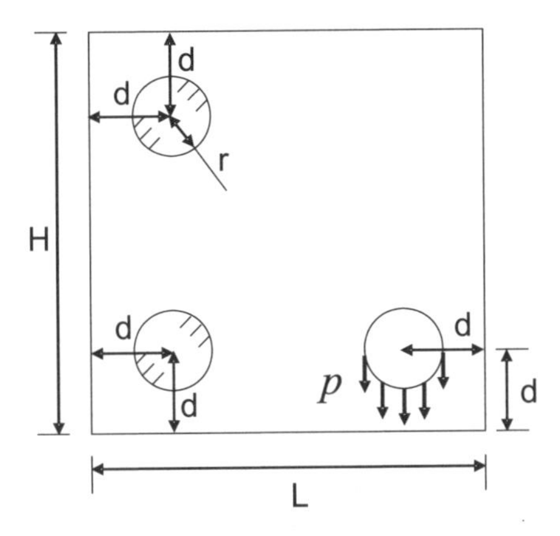

Fig. 5. Problem definition of a 3-hole bracket

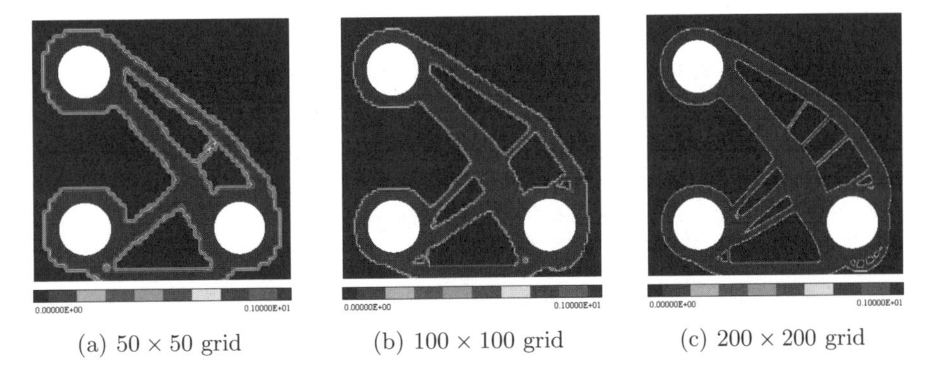

(a) 50×50 grid (b) 100×100 grid (c) 200×200 grid

Fig. 6. The optimal structures of the 3-hole bracket from different grid sizes

Example 4: Cantilever beam with a moving hole

Figure 7 shows the design domain of a short cantilever beam. The design domain has a circular hole whose position coordinates are geometric parameters that are subject to optimization. The length of the rectangle is $L = 0.1$, the height is $H = 0.05$. The thickness of the plate is $t = 0.0025$. The radius of the hole is $r = 0.0075$ and the initial position of the hole is $x_c = 0.03, y_c = 0.0125$. A distributed force $p = 200$ is applied in an interval of 0.005 around the middle point of the right edge of D and the left edge of D is fixed. The volume constraint is $V = 0.025$. Uniformly distributed material $\rho = 1$ is used as the initial design.

In this example, the position of the hole and the material distribution are optimized simutaneously. Figure 8 shows the optimal structures from different

Table 3. The objective and the area of the 3-hole bracket

grid size	area	strain energy
50×50	3.00229e-3	9.16768e+4
100×100	3.00925e-3	7.68538e+4
200×200	2.99114e-3	7.20218e+4

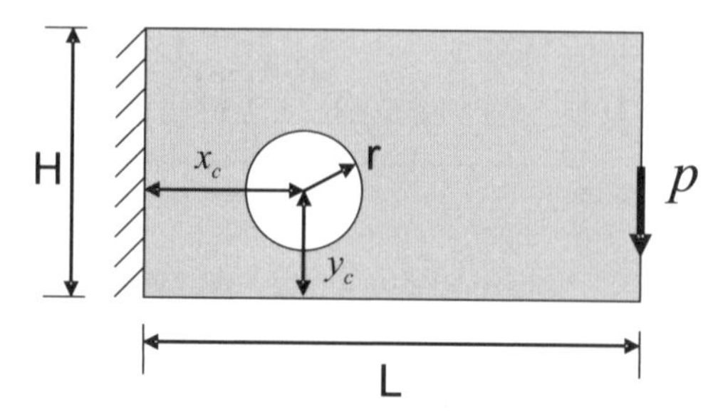

Fig. 7. Problem definition of a cantilever beam with a hole

grid sizes and Table 4 lists values of the objective function, the area of the optimal structures, and the final coordinates of the hole.

Example 5: Stepped cantilever beam

Figure 9(a) shows the design domain of a short cantilever beam. The design domain consists of three segments (steps) whose heights are subject to optimization. The length of the rectangle is $L = 0.1$, the height is $H = 0.05$. $L_1 = 0.03$, $L_2 = 0.04$, $L_3 = 0.03$. The thickness of the plate is $t = 0.0025$. A distributed force $p = 200$ is applied in an interval of 0.005 around the middle point of the right edge of D and the left edge of D is fixed. The volume constraint is $V = 0.025$. The initial heights of the three segments are $h_1 = h_2 = h_3 = 0.025$, and the initial material density is $\rho(x) = 0.5$. Figure 9(b) and 9(c) show the initial design and the optimal structure respectively.

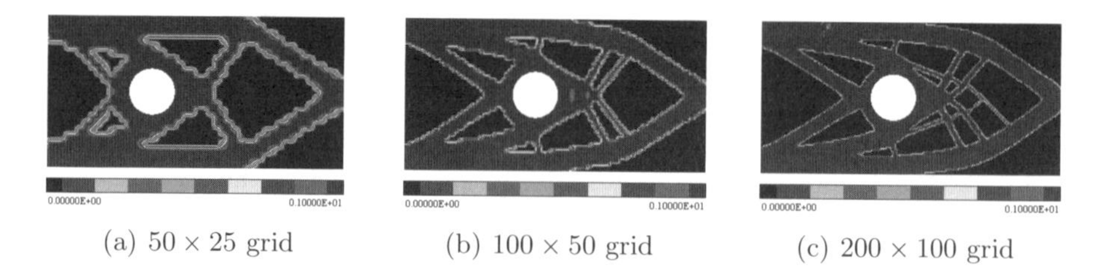

(a) 50×25 grid (b) 100×50 grid (c) 200×100 grid

Fig. 8. The optimal structures of the one hole cantilever beam from different grid sizes

Table 4. The objective, area and the final coordinates of the hole in the one-hole cantilever beam

grid size	area	strain energy	x_c	y_c
50×25	2.50110e-3	1.36148e+4	3.54749e-2	2.49849e-2
100×50	2.50183e-3	1.27441e+4	4.36144e-2	2.49981e-2
200×100	2.49845e-3	1.23955e+4	4.41646e-2	2.51398e-2

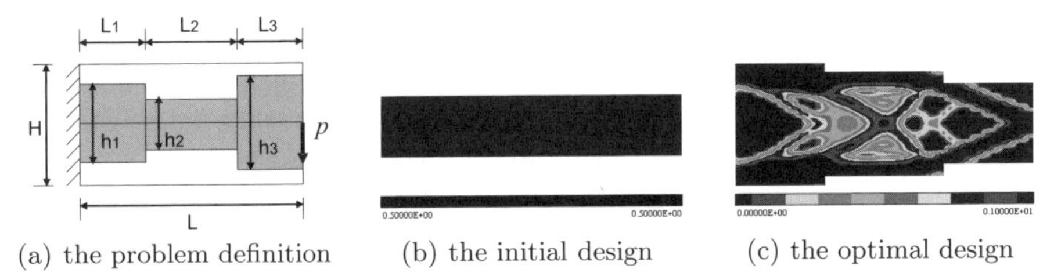

(a) the problem definition (b) the initial design (c) the optimal design

Fig. 9. The optimization of a stepped cantilever beam: (a) the design domain, (b) the initial design with uniform density $\rho = 0.5$, (c) the optimal design

5 Conclusions

5.1 Summary

We proposed a method for representation and optimization of heterogenous models. The key feature of the method is the separation of the material representation and the geometry representation. Representing the continuous material field using B-spline basis functions gives a smooth material field and allows local controls of the material properties. The implicit representation for geometry handles geometrical deformations by changing a few geometric parameters. Due to the separate representations, the material field does not need to conform to the geometry domain, and therefore no particular spatial discretization is required.

We have shown that material sensitivity and shape sensitivity are easily decoupled, supporting simultaneous material and shape optimization. The method was fully implemented in a particular meshfree framework, but it is general enough to be implemented in other computational environments with minimum requirements. Our numerical experiments for the minimum compliance problem and SIMP material model produce results that are at least as good as any published in the literature to date. In particular, we notice the absence of any numerical artifacts, such as checkerboard patterns reported by many others [6,13]. This superior numerical behavior may be attributed to built-in continuity of the material field [21]. Furthermore, to the authors' knowledge, until now SIMP has not been formulated or implemented with simultaneously changing globally parameterized geometric domain.

5.2 Extensions

It should be clear that proposed method can be applied to any material model and extended to other structural design problems. Any gradient-based methods can be used to solve the optimization problem. It is less obvious that the proposed methods can be used with most geometric representations and feature-based heterogeneous models.

The implicit representation of the geometry was used in this paper for two convenience purposes: to define the characteristic function $H(\Phi)$, and to derive the term $\frac{1}{|\nabla\Phi|}\frac{d\Phi}{db}$ in Equation (11). It is clear that the characteristic function computation is supported by any unambiguous representation of a solid through standard point membership classification (PMC) algorithms. Furthermore, we have recently shown [10] that the derivation of sensitivity in section 3.3 only relies on the *existence* of implicit representations, but in fact it does not matter whether the primitive is represented implicitly, parametrically, variationally, or procedurally. Also, we show in [10] that the term $\frac{1}{|\nabla\Phi|}\frac{d\Phi}{db}$ is equivalent to the normal component of the boundary velocity v_n. The proposed approach applies as long as we are able to compute the shape (design) velocity v_n in the direction normal to the primitive's boundary. Once the primitive velocities are computed based on the properties and representations of the individual primitives, they can all be used simultaneously within the framework described in this paper. In this sense, the proposed approach to optimization of heterogeneous models can be used with most geometric representations.

As we already mentioned in the introduction, the separation of material representation from geometry comes at a price: the changes in geometric parameters do not propagate into the material representation, undermining the benefits of the feature-based approach to material modeling. It is proposed in [7] that the material field may be represented as a sum of two independent fields: $F = P + R$, where the $P(\mathbf{b})$ interpolates the material properties prescribed at the features of Ω, while $R(\mathbf{c})$ is a linear combination of B-splines that may be used to control local material properties. Such representation of material field may support both interactive design of heterogeneous models where the material properties follow the geometric parameters and material optimization at the points of the domain that are sufficiently far away from the material features.

Acknowledgements

This works was supported in part by the National Science Foundation grants CMMI-0323514, CMMI-0500380, CMMI-0621116, OCI-0636206, and Wisconsin Industrial & Economic Development Research Program (I&EDR).

References

1. Adzhiev, V., Kartasheva, E., Kunii, T., Pasko, A., Schmitt, B.: Hybrid Cellular-functional Modeling of Heterogeneous Objects. Journal of Computing and Information Science in Engineering 2, 312 (2003)
2. Allaire, G., Jouve, F., Toader, A.M.: Structural optimization using sensitivity analysis and a level-set method. Journal of Computational Physics 194, 363–393 (2004)

3. Belytschko, T., Xiao, S.P., Parimi, C.: Topology optimization with implicit functions and regularization. International Journal for Numerical Methods in Engineering 57, 1177–1196 (2003)
4. Bendsøe, M.P.: Optimal shape design as a material distribution problem. Structural Optimization 1, 193–202 (1989)
5. Bendsøe, M.P., Sigmund, O.: Material interpolations in topology optimization. Archive of Applied Mechanics 69, 635–654 (1999)
6. Bendsøe, M.P., Sigmund, O.: Topology Optimization: Theory, Methods and Applications. Springer, Heidelberg (2003)
7. Biswas, A., Fenves, S., Shapiro, V., Sriram, R.: Representation of Heterogeneous Material Properties in Core Product Model. Engineering with Computers (2007) (in press)
8. Biswas, A., Shapiro, V.: Approximate distance fields with non-vanishing gradients. Graphical Models 66(3), 133–159 (2004)
9. Bloomenthal, J.: Introduction to Implicit Surfaces. Morgan Kaufmann Publishers, San Francisco (1997)
10. Chen, J., Freytag, M., Shapiro, V.: Shape sensitivity of constructive representations. In: Proceedings of the 2007 ACM Symposium on Solid and Physical Modeling, pp. 85–95. ACM Press, New York (2007)
11. Chen, J., Shapiro, V., Suresh, K., Tsukanov, I.: Shape optimization with topological changes and parametric control. International Journal of Numerical Methods in Engineering 71(3), 313–346 (2007)
12. de Boor, C.: A Practical Guide to Splines. Springer, New York (2001)
13. Eschenauer, H.A., Olhoff, N.: Topology optimization of continuum structures: A review. Applied Mechanics Review 54, 331–390 (2001)
14. Frisken, S.F., Perry, R.N., Rockwood, A.P., Jones, T.R.: Adaptively sampled distance fields: A general representation of shape for computer graphics. In: Proceedings of the ACM SIGGRAPH Conference on Computer Graphics, pp. 249–254 (2000)
15. Haug, E.J., Choi, K.K., Komkov, V.: Design Sensitivity Analysis of Structural Systems. Academic Press, New York, NY (1986)
16. Kojekine, N., Hagiwara, I., Savchenko, V.: Software tools using csrbf's for processing scattered data. Computers and Graphics 27(2) (2003)
17. Kou, X.Y., Tan, S.T.: Heterogeneous object modeling: A review. Computer-Aided Design 39(4), 284–301 (2007)
18. Kumar, V., Burns, D., Dutta, D., Hoffmann, C.: A framework for object modeling. Computer-Aided Design 31, 541–556 (1999)
19. Lim, C., Turkiyyah, G.M., Ganter, M.A., Storti, D.W.: Implicit reconstruction of solids from cloud point sets. In: Proceedings of the Third Symposium on Solid Modeling and Applications, pp. 393–402. ACM Press, New York (1995)
20. Liu, H., Cho, W., Jackson, T.R., Patrikalakis, N.M., Sachs, E.M.: Algorithms for design and interrogation of functionally gradient material objects. In: Proceedings of ASME 2000 IDETC/CIE 2000 ASME Design Automation Conference, Baltimore, MD (2000)
21. Matsui, K., Terada, K.: Continuous approximation of material distribution for topology optimization. International Journal of Numerical Methods in Engineering 59, 1925–1944 (2004)
22. Muraki, S.: Volumetric shape description of range data using "Blobby Model. In: Proceedings of the ACM SIGGRAPH Conference on Computer Graphics, vol. 25(4), pp. 227–235 (July 1991)

23. Nocedal, J., Wright, S.J.: Numerical Optimization. Springer, Heidelberg (1999)
24. Ohtake, Y., Belyaev, A., Alexa, M., Turk, G., Seidel, H.-P.: Multi-level partition of unity implicits. ACM Transactions on Graphics (TOG) 22(3), 463–470 (2003)
25. Pegna, J., Safi, A.: CAD modeling of multi-model structures for freeform fabrication. In: In Proceedings of the 1998 Solid Freeform Fabrication Symposium, Austin, TX (August 1998)
26. Press, W.H., Teukolsky, S.A., Vetterling, W.T., Flannery, B.P.: Numerical Recipes in C, 2nd edn. Cambridge University Press, Cambridge (1992)
27. Qian, X., Dutta, D.: Design of heterogeneous turbine blade. Computer-Aided Design 35, 319–329 (2003)
28. Raviv, A., Elber, G.: Three-dimensional freeform sculpting via zero sets of scalar trivariate functions. Computer-Aided Design 32, 513–526 (2000)
29. Ricci, A.: A constructive geometry for computer graphics. Computer Journal 16(3), 157–160 (1973)
30. Rvachev, V.L.: Geometric Applications of Logic Algebra (in Russian), Naukova Dumka (1967)
31. Rvachev, V.L.: Theory of R-functions and Some Applications (in Russian), Naukova Dumka (1982)
32. Rvachev, V.L., Sheiko, T.I.: R-functions in boundary value problems in mechanics. Applied Mechanics Reviews 48(4), 151–188 (1996)
33. Rvachev, V.L., Sheiko, T.I., Shapiro, V., Tsukanov, I.: On completeness of RFM solution structures. Computational Mechanics 25, 305–316 (2000)
34. Savchenko, V.V., Pasko, A.A., Okunev, O.G., Kunii, T.L.: Function Representation of Solids Reconstructed from Scattered Surface Points and Contours. Computer Graphics Forum 14(4), 181–188 (1995)
35. Schmitt, B., Pasko, A., Schlick, C.: Constructive sculpting of heterogeneous volumetric objects using trivariate b-splines. The Visual Computer 20(2), 130–148 (2004)
36. Sethian, J.A.: Level Set Methods and Fast Marching Methods: Evolving Interfaces in Computational Geometry, Fluid Mechanics, Computer Vision, and Materials Science. Cambridge University Press, Cambridge (1999)
37. Sethian, J.A., Wiegmann, A.: Structural boundary design via level set and immersed interface methods. Journal of Computational Physics 163(2), 489–528 (2000)
38. Shapiro, V.: Real functions for representation of rigid solids. Computer-Aided Geometric Design 11(2), 153–175 (1994)
39. Shapiro, V.: Well-formed set representations of solids. International Journal on Computational Geometry and Applications 9(2), 125–150 (1999)
40. Shapiro, V.: A convex deficiency tree algorithm for curved polygons. International Journal of Computational Geometry and Applications 11(2), 215–238 (2001)
41. Shapiro, V.: Semi-analytic geometry with R-functions. Acta Numerica 16, 239–303 (2007)
42. Shapiro, V., Tsukanov, I.: Implicit functions with guaranteed differential properties. In: Fifth ACM Symposium on Solid Modeling and Applications, Ann Arbor, MI, pp. 258–269 (1999)
43. Shapiro, V., Tsukanov, I.: Meshfree simulation of deforming domains. Computer Aided Design 31, 459–471 (1999)
44. Sigmund, O., Petersson, J.: Numerical instabilities in topology optimization: A survey on precedures dealing with checkerboards, mesh-dependencies and local minima. Structural Optimization 16, 68–75 (1998)

45. Tsukanov, I., Shapiro, V.: The architecture of SAGE – a meshfree system based on RFM. Engineering with Computers 18(4), 295–311 (2002)
46. Turk, G., O'Brien, J.: Modeling with implicit surfaces that interpolate. ACM Transactions on Graphics 21(4), 855–873 (1999)
47. Velho, L., Gomes, J., de Figueiredo, L.H.: Implicit Objects in Computer Graphics. Springer, Heidelberg (2002)
48. Wang, M.Y., Wang, X.M., Guo, D.M.: A level set method for structural topology optimization. Computer Methods in Applied Mechanics and Engineering 192(1-2), 227–246 (2003)
49. Wang, S.Y., Wang, M.Y.: Radial basis functions and level set method for structural topology optimization. International Journal for Numerical Methods in Engineering 65(12), 2060–2090 (2005)
50. Washizu, K.: Variational methods in elasticity and plasticity, 3rd edn. Pergamon Press, Oxford (1982)

Automation of the Volumetric Models Construction

Pierre-Alain Fayolle[1], Alexander Pasko[2], Elena Kartasheva[3], Christophe Rosenberger[1], and Christian Toinard[1]

[1] Université d'Orléans, France
p.fayolle@free.fr, christophe.rosenberger@ensi-bourges.fr, christian.toinard@ensi-bourges.fr
[2] Bournemouth University, United Kingdom
apasko@bournemouth.ac.uk
[3] Institute for Mathematical Modeling, Russia
ekart2005@gmail.com

Abstract. The automation of the function-based (FRep) volumetric modeling task is tackled by introducing template parameterized models and a procedure for recovery of constructive models from segmented point-sets. In order to reuse existing models, we propose to parameterize them and to fit the parameters to different point-sets for optimizing and adapting the shape to different objects of the same class of shapes.

The automation of the creation of a constructive FRep model is also considered by creating a recovery procedure for a given segmented point-set and a list of corresponding primitives. A genetic algorithm is used to find the best constructive expression for the object with the given set of primitives in the point cloud segmentation and the set of available operations.

The proposed approach is illustrated by fitting of different models to point clouds and by the automatic generation of constructive trees from segmented point-sets for real mechanical parts.

1 Introduction

Modeling objects in a constructive way by recursively applying set-theoretic operations to primitives is a well-known and powerful paradigm in solid modeling. Combined with the generation of a functional expression for the final solid with the defining function having a distance property, it provides a powerful tool for solid modeling and applications. The construction of objects following this paradigm may however be tedious and sometimes repetitive. The automation of the model construction process is suitable and for that purpose we introduce the notion of template constructive function-based models. The basics of constructive function-based modeling using the function representation (FRep [31]) are described elsewhere in this volume.

A template FRep model is a model where abstract parameters are released and fitted to adapt the shape to different discrete point-sets acquired by scanning

A. Pasko, V. Adzhiev, and P. Comninos (Eds.): HOMA, LNCS 4889, pp. 214–238, 2008.

devices or already available from a different representation of the solid. The defining FRep function can be considered as an algebraic distance measure. Fitting of non-linear parameterized FRep models can be implemented using a combination of meta-heuristics such as simulated annealing or genetic algorithms with local methods such as Newton methods.

However, fitting can be applied in the case the construction of the object is already known, and an analytical expression or a function evaluation procedure with a set of predefined parameters is provided. The next step in the automation of the creation of constructive FRep models is to try to recover a set-theoretic expression for an object from a given segmented point-set and a list of primitives in the segmentation. We introduce a genetic algorithm to find the best constructive expression involving the primitives fitted to the segmented point-set and operations selected from a set of predefined operations. The obtained constructive FRep model can be parameterized and used as a valid template, which can be reused and fitted to different instances of the object.

Parameterized templates modeling and fitting is illustrated in this paper through the processing of different cultural heritage and mechanical objects. Finally, we illustrate the automatic generation of constructive trees from segmented point-sets for real mechanical parts.

The modeling automation methods presented here can be applied to obtain function-based models of heterogeneous objects or their elements. These can be FRep solids, space partitions and material features for spatial attributes modeling, carrier surfaces and trimming objects for trimmed implicit curves and surfaces covered in other chapters of the volume.

2 Previous Works

Modeling requires a lot of skills and can be a difficult and time-consuming task. We look at the automation of the modeling process. We briefly survey the existing methods for the modeling automation for existing shapes using data acquired with scanning devices.

The automation process, called reverse engineering, consists usually of the following steps – not necessary in a linear order:

- data capture; for example using a laser scanner, or others [41]
- preprocessing; like denoising, computing consistent and globally oriented normals, combining multiple views obtained from different data acquisitions
- segmentation and surface fitting; where data points are grouped into sets to which an appropriate surface is fitted
- geometric model creation

Automation is also required when converting data from one format to another, for example from the boundary representation (BRep) to Constructive Solid Geometry (CSG).

Generally, it is possible to distinguish between reverse engineering for computer graphics purposes and CAD purposes, because they have different goals, even if the global framework is the same. Reverse engineering for CAD purposes has stronger requirements for the generated CAD model. The following review of existing works follows this distinction.

2.1 Creation of Triangle Meshes and Implicit Surface Fitting

Creation of triangle meshes: The work by Hoppe [17] reconstructs a surface in three steps: 1) initial surface estimation by computing a signed distance function, 2) mesh simplification, 3) piecewise smooth surface optimization. Amenta's Power Crust method [1] relies on computing an approximation of the medial axis transform with polar balls.

Ohtake et al. [30] compute a mesh approximation of scattered point data by creating an adaptive sparse cover of the point-set, creating auxiliary points corresponding to the spheres, connecting these points, and finally filling the holes and cleaning the mesh.

Implicit surface fitting: The reconstruction method proposed by Muraki [27] consists in fitting blobby models [5] to range data. Savchenko et al. [35] and later Turk et al. [39] proposed to fit a linear combination of radial basis functions to the point-set. Compactly supported radial basis functions were introduced by Morse et al. [26] to decrease the complexity in time and memory of the previous method. Partition of Unity was introduced by Ohtake et al. [28] as an elegant way to partition the point-set in order to decrease drastically the time and memory complexity.

These methods produce verbose models, even in the case of implicit surface fitting which may contain a lot of coefficients for the series items as well as the original point-set. They are also practically useless for inspection and reuse of the structure of the object in contrary to objects built using constructive geometry methods. Furthermore, besides the work of Amenta [1] they lack of theoretical guarantees: existence of cracks and holes, and different topology than the original models.

These methods are however really good for producing nice looking triangle meshes to be used in various applications such as games, virtual museums and others. For complex freeform shapes with lots of small details, they appear also as the only practical methods.

2.2 Reverse Engineering in CAD

The main difference between the reverse engineering for CAD and for computer graphics applications is that the creation of geometric models for CAD purposes concentrate on accurate and consistent models using standard surfaces as found in the common CAD/CAM systems [41]. The reconstructed model should be a valid CAD model for solid modeling and ready to undergo further operations. The standard model representation is usually the boundary representation or BRep.

The problems to be solved include: identifying sharp edges, treatment of blends, providing continuity and smoothness between the patches. Another important problem is the creation of geometric models respecting constraints such as for example: planes are parallel, spheres are concentric, smooth blend between parts, and others [3].

The segmentation part is especially difficult. It consists in clustering the original point-set into subsets that correspond to some common primitives. It is a key step in identifying the logical structure of the final CAD model. The dissertation of Vanco [40] studies the problem of direct segmentation of a point-set, where no intermediate triangulation of the point-set is computed to do the segmentation. It relies on normal and curvature estimations for clustering the points. The drawback of the approach is that computation of normals and curvature is difficult: computing normals with a global consistent orientation is NP [17]. Hoppe [17] proposes a heuristic method to solve it. Computing numerically reliable curvature information is also a difficult problem [29]. Benko and Varady use also a direct segmentation by computing a series of simple tests to split the original point-set [4]. Marshall et al. [22] try to fit common primitives (sphere, torus, plane) using an approximation of the Euclidean distance.

After the point-set is segmented, primitives are associated and fitted to each subset. Sometimes the fitting is part of the segmentation [22].

The final step is the creation of the CAD model. It consists in grouping the fitted primitives to make a valid BRep: most of the difficulties are from the constraints requirements to the final model. Another problem is the creation of a valid BRep model that may be difficult to fulfill.

2.3 Boundary Representation to CSG Conversion

The problem of BRep to CSG conversion is a difficult problem related to reverse engineering of solids. Suppose that we have been successful in generating a BRep model from a point-set, or that we have a segmentation of the point-set, with a primitive fitted to each of the subsets. Then we want to convert the BRep to a CSG representation or to find a CSG representation that uses the fitted primitives.

The question of converting a BRep to a CSG representation has been firstly investigated by Rvachev et al. in the two-dimensional case [34], where an algorithm to convert a shape defined by a two-dimensional linear polygon to a set-theoretic representation is described. In the English literature, a similar algorithm is attributed to Batchelor [2], where the conversion algorithm is based on the concept of convex deficiency tree: any linear polygon can be represented by the difference between its convex hull and a finite number of concavities. Shapiro extended the algorithm to handle curved polygons [36]. Similar algorithms were adapted to three-dimensional polyhedra [44], but unfortunately do not work for some polyhedra. In three-dimensional spaces, the problem has been solved for solids bounded by second degree surfaces [37,6]. These algorithms may require some additional halfspaces not available from the boundary faces information or from the segmentation.

2.4 Discussion

Methods generating a triangle mesh approximation of the surface are unacceptable for several applications involving the reconstructed model. The generated triangle mesh provides no theoretical guarantee of accuracy or consistency. The same fact holds for methods based on fitting implicit surfaces. In most of the cases, there is no proof of accuracy or consistency. For example: can we guarantee that there will be no extra isosurface, or extra isolated point-set of lower dimension corresponding to the isovalue of interest?

Reverse engineering in CAD has stronger requirements than reverse engineering with applications in computer graphics: it aims at providing valid, accurate and consistent models. However, the target model is a boundary representation model, which is known to have various problems. For some applications, it is interesting to obtaine a constructive tree, because it can be used later to inspect or modify the structure of the recovered object.

The same requirement of getting a constructive tree can be made when fitting implicit surfaces. When recovering an FRep model from a given existing solid, a constructive object carries more information than a set of basis functions and their associated coefficients.

Getting a constructive tree gives the possibility to later treat and handle the object in the same way it could be done with any objects modeled by a user. Especially one can inspect the structure of the object, modify primitives or operations, to create new models. Being able to use the constructive tree to evaluate the corresponding real function is another important requirement as the real function can be used for further applications. This function can be used, for example, to generate meshes adapted to the requirements of the finite element methods [18] or it can be used directly in mesh-free methods [13].

Considering these criteria, we look at the problem of reverse engineering objects using constructive FRep models. We introduce for that purpose the idea of parameterized templates. A parameterized template FRep model is a sketch made by the user, where the constructive tree contains only specified operations and types of primitives, while the parameter values of operations and primitives are not defined and should be estimated by fitting as will be discussed below. The creation of template models may still appear to be a difficult task. We also investigate the automation of this task by converting segmented point-sets into constructive FRep models.

3 Automation Methods

3.1 Automation of Modeling Using Template FRep Models

Modeling requires lots of skills and can be a difficult and time-consuming task. We present here the notion of an FRep template to automate the modeling step [9]. A practical case requiring automation consists in modeling existing objects acquired with scanners. The FRep model refers here to the general case functions. However, the use of the distance or approximate distance functions can provide better results in the fitting process of template models [8].

Template FRep Models. An FRep model can be built in a constructive way with abstract parameters. The modification of these parameters can result in various shape modifications including changes of shape topology. Shape parameters can also be estimated to fit some special modeling criteria. In the following, the notation $F(\mathbf{p}, \mathbf{a})$ is used for a parameterized FRep model, where $\mathbf{p} = (x, y, z) \in \mathbb{R}^3$ is a point in the 3D space and $\mathbf{a} = (a_1, \ldots, a_m) \in \mathbb{R}^m$ is a vector of m parameters.

In a given domain of application, objects can have similar shapes that can be parameterized. Template models can exist in specialized libraries for each application domain (mechanical design, human prosthesis design, and others) and may be reused, or need to be created by the user. In the latter case, a modeling work needs to be done by a designer. A parameterized model can be created using measurements or scans of a typical object. The model is required to keep basic ratios of the measured sample object and to proportionally change the dependent parameters according to introduced constraints. In case of scanned data available for a typical object, fitting of the template parameters can be also employed to establish basic ratios and constraints.

An example of a parameterized template FRep model, with different instances of the parameters, is illustrated in Fig. 1. The different model's parameters represent the dimensions of the box, the diameters of the cylinders and their positions.

Fig. 1. Illustration of a parameterized template FRep model: a model is built with FRep in a constructive way with abstract parameters that can be tuned to satisfy some modeling criteria

Fitting Problem Formulation. The problem is to recover a solid from a set of 3D points, $S = \{\mathbf{p_1}, \ldots, \mathbf{p_N}\}$, scattered on the surface of the object. Given S, the task is to find the best configuration for the set of parameters $\mathbf{a}^* = (a_1^*, \ldots, a_m^*)$ so that the parameterized FRep model $F(\mathbf{p}, \mathbf{a}^*)$ closely fits the data points. $F(\mathbf{p}, \mathbf{a})$ is an FRep model, made in a constructive way, which approximates the shape of the solid being reverse engineered. The vector of parameters $\mathbf{a}$ control the final shape of the solid and the best fitted parameters should give the closest possible model according to the information provided by S.

For computing how close a given point is to the surface of the solid with the current set of parameters, a fitness function is needed. The FRep model

$F(\mathbf{p}, \mathbf{a})$ itself can serve for defining such a measure. Note that the better F is an approximation of the Euclidean distance function, the more robust the fitting will be. The error of fit becomes the square of the defining function values at all points (the surface of the solid being the set of points with zero function value):

$$error(\mathbf{a}) = \frac{1}{2} \sum_{i=1}^{N} F^2(\mathbf{p_i}, \mathbf{a}) \tag{1}$$

which can be also rewritten under the following form:

$$error(\mathbf{a}) = \frac{1}{2} \sum_{i=1}^{N} F_i^2(\mathbf{a}) = \frac{1}{2} \mathbf{F}^t(\mathbf{a})\mathbf{F}(\mathbf{a}) \tag{2}$$

where $\mathbf{F}(.)$ is the vector with $F_i(.) = F(p_i, .)$ as the $i - th$ component. Now, we are searching for the vector of parameters $\mathbf{a}^*$ minimizing the error of fit from equation 1. We consider at first local methods for minimizing functions with nonlinear parameters.

Nonlinear Minimization of Least Squares by Local Methods. The best set of parameters $\mathbf{a}^*$ is found by minimization of the least square error (equation 1). This least square error is usually a nonlinear function of the parameters $\mathbf{a}$. Traditional methods for solving such problems are Levenberg-Marquardt methods [33], [25] or Newton methods [7] (full-Newton or quasi-Newton). Such algorithms proceed iteratively from an initial set of parameters and try to converge to a minimum in the parameter space. These methods strongly depend on the initial parameters' estimations, which are used for starting the algorithm.

Such algorithms search in each iteration for a privileged direction to go in the parameter space and for a step to move in that direction. Levenberg-Marquardt and Newton type algorithms differ in the selection of direction and in the ways of computing the step.

The function being minimized, in that case the least square error, is approximated by a Taylor series expansion to the second order:

$$\begin{aligned} error(\mathbf{a}) = {} & \mathbf{F}^t(\mathbf{a_k})\mathbf{F}(\mathbf{a_k}) + (\mathbf{a} - \mathbf{a_k})^t \mathbf{J}(\mathbf{a_k})^t \mathbf{F}(\mathbf{a_k}) + \frac{1}{2}(\mathbf{a} - \mathbf{a_k})^t \\ & (\mathbf{J}(\mathbf{a_k})^t \mathbf{J}(\mathbf{a_k}) + \sum_{i=1}^{N} F_i(\mathbf{a_k}) \nabla^2 F_i(\mathbf{a_k}))(\mathbf{a} - \mathbf{a_k}) \end{aligned} \tag{3}$$

where $\mathbf{a_k}$ is the vector of parameters for the $k - th$ iteration, $\mathbf{J}$ is the Jacobian of $\mathbf{F}$ and $\nabla^2 F_i$ is the Hessian of F_i. Note that $\nabla error = \mathbf{J}^t\mathbf{F}$ and $\nabla^2 error = \mathbf{J}^t\mathbf{J} + \sum_{i=1}^{N} F_i \nabla^2 F_i$.

Newton methods compute the direction as the solution of the following equation:

$$\mathbf{H}.(\mathbf{a_{k+1}} - \mathbf{a_k}) = -\mathbf{J}^t\mathbf{F}(\mathbf{a_k})$$

where $\mathbf{H}$ is the Hessian matrix of the error function, or its numerical approximation when it is not known analytically. Variations of the Newton method differ in whether they use the exact Hessian or an approximation. The Gauss-Newton method, for example, uses $\mathbf{J}^t\mathbf{J}$ as a Hessian approximation.

Levenberg-Marquardt algorithm provides an efficient way for switching between a Newton method and a steepest descent method for selecting a direction. This is done by solving the following equation for the unknown $\mathbf{a_{k+1}}$:

$$(\mathbf{J}^t.\mathbf{J} + \lambda I)(\mathbf{a_{k+1}} - \mathbf{a_k}) = -\mathbf{J}^t\mathbf{F}(\mathbf{a_k})$$

where $\mathbf{a_k}$ is the current vector of parameter, $\mathbf{a_{k+1}}$ is the next vector of parameter. λ is adaptively changed during run-time to allow the use of a Newton method ($\lambda = 0$) or a steepest descent method (big value for λ).

These methods can in general guarantee only a convergence to a local minimum: for parameter spaces with complex topology like, for example, where multiple local minima exist, these methods are likely to be trapped at a local optimum. Good choice of initial parameters is important, because it will determine to which minimum the algorithm may converge. Usually, if the parameters are not in the neighborhood of the global minimum, it is unlikely to converge to it.

It is possible with some further analysis of the model to have some additional information for getting better estimation of the starting parameters. It may also be possible to restart the search algorithm with different starting points. Another method consists in using metaheuristics such as simulated annealing or genetic algorithms to perform the nonlinear optimization.

Simulated Annealing. Simulated annealing is one of the most effective methods for solving combinatorial and continuous global optimization problems [23,19,33]. When trying to minimize an objective function usually only downhills are accepted, but within a simulated annealing algorithm some uphills may be accepted, with a probability $p(T)$, which is initially close to 1, and then decreases to 0, when the temperature T of the system reduces. The procedure, governing the temperature evolution of the system, is called temperature schedule or cooling schedule.

The simulated annealing algorithm has been inspired by the behaviour of some thermodynamical process: in a liquid at high temperature, the molecules move freely with respect to one another; when the liquid is cooled down, the mobility of the molecules decreases, and finally stops. If the cooling is not too fast, then the system will finish in a state of minimum energy. According to the Boltzmann law, a system in thermal equilibrium at temperature T has its energy distributed probabilistically among all different energy states. Even for a low temperature there is a chance for the system to be in a high energy, so that it can escape the local minimum energy and finds a better one.

Convergence to an optimal solution can be theoretically guaranteed, but only after an infinite number of iterations controlled by the cooling schedule. In order to provide a finite time implementation, a proper cooling schedule is needed to simulate the asymptotic convergence behavior of the simulated annealing. For that reason, simulated annealing suffers from slow convergence and may wander around the optimum solution when high accuracy is needed.

A typical simulated annealing algorithm is given below:

Simulated-Annealing

1. Initialization. Choose an initial solution $\mathbf{x_0}$. Fix the parameters for the cooling schedule: the initial temperature T_{max}, the epoch length M, the cooling reduction factor λ, the minimum temperature T_{min}.
2. The main iteration. Repeat M times the following SA search. Generate a trial point randomly $\mathbf{x_{SA}}$ within the feasible domain. Evaluate f at the trial point $\mathbf{x_{SA}}$ and accept the trial point ($\mathbf{x_{k+1}} := \mathbf{x_{SA}}$) if:
   (a) $\Delta f := f(\mathbf{x_{SA}}) - f(\mathbf{x_k}) < 0$ or
   (b) $\Delta f \geq 0$ and $p = exp(\frac{-\Delta f}{T}) \geq r$ where r is a random number in $(0, 1)$.
3. Termination. If the cooling schedule is completed ($T \leq T_{min}$) or the function values of two consecutive improvement trials become close within a tolerance ϵ or when the number of total iterations exceeds some user threshold, then quit. Otherwise, decrease the temperature by setting $T := \lambda T$ and go to step 2.

Genetic Algorithms. Using a genetic algorithm is another way to solve problems of optimization (combinatorial or continuous). A genetic algorithm consists in a population of individuals, also called chromosomes, and operations, inspired by the mechanisms of natural selection and the law of genetics, which act on the elements of the population. An individual encodes a solution of the problem using an appropriate representation for this given problem. In the case of nonlinear optimization of real-valued functions, it consists in a set (an array) of parameters from the parameter space. The different operations acting on the individuals from a population are typically selection, crossover, and mutation [16,14].

Genetic algorithms and simulated annealing are good methods for finding a global optimum among many local optima. Practically it is useful to combine these global methods with local methods to refine the solution and/or accelerate the search.

We have implemented the simulated annealing and direct search methods, such as Levenberg-Marquardt or quasi-Newton type, in C. The two types of algorithms, simulated annealing and the direct search method, are combined in a two-step process. The first step (simulated annealing) should give a configuration in the parameter space being in the vicinity of the global minimum and thus should help avoiding local minima. The second step (the local method) should guarantee a faster convergence by avoiding that the algorithm wanders around the solution and should also improve the quality of the fitted parameters.

Switching from simulated annealing to the local method is done when: a given number of iterations (of simulated annealing) is performed, or the cooling schedule is completed, or the function values of two consecutive improvement trials become close within a given tolerance. If the switch from simulated annealing to the direct search method is done because the given number of iterations of simulated annealing have been finished, then simulated annealing will be run again (continued) after the direct method is completed, starting with the best fitted parameters obtained from the direct method; these two steps – simulated annealing and the direct method – are iterated a given number of times. In other cases, i.e., the cooling schedule is completed or the function values of two consecutive trials are close, the algorithm terminates after the direct search step.

3.2 Constructive Tree Recovery Using a Genetic Algorithm

We propose an algorithm for recovering an FRep model defined by a constructive tree from a segmented point-set and a list of fitted primitives [10]. The goal is to further automate the recovery process by automatically finding an FRep model from a segmented point-set as described in the next section. Once a model is obtained it is possible to parameterize it and further reuse the parameterized model.

The algorithm proposed in the following relies on the existence of a segmentation of a point-set and a list of primitives fitted to the segmented point-set. The same algorithm can be applied to recover a FRep model from a boundary representation (BRep), since the BRep model naturally provides both the point-set and the primitives.

Description of the Algorithm. Let us suppose that we have a set of points $\{\mathbf{p_1}, \ldots, \mathbf{p_n}\}$ on or near the surface of the solid and a set of primitives $\{f_1, \ldots, f_m\}$ fitted to the segmented point-set. Given a finite set of possible operations that can be applied to these primitives $\{\lambda_1, \ldots, \lambda_l\}$, we are searching for an ordering of the primitives with operations acting on them according to the formula:

$$f_{i_1} \lambda_{j_1} \ldots f_{i_m} \tag{4}$$

which is a correct FRep model for the solid defined by the point-set. In the above expression, $j_k \in \{1, \ldots, l\}$, and the set $\{i_1, \ldots, i_m\}$ is obtained from the set $\{1, \ldots, m\}$ by a bijection and is used to order the primitives. A correct FRep model means that the defining function $f = f_{i_1} \lambda_{j_1} \ldots f_{i_m}$ satisfies $f > 0$ inside the solid, $f < 0$ outside the solid and $f = 0$ on the boundary of the solid, defined by the discrete set of points.

If this formula is evaluated from left to right, it is clear that it corresponds to a tree structure (left unbalanced) with operations in the internal nodes and primitives in the leaves. Evaluation from left to right, using an intermediate variable temp, is done as follow:

$$\begin{aligned}
temp &\leftarrow f_{i_1}\ \lambda_{j_1}\ f_{i_2} \\
temp &\leftarrow temp\ \lambda_{j_2}\ f_{i_3} \\
&\ldots \\
temp &\leftarrow temp\ \lambda_{j_{m-1}}\ f_{i_m}
\end{aligned} \tag{5}$$

The representation in Eq. 4 comes from the fact that we want to encode each FRep model in an individual string to be processed by the genetic algorithm. This representation suits well the encoding, and is easy to evaluate.

The question is whether any constructive FRep model can be encoded in that way. If the operations are only set-theoretic, then any FRep model can be represented by a left unbalanced representation. Using DeMorgan transformations: $X \setminus Y \to X \cap \overline{Y}$, $\overline{X \cap Y} \to \overline{X} \cup \overline{Y}$, $\overline{X \cup Y} \to \overline{X} \cap \overline{Y}$, and $\overline{(\overline{X})} = X$, we transform the expression to an equivalent expression containing only $\cup$ and $\cap$. Then exploiting the fact that $\cup$ and $\cap$ are commutative: $X \cup Y = Y \cup X$ and $X \cap Y = Y \cap X$, we can switch internal nodes of the formula to obtain a left unbalanced representation. Using blending operations [32], union or intersection, the representation of a constructive FRep model by a left unbalanced representation is still possible when the blend is symmetric; if the blend is not symmetric, then the operations are not commutative, and a left unbalanced representation may not exist. Unary operations like space deformations: rotations, scaling or any other inverse space mappings, do not pose any problems at all and can be appended to the primitives.

Note that, using DeMorgan laws to transform a constructive tree to a left unbalanced tree introduces the complement of point-sets. Practically it means that the primitives are oriented. When converting BRep models to FRep models it may generally be the case, since the BRep model requires a global and consistent orientation. When fitting primitives to a point-set, the orientation can be obtained from the orientation of the point-set, which is a difficult problem [15].

Because the points $\{\mathbf{p_1}, \ldots, \mathbf{p_n}\}$ belong to the surface of the solid, a correct FRep model should be equal to 0 at each point. With the notation, $f = f_{i_1} \lambda_{j_1} \ldots f_{i_m}$, it means $\forall \mathbf{p_i}$, $f(\mathbf{p_i}) = 0$. The problem can be reformulated as the search of a formula f such that $\sum_{\mathbf{p_i}} f(\mathbf{p_i})^2$ is minimum. A genetic algorithm is used for the minimization problem. This problem is combinatorial and the genetic algorithm is used to perform two tasks: to find the order the primitives and to find the correct binary operations to apply between two primitives.

An individual of the population represents a possible solution to the problem. In this case, it is a left unbalanced expression of a constructive FRep model as illustrated by Eq. 4 (this is the phenotype or what exactly is the individual in the real world). Each individual contains m pairs of integers (op_k, L_k), $1 \leq k \leq m$, corresponding to the type of operations – op_k is an index to one of the operations from the set of possible operations – and L_k the position of the primitive k in the expression. The operation, indexed by op_k, is applied between the primitive k and the preceding primitive, at the position $L_k - 1$, in the reconstructed expression. One pair is encoded by a bit string (a gene). The aggregate of the genes, encoding the pairs, describes the individual's genotype. The expression encoded in an individual contains m operations, but a FRep expression using m primitives has in fact only $m - 1$ operations: one of the encoded operations is not used in the expression but is kept for the representation, i.e. appears in the genotype.

Relation with Genetic Programming. Genetic programming evolves computer programs by applying the Darwinian process of evolution [21] to them. It is similar to a genetic algorithm, with the difference that the data structure used to encode the program (the solution of the problem) is a tree of potentially infinite size. The approach, that we used here, looks like genetic programming, but the size of the representation is of fixed length; consequently there are no advantages in using genetic programming.

In fact, using genetic programming would result in long expressions with redundant information, such as, for example, the sub-expressions: $f \vee f$ (union of the solid with itself), $f \wedge f$ (intersection of the solid with itself). Genetic programming can also lead to sub-expressions like $f \setminus f$ which corresponds only to the boundary of the solid defined by f and is not suitable in solid modeling. However such an expression evaluates to 0 on the subset of points corresponding to the fitted primitive f and will not penalize the fitness function. Tracking such invalid sub-expressions or redundant information can only be done at the end, with algorithms for symbolic simplification.

Description of Solids and Primitives. A necessary condition, to be able to represent the solid by a constructive FRep model, given in the form of Eq. 4, is that the list of primitives is sufficient to describe the solid. Shapiro and Vossler proved in the work [37] that boundary patches on the surface may not be enough to describe the same solid by CSG and that extra halfspaces may be needed to create a CSG representation of the same model. They describe how to detect such cases and how to construct these additional halfspaces.

We illustrate this problem in Fig. 2. The left part shows a boundary representation of a simple two-dimensional object. Building a constructive representation of this object, given the boundary patches, is however not possible. An additional halfspace is required as illustrated in Fig. 2, right.

The final CSG representation of the solid is obtained by taking the union of the disk with the additional halfspace, and then intersecting this solid with the remaining halfspaces:

$$f = a \vee b \wedge c \wedge d \wedge e$$

The algorithm proposed here does not automatically induce additional halfspaces. They need to be detected and constructed by the algorithm of Shapiro and Vossler [37] and inserted in the list of primitives. In some cases, however, circular or spherical shapes are used to smoothly blend surfaces together; in these cases, additional halfspaces are not required by our algorithm, because the circular/spherical shapes are incorporated in the blending operations, which can be used as valid operations. We also noticed that when the algorithm is applied to the recovery from a point-set the need of extra halfspaces is dependent on the segmentation and fitting algorithms. In the example used above, if a rectangle is fitted instead of a collection of segments, then the extra halfplane is not needed because it is already embedded in the rectangle.

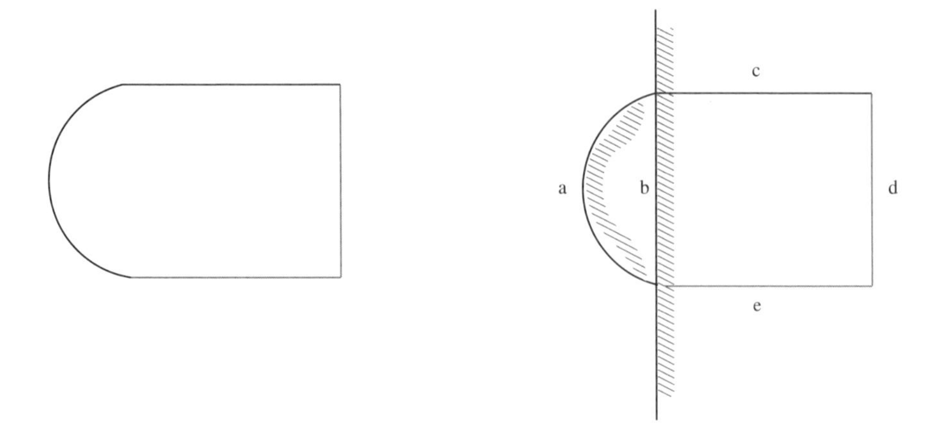

Fig. 2. Additional halfspaces may be required to build a constructive representation from a boundary representation. The left object is a boundary representation of a simple two-dimensional solid. The right object contains an additional halfspace, which is needed to build a constructive representation of the solid.

Implementation of the Constructive Tree Recovery Algorithm. The genetic algorithm is implemented in C++ language using the GALib library [43]. A simple genetic algorithm, as introduced and discussed by Goldberg [14], is used. This genetic algorithm uses non-overlapping populations. For each generation, a new population is created by selecting from the previous population, and mating to produce the offsprings for the new population. These two steps are repeated until the termination criterion is met.

The fitness function is defined by:

$$\phi(g) = \sum_{\mathbf{p_i} \in S} F_g(\mathbf{p_i})^2 \tag{6}$$

where g is an individual in the current population; S is the point-set, with points on or near the surface of the solid; F_g is the FRep encoded by the individual g and corresponds to a left unbalanced tree (Eq. 4).

We use a one dimensional binary string ('GA1DBinaryString') as the data structure to encode an individual's genotype in the genetic algorithm. As explained above, an individual corresponds to an array of m pairs (op_k, L_k), $1 \leq k \leq m$, where op_k is an index to one of the operations, and L_k is the priority level – or position – of the primitive k in the expression. Each pair is encoded as a sequence of bits (a gene).

We use an example to explain in more details how the decoding and encoding of the individuals are performed; this example consists of 16 primitives and the possible operations are union, intersection and blending intersection. Each pair (op_k, L_k) can then be encoded by a gene of size $2 + 4 = 6$ bits. An operation op_k can be encoded in 2 bits, since three operations are considered. The last 4 bits are used to encode the index of the kth primitive (priority level) L_k in the expression construction. An expression (phenotype) is thus encoded by a string of size $6 * 16$=96 bits, i.e. an array of 16 genes of size 6 bits.

For a given binary string, we need also to be able to reconstruct the expression to evaluate it, i.e. finding the phenotype of an individual from its genotype. First, each gene, corresponding to a string of bits, is decoded to a pair of integers (op_k, L_k). Then we use the following procedure:

- select the next pair (op_k, L_k) with the lower L_k value
- in case of several pairs with same L_k value, take the one with the minimal index k in the initial list

This procedure sorts the pairs within the array with respect of the priority level L_k of the primitive in the expression. It orders the primitives within the expression. This procedure is illustrated by the following example. Let a string be encoded by 16 pairs (op_k, L_k), where $op_k \in \{0, 1, 2\}$, and $L_k \in \{0, .., 15\}$. $k = 1 : (0, 15)$, $k = 2 : (0, 1)$, $k = 3 : (1, 0)$, $k = 4 : (1, 2)$, $k = 5 : (1, 1)$, $k = 6 : (1, 0)$, $\ldots$, $k = 16 : (1, 7)$.

Following the procedure described above, we select the pairs in the following order: 3, 6, 2, 5, 4 $\ldots$ 1 and then the reconstructed expression becomes: $F = (op_3) f_3 op_6 f_6 op_2 f_2 op_5 f_5 op_4 f_4 ... op_1 f_1$. The first operation is in parenthesis because it is not taken into account in the evaluation and is here only for the symmetry of the expression. Finally each operation can be replaced. Let us suppose that $op_i == 0$ corresponds to union (|), $op_i == 1$ for intersection (&), and $op_i == 2$ for blending intersection (&&); we get the following expression: $F = f_3 \& f_6 | f_2 \& f_5 \ldots | f_1$. This expression is evaluated from left to right using R-functions or SARDF operations. The evaluation of this expression at each point of the point-set serves to define the fitness function for the given individual (Eq. 6).

We need to define now the genetic operations on the individuals. There are at least three operations that need to be defined for a genetic algorithm:

- the selection (selection of individuals in the previous population),
- the crossover (mating selected individuals to create new offspring),
- the mutation (mutate the selected individual)

Selection: The selection is implemented by the roulette wheel algorithm. The idea of the roulette wheel is to choose randomly between all the individuals of the current population, with a higher probability to select an individual with a good fitness value.

Mutation: The first algorithm used for the mutation is a random flip of bits. If an individual is selected for a mutation, then 1 bit of the individual is chosen randomly and flipped (meaning that if the value of the chosen bit were 1 then it would become 0 and 0 would become 1). This algorithm is the default one. It is however not aware of the problem representation.

We have also implemented a different algorithm taking into account the structure of the problem. Let us suppose, that there are m pairs, and that a pair is coded using p bits as a gene. Note that with the example above $m = 16$ and $p = 6$. If an individual is selected for a mutation, then the mutation is done as follows:

1. choose randomly a position in the string
2. find the beginning of the p-bit sequence it belongs to
3. flip all the p bits of the sequence

Crossover: The default crossover operation is a one point crossover. Two individuals are selected from the previous population to be mated, then one point is chosen randomly in both individuals and the different parts are swapped. By default, it is not aware of the structure of the problem, therefore the following crossover can be used:

1. choose randomly a position in the string of parent 1
2. find the beginning of the p-bit sequence it belongs to
3. find the symmetric position in the parent 2
4. swap the subparts of the parents to generate the two children

The experimental results on the constructive tree recovery are described in the next section.

4 Experimental Results and Applications

4.1 Experiments with Fitting Template FRep Models

Fitting template models is illustrated in the following through several examples. At first template models for a mechanical part and a sake pot are fitted to point-sets. Then, the application of template models fitting for finite element remeshing is described.

Simple CAD Part. The first test part contains 10714 points scattered on the surface. The FRep defining function F shown below is used as a parameterized model for the recovery process:

$$f(\mathbf{x}, \mathbf{a}) := (box(\mathbf{x}, \mathbf{a}) \backslash cylinderZ(\mathbf{x}, \mathbf{a})) \\ \backslash cylinderZ(\mathbf{x}, \mathbf{a}); \tag{7}$$

This FRep model consists of three simple primitives: one box and two infinite cylinders oriented along the Z axis; each primitive is defined by its parameterized model. For example, in the case of the cylinder, the defining function is: $cylinderZ(x, a) := a[1] - sqrt((x[1] - a[2])^2 - (x[2] - a[3])^2)$, where $a[1]$, $a[2]$, and $a[3]$ are parameters meaning the radius and a point on the $x - y$ plane, through which the axis of the cylinder passes. All the primitives are combined together using the subtraction operator \, which is itself defined analytically by an R-function as discussed in [31].

Twelve parameters are released in the model, corresponding to the lower-left corner of the box, the three dimensions of the box, and the center and radius for each of the cylinders. The fitting algorithms need an initial estimation for the parameters, in the tests we use two sets of initial values configurations: one is close to the best fit (set1), in contrary to the second one (set2).

Table 1. Time (in seconds) taken by each method to converge to the best fit and the least-square error (sum of the deviations squared) of the best fit for each set of initial values. QN stands for Quasi-Newton, SA for simulated annealing, and SAQN is an hybrid algorithm consisting in a combination of the simulated annealing and Quasi-Newton algorithms.

	Time in sec		Least square error	
	set1	set2	set1	set2
QN	1.852	9.643	5.47	595.04
SA	1635.09	1773.109	5.49	5.49
SAQN	72.042	144.177	5.47	5.47

The results of the tests are given in terms of the following: least square error of the reconstructed model for the three methods: Quasi-Newton (QN), simulated annealing (SA), and hybrid method simulated annealing – Quasi-Newton (SAQN) (see Table 1), time given in seconds taken to converge to the best fit for each of these methods (see Table 1), and the visual shape of the best fit (Fig. 3).

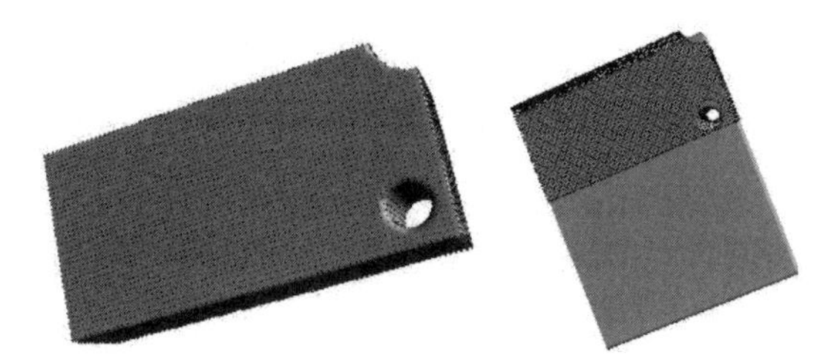

Fig. 3. Shapes for the best fitted FRep in two cases: (right) the best fitted object does not correspond to the real object when starting with set2 and using the QN method; (left) the best fitted object corresponds to the real object when starting with the set2 and using the hybrid method.

Table 1 shows that the local method stops at a local minimum for the set2 of initial parameters, resulting in a wrong shape (Fig. 3, right shape), whereas with simulated annealing, it always converges to the global minimum. Unfortunately, the counterpart is the slow rate of convergence for the sampling method (Table 1).

When using a combination of SA and QN, we can recover correct parameters and shape (Fig. 3, left shape, and Table 1, last line). The steps of the shape evolution during the hybrid method search are shown in Fig. 4. Experimentally, a combination of SA and QN avoids local optima with a better convergence rate than SA alone.

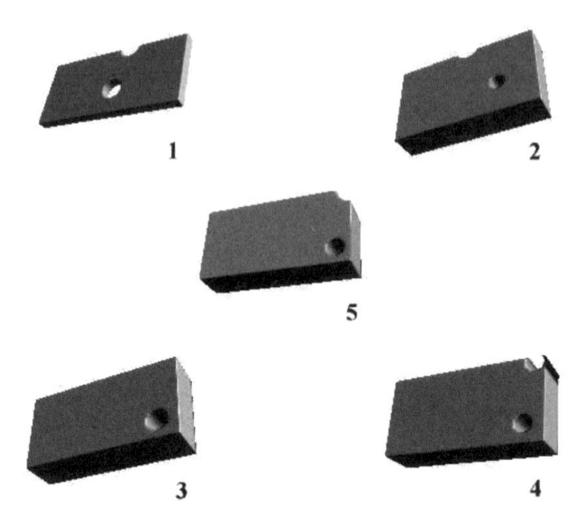

Fig. 4. Evolution of the shape during the fitting process using the hybrid method

Lacquer Ware Sake Pot. The next example is the fitting of a model of a hand-crafted lacquer ware pot, which is used for pouring sake (Japanese rice wine). The discrete data set of the sake pot includes 27048 3D points, scattered on the surface of the object. The parameterized model of the sake pot sketched and discussed in the work on cultural heritage [42] is reused in our experiment. The parameterized model was created using hand measurements of a typical sake pot. The major parameters are the coordinates of the origin (position), the basic radius of the pot body, and the height of the pot handle. The model is required to keep basic ratios of the measured sample object and to proportionally change the dependent parameters like those of the blend area between the spout and the body, and the shape of the lid holder (note non-linear changes of these shapes in Fig. 6).

The first fitting test is made using a Quasi-Newton algorithm. At the end of the algorithm, the value of the fitness function is big enough to indicate that the method stopped at a local minimum. A comparison of the discrete model and the fitted parameterized FRep, illustrated by Fig. 5, indicates that the best fitted parameters correspond to a local minimum of the fitness function.

A hybrid algorithm, combining Simulated-Annealing and Quasi-Newton is used to go to the vicinity of the global optimum. The initial vector of parameters at the start of the process is the same as the one used before. In our

Fig. 5. Local minimum effect: result of the fitting with a local method starting with initial parameters far from the global optimum. The point-set for the sake pot is also displayed for comparison.

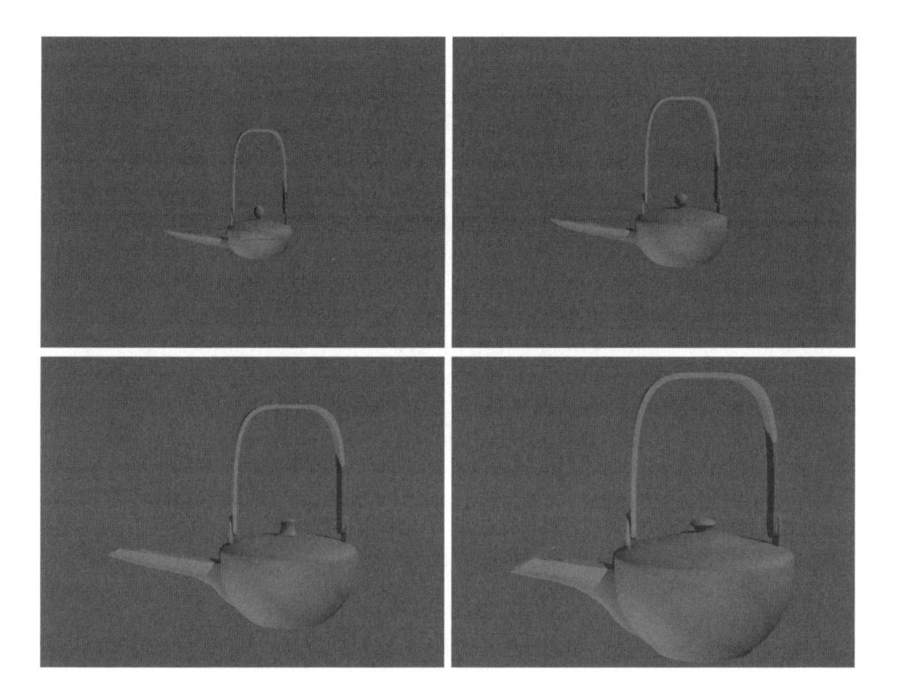

Fig. 6. Evolution of the shape of the sake pot during the fitting process

experiments, the Simulated-Annealing algorithm is stopped after the value of the fitness function goes below a threshold given by the user in order to determine the final switch to a Quasi-Newton method. The current shape is confirmed by visual feedback. Then, the obtained parameters are reused as initial values for a final call to the Quasi-Newton method. The steps of the evolution of the shape during the hybrid fitting of the FRep model can be seen in Fig. 6.

Application in Finite Element Meshes (FEM)

Approaches to FEM generation. Surface remeshing is very important for applications associated with numerical simulation procedures, in particular with finite element analysis (FEA). These applications impose strict constraints on the quality of the surface approximation and on the shapes and sizes of mesh elements. Moreover, finite element meshes have to be adapted both to physical and geometric features of computational tasks. Changes in the boundary or initial conditions of the simulated process may cause remeshing even if the computational domain remains the same.

In many cases the initial description of computational domains in FEA is represented by their boundary surface triangulations. These triangulations can be exported from various modeling systems, produced by 3D scanning, or be a result of previous FE computations. Usually these initial triangulations consist of badly shaped triangles and are not adapted to physical conditions and an appropriate remeshing is required. Mesh refinement and optimization procedures need accurate information about the geometry of the computational domain. Therefore, the creation of an adequate description of a solid based on the initial triangulation of its boundary surface is an important problem for the FE mesh generation and optimization. Different approaches were considered to solve this problem. In [12], finite element adaptation is based on the local approximation of the underlying surface geometry by a quadric surface. The authors of [20] convert a CAD model into a volume representation by sampling its distance field on a uniform grid and then applying the extended marching cubes algorithm to this volume.

Taking into account that many mechanical parts can be represented as constructive solids, we propose to apply FRep recovery to support FE mesh generation for objects whose initial geometry is represented by boundary surface triangulations. The initial mesh is used for the selection or creation of a parameterized FRep model. Then, the parameters of the FRep model are fitted to the vertices of the mesh. The final model can be used for the surface and volume finite element adaptation by the methods described in [18].

Fitting to a CAD mesh. As an example of application of the FRep shape recovery for the FEM generation, a parameterized FRep model corresponding to the CAD mesh Fig. 7 (top, left) is created and fitted using the previously proposed techniques.

The FRep model including 14 parameters is sketched corresponding to the shape shown in Fig. 7, top, left; the initial values for the parameters are chosen randomly. The convergence is obtained using an hybrid simulated annealing/ Quasi-Newton scheme. The FRep shape corresponding to the best set of parameters is shown in Fig. 7, top, right.

Starting with the acquired FRep model, it is then possible to apply the mesh adaptation methods from [18]. The results of such methods are shown in Fig. 7, bottom. The left picture shows an optimized surface mesh, which was then used for the 3D tetrahedral mesh generation (right) using the extended advancing front method [18].

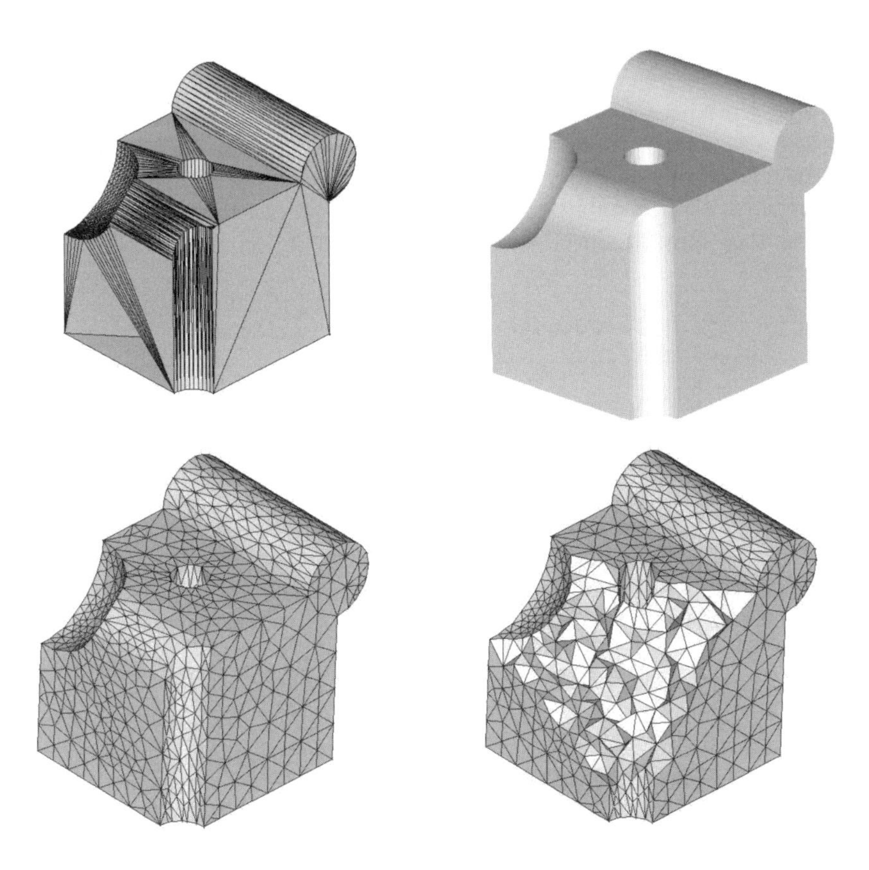

Fig. 7. A surface mesh, generated by a CAD system (top, left), the recovered shape (top, right), the associated optimized mesh (bottom, left), and the 3D tetrahedral mesh generated from it (bottom, right). The original BRep model and the generated optimized and 3D tetrahedral meshes are courtesy of Elena Kartasheva.

4.2 Constructive Tree Recovery Using a Genetic Algorithm

The genetic algorithm is applied to recover constructive trees for some mechanical parts. The input of the algorithm is a set of points scattered on the surface of the solid and the primitives associated with the segmentation of the point-set. Point-set segmentation and fitting of primitives to the different subsets of the segmented point-set are obtained by a brute force algorithm. A genetic algorithm is run on a list of primitives (cube, box, sphere, torus, ellipsoid), the best fitted primitive is selected and the corresponding points from the point-set are removed, then we loop to the first step until the point-set contains only a few points. Primitives are defined by Euclidean distance functions and set-theoretic operations are implemented using the SARDF operations (described elsewhere in this volume).

Example 1. The first example involves a point-set of 9530 points and 10 primitives. These primitives include planes, spheres and cylinders and were recovered using a brute force genetic algorithm as described above. In fact, a box was fitted instead of planes, but we decided to use the planes to increase the number of primitives. It should be noticed that the use of a different segmentation algorithm may have produced such planes instead of a box.

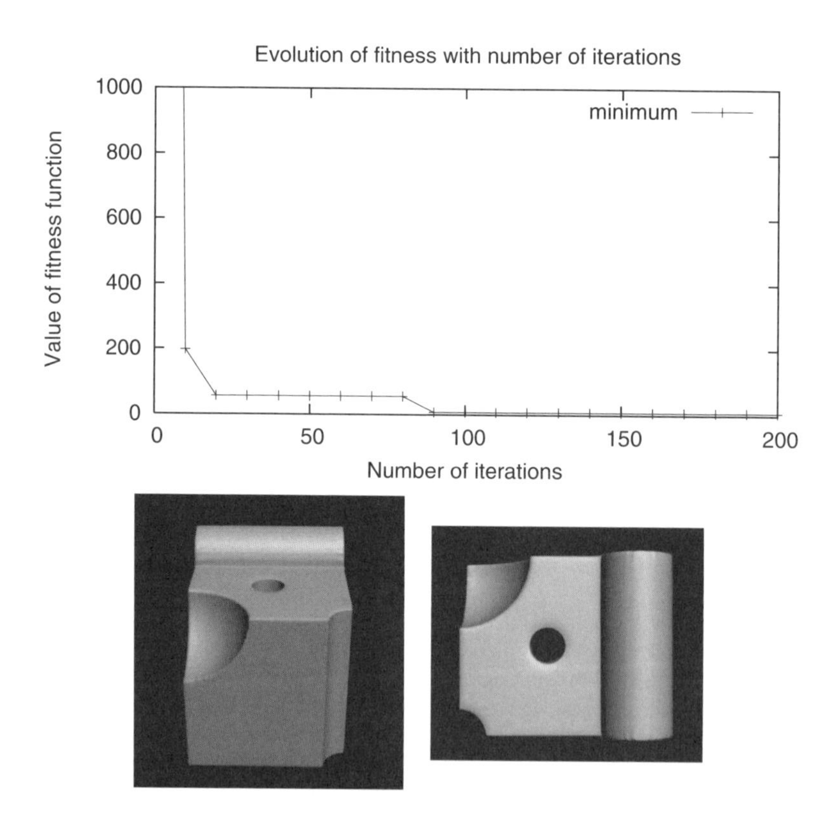

Fig. 8. Top: Evolution of the fitness of the best individual for a population of size 1000; Bottom: Result for the best individual for the first mechanical part

We used the genetic algorithm and the genetic operations described above to recover the constructive tree of the object. We used a probability of 0.1 for the mutation and of 0.6 for the crossover. Figure 8, top, illustrates the convergence after 200 generations of an initial population of size 1000.

The solid resulting from the best individual is given in Fig. 8, bottom.

Example 2. The second example involves a point-set of 49388 points segmented into 10 primitives. The same parameters as above are used for the size of population, probability of crossover and probability of mutation. Figure 9, top, illustrates the convergence after 200 generations of a population of size 1000. The solid obtained from the best individual is illustrated by Fig. 9, bottom.

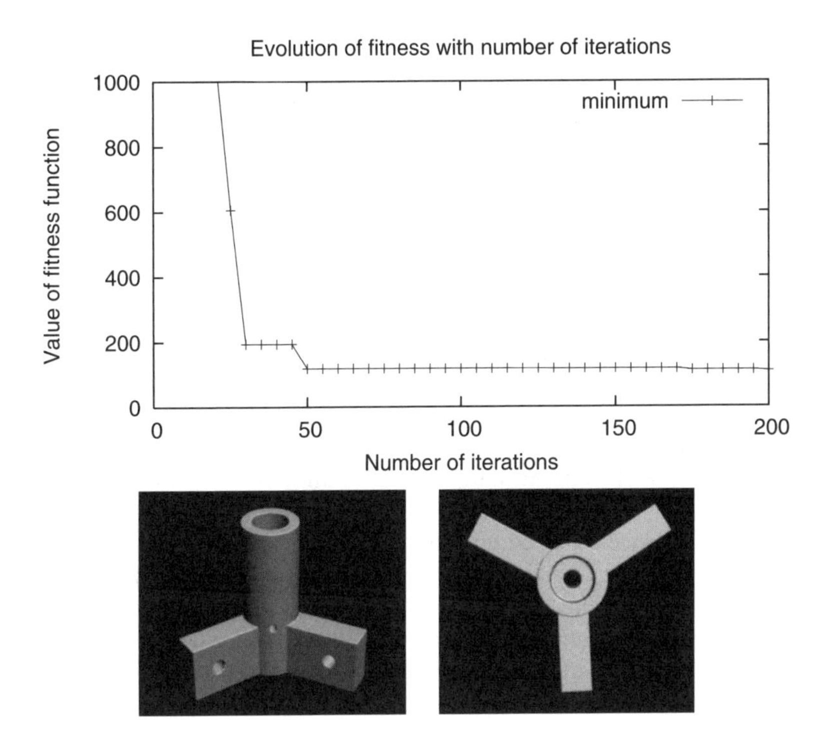

Fig. 9. Top: Evolution of the fitness of the best individual for a population of size 1000; Bottom: Result for the best individual for the second mechanical part

Comparing figures 8 and 9, we find a higher fitness value for the second example. It can be explained by a bigger number of points in the point-set used: the second point-set is approximately 5 times bigger. It can also be explained by a different approximation in fitting the primitives. The number of points in the point-set has some importance. If the number is too big, it slows down the evaluation of the fitness function and the speed of the algorithm convergence.

5 Conclusion

The automation of function-based volumetric modeling is approached in this work with involvement of some additional information on the object shape, for example, an existing model for a family of similar objects or a scanned surface data of a physical instance of the object to be modeled. Some objects do indeed belong to more general families of similar objects: for example several vases may possess the same structure and overall shape. Another example is human body, which is unlikely to radically change its shape in the near future. Once a model has been made for an object, it can be abstracted by a template model, where each of the most important parameters have no specific values and will be fitted to adapt the model to other instances of the same object type.

We have introduced new algorithms for template FRep models fitting, which rely on the combination of heuristic methods – such as simulated annealing or genetic algorithms – with gradient based methods – such as Quasi-Newton or Levenberg-Marquardt. Template models and the proposed fitting algorithms have been illustrated by fitting different examples of template models for mechanical parts and a more complex sake pot to discrete point-sets. A different mechanical part was used to illustrate an application of fitting template models in remeshing for finite element applications.

We have also introduced a new algorithm for the recovery of constructive FRep models from segmented point-clouds and the associated set of primitives. The search is performed by a genetic algorithm which looks for the best expression involving the primitives. The proposed algorithm is not restricted to the set-theoretic operations, and can be extended, by using blending operations. From the recovered constructive FRep models, it is possible to make a parameterized FRep template model that can be reused in fitting, mesh optimization or other operations.

The modeling process can be greatly automated by reuse of template models and their optimization. The idea of the template model can be extended in some directions. We studied only the optimization of the shape minimizing the algebraic distance to a point-set, but different optimizations and objective functions can be experimented leading to more general shape optimization. The idea of parameterized templates leads to a more general idea of the automation of the properties modeling for heterogeneous objects. Similarly to the parameterization of object shapes, the geometry of space partitions associated with attributes can be parameterized and fitted to satisfy some modeling constraints.

References

1. Amenta, N., Choi, S., Kolluri, R.K.: The power crust. In: SMA 2001: Proceedings of the sixth ACM symposium on Solid modeling and applications, pp. 249–266. ACM Press, New York (2001)
2. Batchelor, B.G.: Hierarchical shape description based upon convex hulls of concavities. Journal of Cybernetics 10, 205–210 (1980)

3. Benko, P., Kos, G., Varady, T., Andor, L., Martin, R.: Constrained fitting in reverse engineering. Computer Aided Geometric Design 19(3), 173–205 (2002)
4. Benko, P., Varady, T.: Direct segmentation of smooth, multiple point regions. In: Proceedings of GMP, pp. 169–178 (2002)
5. Blinn, J.: A generalization of algebraic surface drawing. ACM Trans. Graph 1(3), 235–256 (1982)
6. Buchele, S.F., Crawford, R.H.: Three-dimensional halfspace constructive solid geometry tree construction from implicit boundary representations. Computer-Aided Design 36(11), 1063–1073 (2004)
7. Dennis, J.E., Gay, D.M., Welsch, R.E.: An adaptative nonlinear least-squares algorithm. ACM Transaction on mathematical software 7, 348–368 (1981)
8. Faber, P., Fisher, R.B.: Pros and cons of Euclidean fitting. In: Radig, B., Florczyk, S. (eds.) DAGM 2001. LNCS, vol. 2191, pp. 414–420. Springer, Heidelberg (2001)
9. Fayolle, P.-A., Pasko, A., Kartasheva, E., Mirenkov, N.: Shape recovery using functionally represented constructive models. In: Proceedings of International Conference on Shape Modeling and Applications 2004 (SMI 2004), pp. 375–378 (2004)
10. Fayolle, P.-A., Pasko, A., Mirenkov, N., Rosenberger, C., Toinard, C.: Constructive tree recovery using genetic algorithms. In: Procedings of the International Conference on Visualization, Imaging and Image Processing (2006)
11. Fayolle, P.-A., Rosenberger, C., Toinard, C.: 3d shape reconstruction of template models using genetic algorithms. In: Proceedings of 17th International Conference on Pattern Recognition (ICPR 2004), pp. 269–272 (2004)
12. Frey, P.J., Borouchaki, H.: Geometric surface mesh optimization. Computing and visualization in science 1(3), 113–121 (1998)
13. Freytag, M., Shapiro, V., Tsukanov, I.: Field modeling with sampled distances. Computer Aided Design 38(2), 87–100 (2006)
14. Goldberg, D.: Genetic algorithms in search, optimization and machine learning. Addison-Wesley, Reading (1989)
15. Hart, J.C.: Distance to an ellipsoid. In: Heckbert, P. (ed.) Graphics Gems IV, pp. 113–119. Academic Press, Boston (1994)
16. Holland, J.H.: Adaptation in natural and artificial systems. The University of Michigan Press, Ann Arbor (1975)
17. Hoppe, H.: Surface reconstruction from unorganized points, Ph.D. thesis, University of Washington (June 1994)
18. Kartasheva, E., Adzhiev, V., Pasko, A., Fryazinov, O., Gasilov, V.: Surface and volume discretization of functionally based heterogeneous objects. Journal of Computing and Information Science in Engineering, Transactions of the ASME 3(4), 285–294 (2003)
19. Kirkpatrick, S., Gelatt, C., Vecchi, M.: Optimization by simulated annealing. Science 220, 671–680 (1983)
20. Kobbelt, L., Botsch, M., Schwanecke, U., Seidel, H.-P.: Feature sensitive surface extraction from volume data. In: Procedings of SIGGRAPH 2001, pp. 57–66. ACM, New York (2001)
21. Koza, J.: Genetic programming. MIT Press, Cambridge (1992)
22. Marshall, D., Lukacs, G., Martin, R.: Robust segmentation of primitives from range data in the presence of geometry degeneracy. IEEE Transactions on pattern analysis and machine intelligence 23(3) (2001)
23. Metropolis, N., Rosenbluth, A., Rosenbluth, M., Teller, A., Teller, E.: Equations of state calculations by fast computing machine. J. Chem. Phys. 21, 1087–1092 (1953)

24. Michalewicz, Z.: Genetic algorithms + data structures = evolution programs. Springer, Heidelberg (1996)
25. More, J.: The levenberg-marquardt algorithm implementation and theory. Lecture notes in mathematics No630 Numerical analysis 630, 105–116 (1978)
26. Morse, B., Yoo, T., Chen, D., Rheingans, P., Subramanian, K.: Interpolating implicit surfaces from scattered surface data using compactly supported radial basis functions. In: Proceedings of Shape modeling international, pp. 89–98 (2001)
27. Muraki, S.: Volumetric shape description of range data using "blobby model". In: Proceedings of SIGGRAPH, pp. 227–235 (1991)
28. Ohtake, Y., Belyaev, A., Alexa, M., Turk, G., Seidel, H.-P.: Multi-level partition of unity implicits. ACM Trans. Graph 22(3), 463–470 (2003)
29. Ohtake, Y., Belyaev, A., Seidel, H.-P.: Ridge-valley lines on meshes via implicit surface fitting. ACM Trans. Graph 23(3), 609–612 (2004)
30. Ohtake, Y., Belyaev, A., Seidel, H.-P.: An integrating approach to meshing scattered point data. In: SPM 2005: Proceedings of the 2005 ACM symposium on Solid and physical modeling, pp. 61–69. ACM Press, New York (2005)
31. Pasko, A., Adzhiev, V., Sourin, A., Savchenko, V.: Function representation in geometric modeling: concept, implementation and applications. The Visual Computer 11(8), 429–446 (1995)
32. Pasko, A., Savchenko, V.: Blending operations for the functionally based constructive geometry. In: set-theoretic Solid Modeling: Techniques and Applications, CSG 1994 Conference Proceedings, pp. 151–161. Information Geometers (1994)
33. Press, W., Flannery, B., Teukolsky, S., Vatterling, W.: Numerical recipes in c - the art of scientific computing. Cambridge University Press, Cambridge (1992)
34. Rvachev, V.L., Kurpa, L.V., Sklepus, N.G., Uchishvili, L.A.: Method of r-functions in problems on bending and vibrations of plates of complex shape (in Russian) (1973)
35. Savchenko, V., Pasko, A., Okunev, O., Kunii, T.: Function representation of solids reconstructed from scattered surface points and contours. Comput. Graph. Forum 14(4), 181–188 (1995)
36. Shapiro, V.: A convex deficiency tree algorithm for curved polygons. International Journal of Computational Geometry and Applications 11(2), 215–238 (2001)
37. Shapiro, V., Vossler, D.L.: Separation for boundary to csg conversion. ACM Trans. Graph 12(1), 35–55 (1993)
38. Shepard, D.: A two-dimensional interpolation function for irregularly spaced data. In: Proceeding 23 National Conference, vol. 23, pp. 517–524. ACM, New York (1968)
39. Turk, G., OBrien, J.: Shape transformation using variational implicit functions. In: Proceedings of SIGGRAPH, pp. 335–342 (1999)
40. Vanco, M.: A direct approach for the segmentation of unorganized points and recognition of simple algebraic surfaces, Ph.D. thesis, Technische Universität Chemnitz (2003)
41. Varady, T., Martin, R.R., Cox, J.: Reverse engineering of geometric models – an introduction. Computer Aided Design 29(4), 255–268 (1997)
42. Vilbrandt, C., Pasko, G., Pasko, A., Fayolle, P.-A., Vilbrandt, T., Goodwin, J., Goodwin, J., Kunii, T.: Cultural heritage preservation using constructive shape modeling. Comp. Graph. Forum 23(1), 25–41 (2004)
43. Wall, M.: A c++ library of genetic algorithm components (1996), http://lancet.mit.edu/ga
44. Woo, T.C.: Feature extraction by volume decomposition. In: Proc. Conference on CAD/CAM Technology in Mechanical Engineering, Cambridge, MA (1982)

Heterogeneous Modeling of Biological Organs and Organ Growth

Roman Ďurikovič[1,2], Silvester Czanner[3], Július Parulek[2,5], and Miloš Šrámek[2,4]

[1] University of Saint Cyril and Metod, Trnava, Slovakia,
[2] Faculty of Mathematics, Physics and Informatics, Comenius University, Slovakia
roman.durikovic@fmph.uniba.sk
[3] Warwick Manufacturing Group, University of Warwick, UK
S.Czanner@warwick.ac.uk
[4] Austrian Academy of Sciences, Austria
[5] Institute of Molecular Physiology and Genetics, Slovak Academy of Sciences, Slovakia

Abstract. The growth of the organs of human embryo is changing significantly over a short period of time in the mother body. The shape of the human organs is organic and has many folds that are difficult to model or animate with conventional techniques. Convolution surface and function representation are a good choice in modelling such organs as human embryo stomach and brain. Two approaches are proposed for animating the organ growth: First, uses a simple line segment skeleton demonstrated on a stomach model and the other method uses a tubular skeleton calculated automatically from a 2D object outline. The growth speed varies with the position within the organ and thus the model is divided into multiple geometric primitives that are later glued by a blending operation. Animation of both the embryo stomach and brain organs is shown.

1 Introduction and Previous Works

The purpose of this manuscript is to model the outer shape and the shape metamorphosis during the growth of some human embryo organs, particularly brain and digestive system. Popular methods like 3D shape reconstruction from Computer Tomography (CT) sections or ultrasound data can not be used for this type of modelling because the resolution of the devices used in those methods are much higher comparing to the size of human embryo. Four weeks old embryo is approximately 3 *mm* tall while the CT resolution is 1 *mm* giving us only three sections for a reconstruction process. Usually, the microscopic cross-sections are used to reconstruct the polygonal representation of an embryo, which is exact but complicated process. In case of such destructive approach often a mouse embryo is used instead of the human embryo [1]. To control the shape metamorphosis between two mesh objects become a problem when they have different topology and geometry. To create the realistically looking human organ models and to

A. Pasko, V. Adzhiev, and P. Comninos (Eds.): HOMA, LNCS 4889, pp. 239–258, 2008.

generate the animations demonstrating the growth process requires a proposed methodology.

Growing human organs can be described as dynamical systems with a dynamical structure [2]. In such systems not only the values of variables characterizing system components, but also the number of components and the connections between them, may change over time. There is a need to construct a mathematical description of a system. The model can then be used for simulation or optimization. All models are predictive in the meaning that simulation output predict what could occur in the real world where the system is operating. Numerous interdisciplinary research initiatives are generating excellent research results with regards to modelling, simulation and visualization of human anatomy and physiology. These research initiatives focus on different (biological) levels [3]; molecular and cellular levels, tissue [4] and organ levels [5], [6], [7], and system and human (organism) levels.

The simulation of human organs growing can be seen as an imitation of the reality for studying the effect of changing parameters in a model as a means of preparing a decision or predicting experiment results. Since the human body is mainly made up of a variety of organs, the medical consequence of organ modelling is very important, ranging from heart surgery to minimally-invasive surgery.

In the area of modelling and simulation of human organs many research works have been carried out. One class of reconstruction methodologies uses implicit functions. They allow extracting an iso-surface either by a procedural method, or skeletal implicit surfaces (surfaces generated by a field function and a skeleton). Amrani introduced a method using the skeleton-based implicit surface for implicit reconstruction [7].

Another construction method is presented by Leymarie. His approach is based on propagation along the scaffold from initial sources of flow as a means to efficiently construct it. The detection of these sources can be shown to be reduced to considering pairs of input points, which then constitutes the computational bottleneck of this method [8].

A semi-automatic reconstruction method that can be used on noisy scattered points of a medical organ is presented by Tsingos [5]. The method is based on implicit iso-surfaces generated by skeletons that provide a smooth and compact representation of the surface. The user can guide the reconstruction by initializing some skeletons and their reconstruction windows, thus taking benefits of his initial knowledge of the data.

The method developed by Attali *et al.* [9] computes the Voronoi graph of the point set to build the skeleton of the object and reconstructs its surface. The surface thus reconstructed has only fixed topological type.

Even though the convolution surfaces provide nice blending between several parts of organs, the control of the blend shape is very limited. The functional representation [10] is a tool that generalize the set theoretic operations and generates full range of shapes from simple object union to smooth blend. The animation of such surfaces follow the changes smoothly, even if the topology changes. Because of this advantage the functional representation become a popular tool where the shapes to be modelled are from the natural world. We explain here our modelling experience that can be useful for others.

2 Shape Representation for Growth Animation

The polygonal models can not capture the development of such complex process as the growth of the digestive system. So far, we have created the skeletons of different physiological parts, we need to blend them together to get the smooth shapes. Even though, the convolution surfaces provide nice blending between several parts of organs, the control of the blend shape is very limited. The functional representation is a tool that generalizes the set theoretical operations and generates full range of shapes from simple object union to smooth blending. The animation of such surfaces follow the changes smoothly, even if the topology changes. Because of this advantage the *functional representation* is an excellent tool when the shapes to be modelled are from the natural world. We discuss herein the shape modeling based on skeleton calculated from dynamic simulation and L-system growth.

An *implicit surface* is defined by an isosurface of some potential field $F : \mathcal{R}^3 \rightarrow \mathcal{R}$ at threshold level T: $S = \{p \in \mathcal{R}^3 : F(p) - T = 0\}$. The function $F(p)$ is also called an implicit function. A convolution surface is implicitly defined by a potential function F obtained via convolution operator between a kernel and all the points of a skeleton. The convolution surface thus obtained is a smoothed skeleton. The skeleton is a collection of geometric primitives such as point, line segment, arc and plane that outline the structure of an object being modelled. Convolution surface build from complex skeletons can be evaluated individually by adding the local potentials for each primitive, because convolution operator is linear.[3] Let us have N skeleton primitives the above statements can be written as the following modelling equation in an implicit form:

$$\sum_{i=1}^{N} F_i(x_1, x_2, x_3) - T = 0, \tag{1}$$

where F_i is the source potential of i-th skeleton primitive and T is the iso-potential threshold value.

3 Function Representation

Let us consider closed subsets of n-dimensional Euclidian space E^n with the definition:

$$f(x_1, x_2, ..., x_n) \geq 0,$$

where f is a real continuous function defined on E^n. The above inequality is called a function representation (F-rep) of a geometric object and function f is called the defining function. In three-dimensional case the boundary of such a geometric object is called implicit surface. The major requirement on the function is to have at least C^0 continuity. The set of points $X_i(x_1, x_2, ..., x_n) \in E^n$, $i = 0, ..., N$ associated with Eq. 3 can be classified as follows:

$$f(X_i) > 0 \text{ if } X_i \text{ is inside the object,}$$
$$f(X_i) = 0 \text{ if } X_i \text{ is on the boundary of the object,}$$
$$f(X_i) < 0 \text{ if } X_i \text{ is outside the object.}$$

Let us consider from now on the defining function given by the convolution operator between a kernel and all the points of a skeleton, i.e. function F as defined in the last equation of the previous section.

3.1 Set-Theoretic Operations

The binary operations on geometric objects represented by functions can be also defined in the form of function representation by

$$\mathcal{F}(f_1(X), f_2(X)) \geq 0, \tag{2}$$

where $\mathcal{F}$ is a continuous real function of two variables .[10] Such operations are closed on the set of function representations. After set theoretic operation between two subjects defined by functions f_1 and f_2 the resulting object has the defining function as follows:

– For object union

$$f_3 = f_1 | f_2 \equiv \frac{1}{1+a}(f_1 + f_2 + \sqrt{f_1^2 + f_2^2 - 2a f_1 f_2}),$$

– for object intersection

$$f_3 = f_1 \& f_2 \equiv \frac{1}{1+a}(f_1 + f_2 - \sqrt{f_1^2 + f_2^2 - 2a f_1 f_2}),$$

– for object subtraction

$$f_3 = f_1 \backslash f_2 \equiv f_1 \& (-f_2),$$

where $|, \&, \backslash$ are notations of so-called R-functions and parameter $a = a(f_1, f_2)$ is the arbitrary continuous function satisfying the conditions

$$-1 < a(f_1, f_2) \leq 1$$
$$a(f_1, f_2) = a(f_2, f_1) = a(-f_1, f_2) = a(f_1, -f_2).$$

Please, note that even thought the resulting defining function for set above theoretic operations is continuous, the resulting object is not continuous in general.

3.2 Blending Union Operation

Intuitively the blending union operation between two initial objects from the set of function representations is a gluing operation. It allows us to control the gluing type in the wide range of shapes from pure set union to convolution like summation of terms. Mathematically the blending union operation is defined by

$$\mathcal{F}(f_1, f_2) = f_1 + f_2 + \sqrt{f_1^2 + f_2^2} + \frac{a_0}{1 + (\frac{f_1}{a_1})^2 + (\frac{f_2}{a_2})^2},$$

where f_1 and f_2 are functions representing objects that are blended. The absolute value a_0 defines the total displacement of the bending surface from two initial surfaces. The values $a_0 > 0$ and $a_1 > 0$ are proportional to the distance between blending surface and the original surface defined by f_1 and f_2, respectively. The effect of this operation compared to other possible object connections is demonstrated on two object primitives whose skeleton consists of two line segments one vertical and the other one diagonal, see Figure 1 top-left. Simple plus operation between convolution functions deforms the thickness of vertical convolution cylinders as shown in top-right image. Considering four line segments as a single skeleton of geometric primitive results in the shape shown in top-center image. The sequence of shapes shown on bottom of Figure 1 are the blending union operations between two parallel geometric primitives. The geometric primitives and their skeletons do not change but the blending parameters used to blend them are different for each image. In orderer from left side the used parameters are $a_i = 0.01$, $a_i = 0.07$, $a_i = 0.3$, $a_i = 0.5$, and $a_i = 0.7$, respectively. We can conclude that in the case when the shape and size of geometric primitives must be preserved the blending union operation with different parameters a_0, a_1, and a_2 is a good choice. On the other hand when the blending shape is main concern the convolution plus operation should be used. When both the shape of geometric primitives and that of blending are important the small values of blending union parameters is a choice. The F-rep blending union operation has similar advantages as simple convolution union with respect to minimizing unwanted bulges.

4 Shape of Organic Models

In previous sections have been discussed the theory of F-rep and convolution surfaces. As next, we will show a method to model the organic shapes by F-rep, where each of the geometric primitives is defined by

$$\sum_{i=1}^{N} F_i(x_1, x_2, x_3) - T = 0,$$

where F_i are the source potentials of skeleton primitives i.e. points, lines or triangles and T is a threshold value. Therefore, what we need to design next are the skeletons for different organs.

Fig. 1. Blending union operation. top: standard and bottom: Blending union operation.

4.1 Human Brain Model

First step in the model creation process is to obtain the size measurements of brain and stomach stages from atlas of embryology. Embryological atlas contains hand-drawing pictures and photographs of human embryo organs ordered by age. For the purpose of this study the models from 28 - 56 days old brain were used. The brain pictures has been scanned, stored in binary form and measured by ruler. The model, at this stage of precessing, was divided into physiological parts to suite the animation purposes. The outlines of physiological parts were drawn over the pictures and photographs, see Figure 2.

4.2 Brain: Central Skeleton

The result of the measurements is a 2D planar contour, call *the central skeleton*, nearly outlining the outer contour of the shape. Interior of central skeleton is triangulated such that it crates a triangular strip. One can observe different growth speed for different pars of embryo brain. It is therefore natural to divide the central skeleton into those parts. Additional parts could be necessary to model the folds and control the unwanted blending problem near the folding areas. Figure 3, shows namely the part I corresponding to the part of brain called rhombencephalon, part II will develop to mesencephalon and part III is a

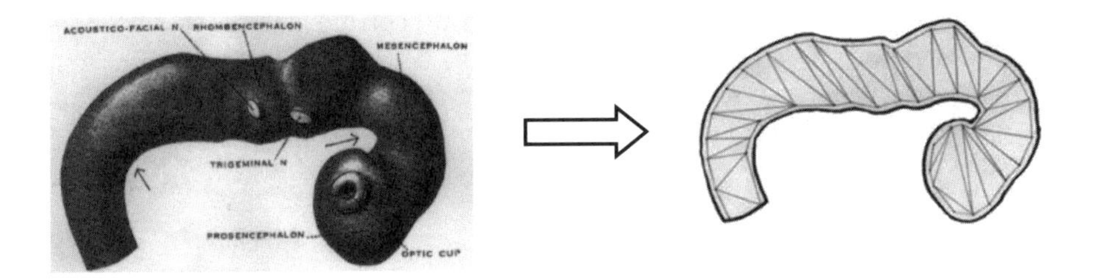

Fig. 2. The conversion of drawing human embryo brain to central skeleton

prosencephalon. The next step is to calculate the central line that will be used as a base to define the thickness of the model along the line forming the tubular object. Central line passes through the center of central skeleton, connecting the mid points of vertical edges of a triangular strip.

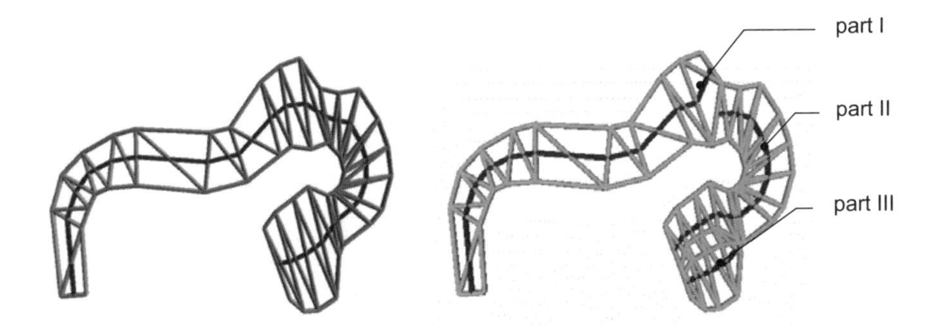

Fig. 3. Dividing the central skeleton to 3 parts. The line in the middle of the central skeleton is called central line.

4.3 Brain: Skeleton

By adding the thickness to 2D central skeleton the 3D skeleton of the model is obtained. Multiple number of copies of central skeleton are slightly scaled and shifted to left and right sides of central skeleton. By this way the cross sections are produced which are then connected to form the tubular skeleton, see Figure 4:

- Each of side skeletons is scaled to fit the ellipses whose center is on the central line. Radius a of the ellipse is a distance to the central line from the border of the central skeleton. Radius b follows the equation, $b = \alpha a$, where α is a ratio parameter.
- As next step, for a given θ the side skeletons are translated by distance $t = c \cos \theta$, where c is known from parametric equation of ellipse shown in Figure 5.

- Finally, side skeletons are connected with a central skeleton or with other side skeletons by a triangular mesh.
- After erasing all interior triangular patches we obtain multiple tubular shapes forming together the entire skeleton of the brain.

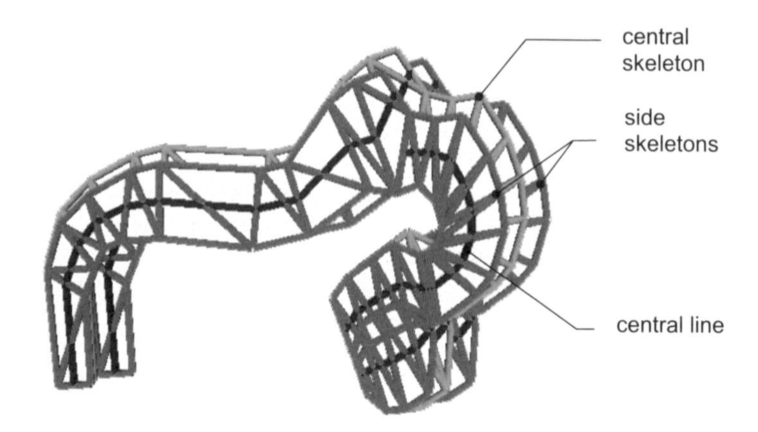

Fig. 4. Adding the thickness by scaling and shifting the central skeleton

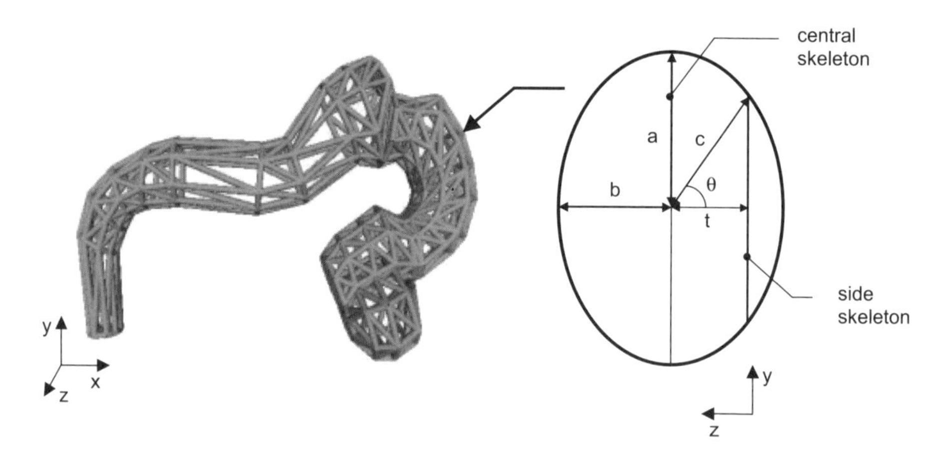

Fig. 5. A 3D skeleton for 36 days old human embryo brain

4.4 Model of the Human Digestive System

To approximate the shape of an organ while considering the speed and direction of cell growth at the same time, we group the entire set of cells into a number of cylindrical bunches (clusters). Thus, the skeleton of the organ is defined by a chain of linear segments passing through the cluster centers, see Fig. 6. Organ growth can then be modeled by the growth of the line skeleton, and variations in shape thickness during the growth process can be captured by variations in cylinder size. When a cylinder changed in size, it was understood that the organ cells grew in the directions emanating from the cluster center. Similarly, when the skeleton segment underwent changes in length, it was understood that the cells included in two adjoined clusters grew in directions parallel to this segment.

Taking into account the development, the organs were divided into physiological parts having different speed and direction of growth to suite the animation purposes. The physiological parts of the intestine system are shown in Fig. 7 and marked I, II and III for stomach, marked IV for small intestine, marked VII for large intestine, marked V for appendix, and marked VI for vitteline duct. While refering to Langman's embryology [11] we collected data that are shown in Table 1. For each available embryo age (developmental stage) of large intestine its mean thickness and skeleton length are listed. Statistical data for a human embryo stomach have already been summarized by Ďurikovič *et al.*[2].

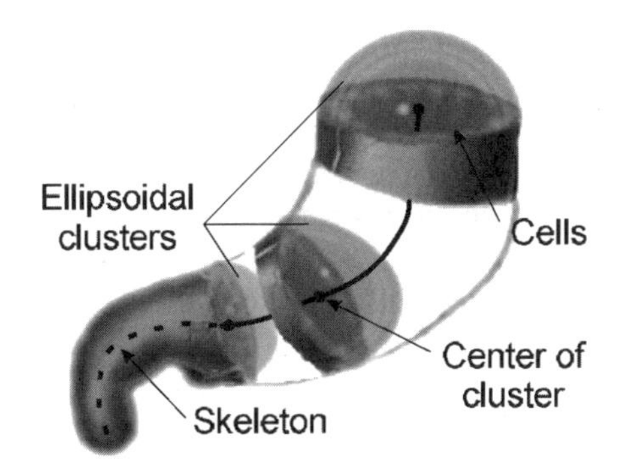

Fig. 6. Skeleton of the organ and the clusters

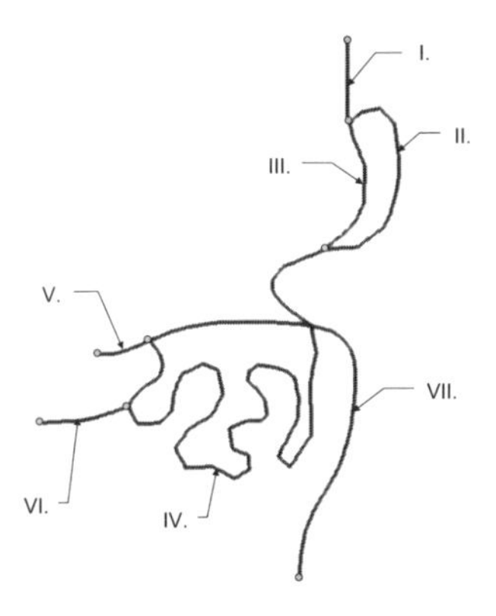

Fig. 7. Physiological parts represented with line skeleton

Table 1. Shape measurements of large intestine, physiological part no. IV

Embryo age (day)	Length (mm)	Thickness (mm)	
28	2.76	0.30	herniation
49	15.58	0.45	
58	19.49	0.52	
70	21.77	0.61	reduction
83	24.04	0.82	
113	28.62	1.00	fixation

The organs, at this preprocessing stage, were divided into physiological parts having different speeds and directions of growth to suite the animation purposes.

4.5 Digestive System: Skeleton

The topology of the digestive system is expressed by a tree structure and the development of the tree-like structure can be easily modeled with an algebraic L-system [12,13]. An L-system formalism was proposed by Lindenmayer [14], and the method has been used as a general framework for plant modeling. The L-systems are extended to by introducing continuous global time control over the productions, stochastic rules for the capture of small variations, and explicit functions of time used to describe continuous aspects of model behavior, in addition to differential equations.

In some cases it is convenient to describe continuous behavior of the model using explicit functions of time rather then differential equations. For example, global shape transformations varying over time require a large and complicated system of differential equations, while only few explicit functions of time are sufficient for the description of these transformations.

4.6 Cell Model

Let's move to a micro structure of muscle cell structures on the organelle level. We present a modeling concept based on the theory of implicit surfaces that allows for creation of a realistic infrastructure of the micro-world of muscle cells. From the viewpoint of geometry, the structure of living cells is given by the three-dimensional organization of their numerous intracellular organelles of various sizes, shapes and locations.

4.7 Cell: Central Skeleton

The initial step involves creation of the central skeleton of the cell, which is represented by a system of parallel cross-sectional graphs (c-graphs) distributed along the longitudinal axis. We define the c-graph as a continuous planar graph which divides the plane in a finite number of closed non-intersecting polygons. Then we exploit the two-dimensional c-graphs to create the myofibrillar system

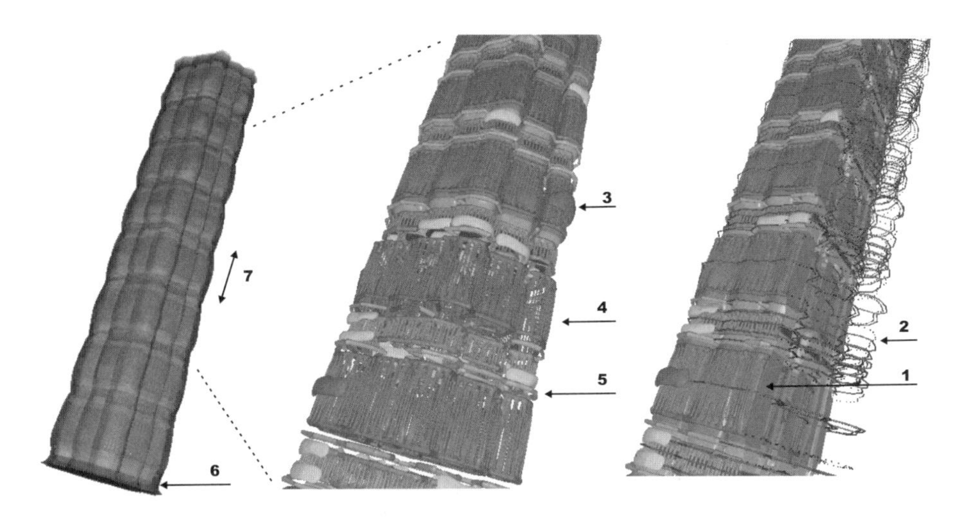

Fig. 8. An example demonstrating eight consecutive sarcomeres of a muscle cell (left). For better clarity, the sarcolemma is hidden and, also, the bottom part of the myofibrillar system is clipped of by a transversal plane (middle). The complex system of underlying skeletons is made visible by clipping with a longitudinal plane (right). The myofibrillar system (1) is defined by means of c-graphs (2). The remaining organelles include mitochondria (3), sarcoplasmic reticulum (4), t-tubules (5) and sarcolemma (6); given in the basic repetitive unit, sarcomere (7).

by means of the F-rep representation of polygons and interpolation. For better clarity, this concept is demonstrated in Figure 8.

In the following subsections we propose approaches for creation of the most complex structures of muscle cells, reticulum and mitochondria.

4.8 Cell: Skeleton

The basic modeling object at this step is a set of seed points distributed in a system of several cross-sectional planes as shown in Figure 9a. Let $\mathbf{S} = \{s_1, s_2, \ldots, s_n\}$ stand for the set of seeds in one crossection. Each seed produces an implicit circle f_i with an appropriate radius. The whole contribution of $\mathbf{S}$ is represented by CSG union:

$$f(\mathbf{S}) = \bigcup_{s \in \mathbf{S}} f_s. \tag{3}$$

Similarly, we define the set $\mathbf{R}$ of seeds in a neighboring crossection. To create a smooth junction between shapes $f(\mathbf{S})$ and $f(\mathbf{R})$ we apply an interpolation method.

Assume that both shapes (sets of implicit circles) contain a set of control points, $\mathbf{P} = \{p_1, \ldots, p_n\}$ for the function $f(\mathbf{S})$ and $\mathbf{Q} = \{q_1, \ldots, q_n\}$ for the function $f(\mathbf{R})$. Moreover, vectors of correspondence are specified between these

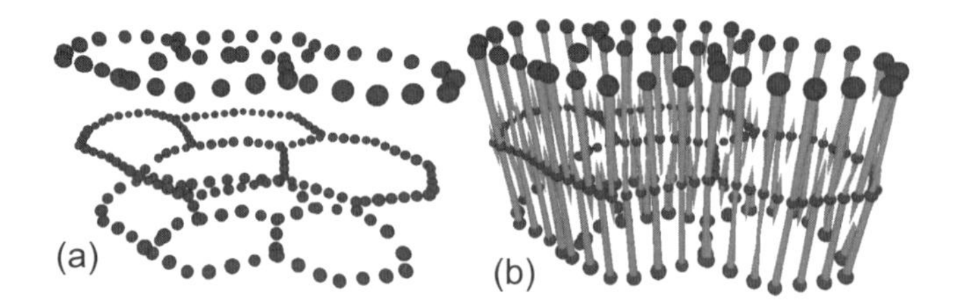

Fig. 9. (a) Seeds (here represented by the red spheres) are distributed in sets of parallel planes. (b) A classical interpolation technique results in non-interconnected segments.

points, $\mathbf{C}_P = \{q_1 - p_1, \ldots, q_n - p_n\}$ and $\mathbf{C}_Q = \{p_1 - q_1, \ldots, p_n - q_n\}$. The set $\mathbf{C}_P$ is attached to the set $\mathbf{P}$, and the set $\mathbf{C}_Q$ is attached to the set $\mathbf{Q}$. Now, we create two weighting displacement functions ϕ^p and ϕ^q, which represent transformations of the given shapes in the directions defined by the vectors of $\mathbf{C}_P$ and $\mathbf{C}_Q$. The weighting displacement functions are defined by

$$\begin{aligned} \phi^p(\mathbf{x}) &= \mathbf{x} + h_1(t)d_1(\mathbf{x}) \\ \phi^q(\mathbf{x}) &= \mathbf{x} + h_2(t)d_2(\mathbf{x}), \end{aligned} \tag{4}$$

where $h_1(t)$, $h_2(t)$ represent weighting proportions within the interval $< 0, 1 >$, and $d_1(x)$, $d_2(x)$ represent interpolation of control points given by vectors of $\mathbf{C}_P$ and $\mathbf{C}_Q$. To interpolate the displacement d_1, d_2 we adopt volume splines—the so-called thin-plate function [15,16]. The weighting factors h_1 and h_2, i. e. functions that specify the size of control point displacements, are defined as

$$\begin{aligned} h_1(t) &= (1 - t^a)^b \\ h_2(t) &= 1 - h_1(t), \end{aligned} \tag{5}$$

where the parameters a, b modify the slope and curvature of the transition (Fig. 10a).

To create the required smooth transformation without gaps, the linear interpolation is modified by the displacement functions, Eqs. 4:

$$F_{lt} = (1 - t)f(\mathbf{S})(\phi^p(\mathbf{x})) + tf(\mathbf{R})(\phi^q(\mathbf{x})) + aw_3(t), \tag{6}$$

where $aw_3(t)$ is the additional blending term used to fine-tune interconnection of shapes by adding material primarily in the central part of the interpolation region. The parameter a stands for the amount of blending and the weighting function $w_3(t)$ is defined as

$$w_3(t) = (1 - (2t - 1)^c)^d, \tag{7}$$

where the parameters c, d modify the slope and the curvature (Fig. 10b).

A result of this approach with three sets of seeds defined in three parallel planes is depicted in Figure 11a.

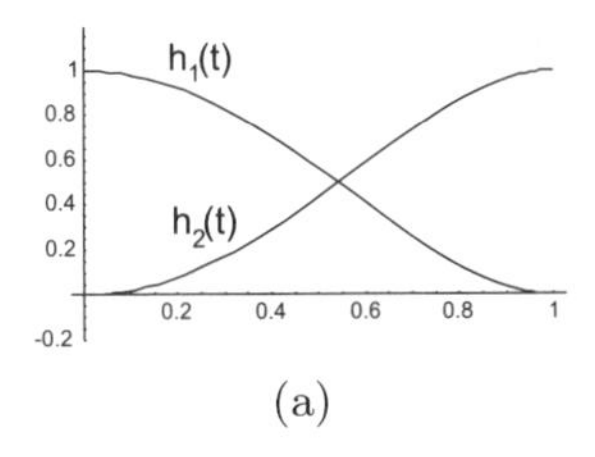

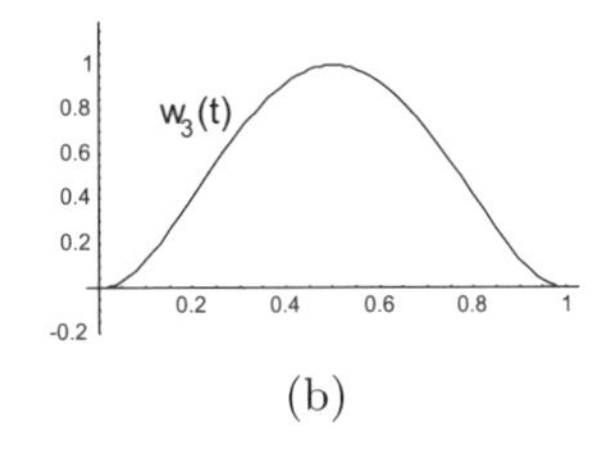

Fig. 10. (a) Weighting functions h_1 and h_2. (b) The weighting function w_3 has the maximum in the middle.

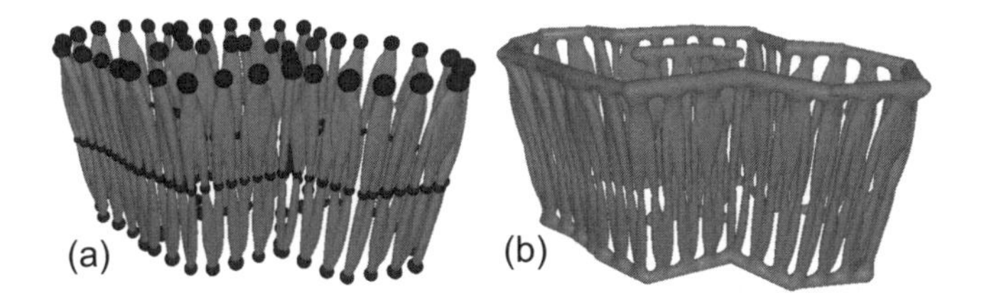

Fig. 11. Modeling of sarcoplasmic reticulum. (a) The final warping interpolation provides gap free interconnections. (b) The smooth junction between terminal cisterns and tubes is obtained by the blended union.

The second step in the building process is formation of terminal cisterns. These cylindrical shaped objects form a smooth junction to systems of longitudinal tubules. Terminal cisterns are created as blended union of implicit cylinders. Their underlying line segments are obtained by connecting the seeds in the bottom most and the top most plane of the system of crossectional planes, see Figure 11b.

4.9 Cell: Mitochondria

In order to capture the varying elliptical shape of mitochondria, we use implicit sweep objects. The basic components of sweep objects are a 2D sweep template and a 3D sweep trajectory. Here, the 2D template is a 2D implicit ellipse with variable dimensions. Figure 12 demonstrates such a mitochondrion defined by a trajectory specified by means of spline control points.

5 Organ Growth

Continuous processes such as the elongation of skeleton segments, and growth of cell clusters, over time can easily be described by the growth functions. Growth functions can be then included into algebraic L-systems as explicit functions or differential equations. Growth is often slow initially, accelerating near the maximum stage, slowing again and eventually terminating. A popular example

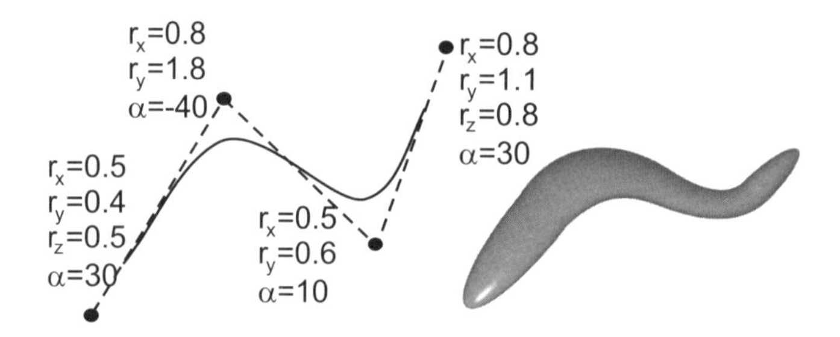

Fig. 12. The curve, represented as a quadratic B-spline, is created from the control points, where each has assigned corresponding radii and rotation angles (left). Note end control points have specified also z radius for 3D ellipsoid. The resultant sweep object is depicted on the right.

of the growth function [17] is the *logistic* function which is a solution to the following differential equation

$$\frac{\partial r}{\partial t} = p\left(1 - \frac{r}{r_{\max}}\right) r \equiv g_{r_{\max},p}(r). \tag{8}$$

Logistic function monotonically increases from initial value r_0 to r_{max} with growth rates of zero at start and end of time interval $[T_0, T]$. It is an S shape function with a steep controlled by a parameter p.

The details of the L-system tables have been described by Durikovic [13]. He has described the skeleton elongation, the global bending of the skeleton parts, dynamics of skeleton structures, and growth functions.

6 Shape from Skeleton

The measured organ models discussed in previous section were divided into physiological parts, at preprocessing stage, having different speed and direction of growth to suit the animation purposes. A single physiological part has the shape defined by the skeleton based F-rep. The skeleton of the physiological part can be animated directly by a key-frame animation or we can use a sophisticated methods to simulate the skeleton growth based on L-system or the dynamic L-system.

6.1 Brain Shape

A smooth convolution surface defined over the triangular mesh of tubular skeleton creates the model of embryo brain. In order to create brain model with convolution surfaces, we use HyperFun[1,9] as modelling library and POV-Ray[8] as rendering software. HyperFun command hfConvTriangle generates convolution surface over the triangles which suites our problem. Let us discuss all parameter settings for one particular example, the *stage3* human embryo brain shown in Figure 13. The convolution kernel width is set to $s = 0.5$ and iso-potential threshold value is $T = 0.6$. The ration parameters of brain thickness have been set to $\alpha = 1.0$ at parts I and II and to $\alpha = 1.2$ at part III. Nice blending during the animation can be guarantied by blend-union operation between three parts of this model using the HyperFun command hfBlendUni. The blending parameters $a_1 = a_2 = a_3 = 0.2$ are used for both gluing parts I, II and parts II and III, respectively.

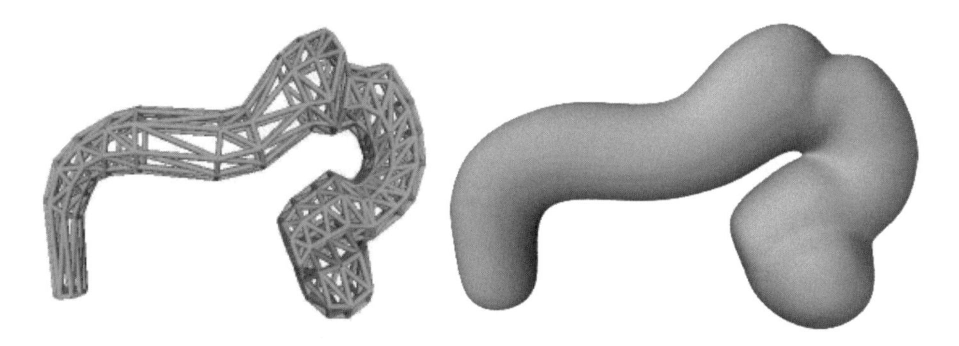

Fig. 13. Stage3 human embryo brain. Left: 3D tubular skeleton, right: entire brain model, defined by function representation.

6.2 Shape of the Digestive System

We represent the smooth shape of the digestive system in a compact way by piecewise linear skeleton and locally defined convolution cylinders along each linear segment of a skeleton. Thus, the resulting smooth tubular surface is represented by a real function as the blend union operation between many convolution cylinders. The shape of a convolution surface can be varied in several ways: by varying the skeleton, by varying the thickness of convolution cylinders with parameter s from Eq. 9, and by the iso-potential threshold value T:

$$f_i(X) = \int_{V_i} \frac{1}{(1 + s^2 r^2(v))^2} dv - T. \tag{9}$$

For example, the small and large intestines monotonically increase their thickness which can be modeled with the monotonically decreasing parameter s as seen for the six developmental stages of intestine in Table 2.

Thickness. As was already mentioned, the increasing thickness of convolution cylinders distributed along the skeleton segments is given by monotonically decreasing the width parameter s in time as shown in Table 2. We will transform the solution of Eq. 10, $\hat{s}$, that monotonically increases from 0.01 to 0.16 over the time interval $[28, 120]$ into a monotonically decreasing function s by Eq. 11,

Table 2. Convolution parameters for thickness of small and large intestine

Embryo age (day)	Intestine Small s	T	Large s	T
28	0.45	0.2	0.69	0.25
49	0.42	0.2	0.67	0.25
58	0.40	0.2	0.63	0.25
70	0.35	0.2	0.60	0.25
83	0.32	0.2	0.57	0.25
113	0.29	0.2	0.54	0.25

where $s_{max} = 0.7$:

$$\frac{\partial \hat{s}(t)}{\partial t} = g_{0.16,0.003}(\hat{s}), \quad \hat{s}(28) = 0.01 \tag{10}$$

$$s(t) = s_{max} - \hat{s}(t). \tag{11}$$

Function $s(t)$ is the growth function controlling the thickness of large intestine with a good approximation of data from Table 1. The graph of the growth function over the time is shown on right of Fig. 14.

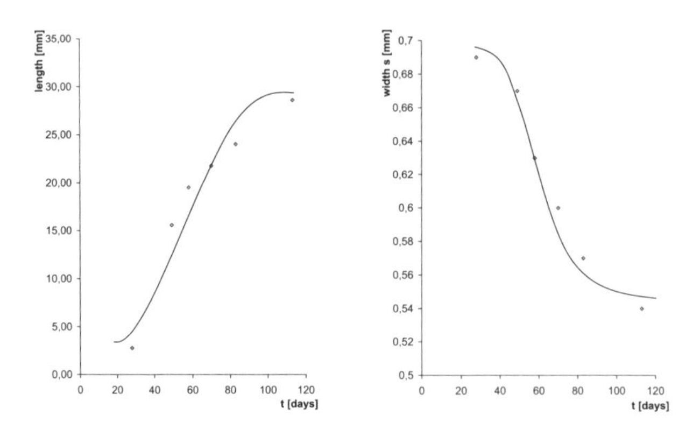

Fig. 14. Graphs of growth functions. Left) Total length of large intestine in time. Right) Change of width parameter s in time, see Eq. 9.

7 Results

Few frames from animation of *Organ growth* show the embryo stomach and brain described by embryo age and the real size scale bar, see Figures 15, 16.

Shown in Fig. 17 are several frames from a generated animation simulating digestive growth based on the proposed L-system using the above growth functions. The environment forces and self collision were handled by the spring

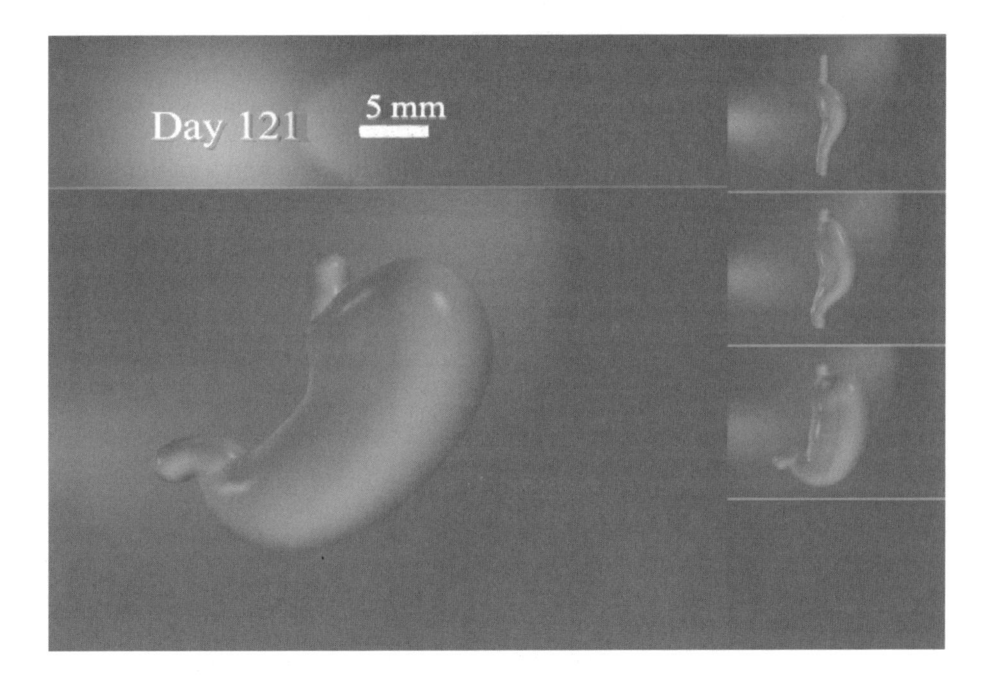

Fig. 15. A single frame from the human embryo stomach animation

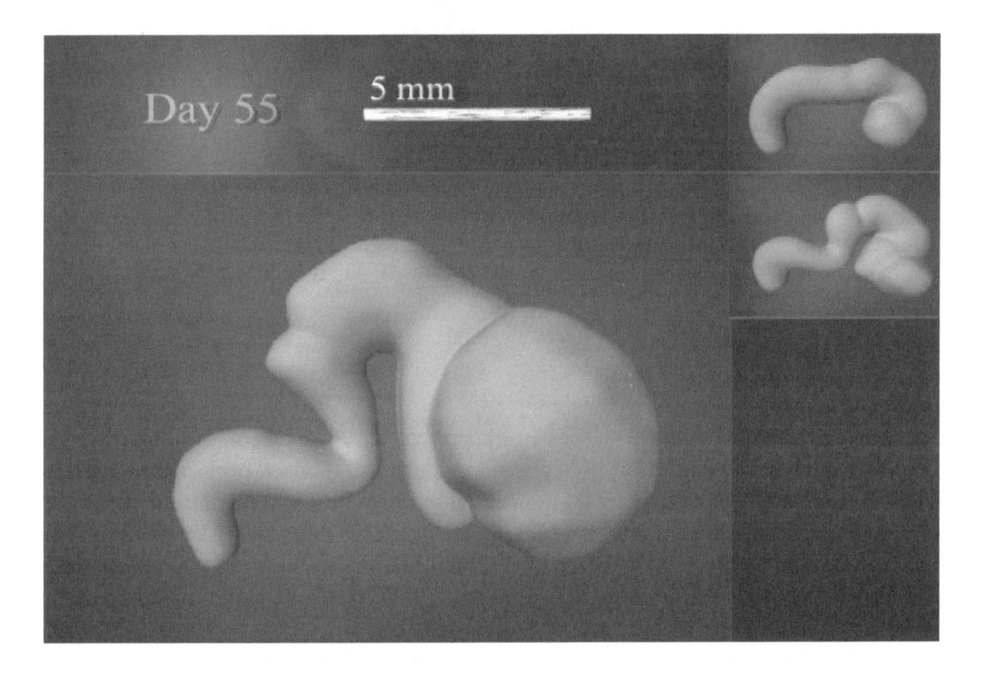

Fig. 16. A single frame from the human embryo brain animation

representation of results obtained from L-system. The shape of the digestive system shown in this figure undergoes global bending transformation and deformations resulting from gravity, animator intervention (looping process), and collision. Some of the intermediate shapes in Fig. 17 have disjoined elements due to aliasing in the implicit polygonizer that has difficulties to find a mesh for long thin structures.

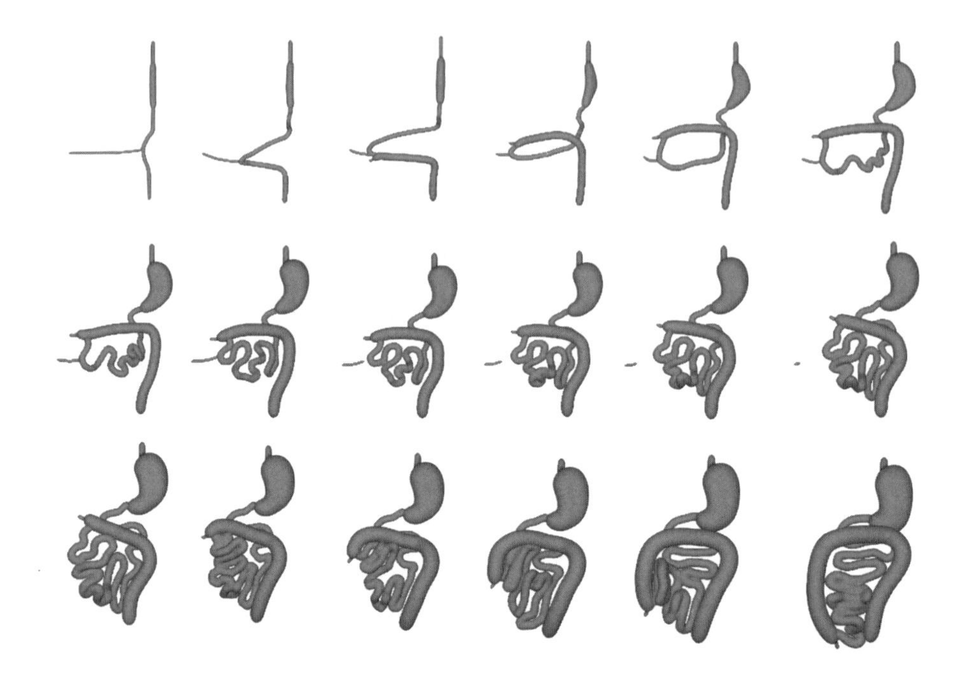

Fig. 17. Development of a human embryo digestive system with proposed method taking into account the skeleton dynamics and growth functions in algebraic L-system. Function representation is used to define a smooth shape. Indicated stages from left to right represent 28, 34, 40, 49, 52, 58, 64, 70, 76, 79, 83, 94, 101, 107, 110, and 113 days of animation sequence.

8 Conclusions

We have presented a method for simulation of the growth of human embryo digestive system. The method uses the shape calculated based on F-rep using isosurfaces generated by skeleton segments, which provides a smooth and compact representation of the surface usable for complex animations. We proposed a method in which the organ growth and global bends are separate processes. The differential growth functions are introduced for an algebraic L-system which efficiently control the elongation of skeleton segments.

We succeeded to model structure of living cells, virtual human embryo organs, namely brain, stomach and digestive system using convolution surfaces and functional representation. The growth animation of a stomach was generated for all 9 months of development while the brain growth animation was generated for first 4 months of embryo development. The advantage of skeleton based approach is that it avoids the the topology artifacts that can occur when using the nonlinear interpolation between two defining functions of F-rep models. Variable speed of growth and shape thickness is successfully modelled by convolution plus or blending union between model parts.

We have proposed the skeletons consisting of triangular patches which gives us the opportunity to define the flat shapes like pillow, refer to the brain model.

Acknowledgements

Authors wishes to thank Mineo Yasuda from Medical School of Hiroshima University for sharing the knowledge as an embryologist. The authors are grateful to Laboratory of Cell Morphology at the Institute of Molecular Physiology and Genetics, Slovak Academy of Sciences. The images were rendered by the POVRay ray-tracing program programmed by POVRay Team. This research was supported by a Marie Curie International Reintegration Grant within the 6th European Community Framework Programme EU-FP6-MC-040681-APCOCOS.

References

1. Ďurikovič, R., Kaneda, K., Yamashita, H.: Imaging and modelling from serial microscopic sections for the study of anatomy. IEEE Medical & Biological Engineering & Computing 36(3), 276–284 (1998)
2. Ďurikovič, R., Kaneda, K., Yamashita, H.: Animation of biological organ growth based on l-systems. Computer Graphics Forum (EUROGRAPHICS 1998) 17(3), 1–13 (1998)
3. Smith, C., Prusinkiewicz, P.: Simulation modeling of growing tissues. In: Proceedings of the 4th International Workshop on Functional-Structural Plant Models, pp. 365–370 (2004)
4. Felkel, P., Fuhrmann, A., Kanitsar, A., Wegenkittl, R.: Surface reconstruction of the branching vessels for augmented reality aided surgery. In: Proceedings of the Biosignal, pp. 252–2549 (2002)
5. Tsingos, N., Bittar, E., Gascuel, M.P.: Implicit surfaces for semi-automatic medical organ reconstruction. In: Earnshaw, R., Vince, J. (eds.) Computer Graphics, development in virtual environments, proceedings of Computer Graphics International 1995, Leeds, UK, May 1995, Academic Press, London (1995)
6. Tait, R., Schaefer, G., Kühnapfel, U., Çakmak, H.K.: Interactive spline modelling of human organs for surgical simulators. In: 17th European Simulation Multiconference, Nottingham Trent University, Nottingham, England, pp. 355–359 (2003) ISBN 3-936150-25-7
7. Amrani, M., Jaillet, F., Melkemi, M., Shariat, B.: Simulation of deformable organs with a hybrid approach. In: In Proceedings of the Revue Int. d'infographie et de la CFAO (2001)
8. Leymarie, F.F., Kimia, B.B.: Shock scaffolds for 3d shapes in medical applications. In: DIMACS Workshop on Medical Applications in Computational Geometry, DIMACS Center, Rutgers University, Piscataway, NJ (2003)

9. Attali, D., Montanvert, A.: Semi-continuous skeletons of 2D and 3D shapes in C. World Scientific, New York (1994)
10. Pasko, A., Adzhiev, V., Sourin, A., Savchenko, V.: Function representation in geometric modeling: concepts, implementation and applications. The Visual Computer 11(8), 429–446 (1995)
11. Sadler, T.W.: Langman's medical embryology, 7th edn. Williams & Wilkins, New York (1995)
12. Mech, R., Prusinkiewicz, P.: Visual models of plants interacting with their environment. In: SIGGRAPH, pp. 397–410 (1996)
13. Ďurikovič, R.: Growth simulation of digestive system using function representation and skeleton dynamics. International Journal on Shape Modeling 10(1), 31–49 (2004)
14. Prusinkiewicz, P., Lindenmayer, A., Hanan, J.: Development models of herbaceous plants for computer imagery purposes. In: SIGGRAPH 1988: Proceedings of the 15th annual conference on Computer graphics and interactive techniques, pp. 141–150. ACM Press, New York (1988)
15. Duchon, J.: Splines minimizing rotation-invariant semi-norms in Sobolev spaces. In: Schempp, W., Zeller, K. (eds.) Constructive Theory of Functions of Several Variables. Lecture Notes in Mathematics, vol. 571, pp. 85–100. Springer, Heidelberg (1977)
16. Savchenko, V., Pasko, A.: Transformation of functionally defined shapes by extended space mappings. The Visual Computer 14(5-6), 257–270 (1998)
17. Edelstein-Keshet, L.: Mathematical models in biology. McGraw-Hill, New York (1988)

Universal Desktop Fabrication

T. Vilbrandt[1], E. Malone[2], H. Lipson[2], and A. Pasko[3]

[1] Digital Materialization Group, Japan, and MIT-FabLab, Norway
turlif@turlif.org
[2] Cornell University, USA
{em224, Hod.Lipson}@cornell.edu
[3] Bournemouth University, United Kingdom
apasko@bournemouth.ac.uk

Abstract. Advances in digital design and fabrication technologies are leading toward single fabrication systems capable of producing almost any complete functional object. We are proposing a new paradigm for manufacturing, which we call Universal Desktop Fabrication (UDF), and a framework for its development. UDF will be a coherent system of volumetric digital design software able to handle infinite complexity at any spatial resolution and compact, automated, multi-material digital fabrication hardware. This system aims to be inexpensive, simple, safe and intuitive to operate, open to user modification and experimentation, and capable of rapidly manufacturing almost any arbitrary, complete, high-quality, functional object. Through the broad accessibility and generality of digital technology, UDF will enable vastly more individuals to become innovators of technology, and will catalyze a shift from specialized mass production and global transportation of products to personal customization and point-of-use manufacturing. Likewise, the inherent accuracy and speed of digital computation will allow processes that significantly surpass the practical complexity of the current design and manufacturing systems. This transformation of manufacturing will allow for entirely new classes of human-made, peer-produced, micro-engineered objects, resulting in more dynamic and natural interactions with the world. We describe and illustrate our current results in UDF hardware and software, and describe future development directions.

1 Introduction

Humans and animals have evolved and live in an enormously complex dynamic system, the natural world. Lacking the vast computational resources necessary to explicitly represent and manipulate the complexity of the world, the animal and human mind developed the ability to represent objects implicitly, as simple, clearly delineated boundaries of space [1, 2]. It is hardly a surprise then, that traditional manufacturing and design processes assume that any given object or an independent part of a larger object is made from a single, homogeneous material. Raw materials extracted from nature are separated and purified so that they can easily be utilized in this framework. The lack of explicit computation and thus the homogenization of nature results in 'man-made' objects that clearly stand apart from nature.

A. Pasko, V. Adzhiev, and P. Comninos (Eds.): HOMA, LNCS 4889, pp. 259–284, 2008.

Over the past two decades, advances in digital computational power and the development of inexpensive and interactive three-dimensional modeling and visualization systems have extended the human capacity to conceive of and represent increasingly complex and optimal—more “natural”—objects. This has lead to the design of objects and software tools that do not respect the constraints of traditional manufacturing. At the same time, it has also instigated a family of technologies known as Rapid Prototyping (RP) or Solid Freeform Fabrication (SFF), better equipped to handle these new, “natural”, digital objects. SFF builds up complex three-dimensional objects directly from digital design data by depositing or solidifying material, layer by layer, under computer control. For the designer, the ability to simply “print-out” extremely complex and otherwise impossible to fabricate designs has driven the demand for RP/SFF technologies to produce not merely prototypes, but parts accurate and durable enough to obviate traditional manufacturing [3]. This general category of technology is referred to as Digital Fabrication (DF) [4]. The current state-of-the-art commercial DF systems allow net- or near net-shape mechanical parts with very complex geometry to be produced in a variety of engineering materials, ranging from thermoplastics to ceramics to high-performance metal alloys. Researchers are extending the range of what can be produced with DF processes to include sensors, actuators, electronics, power sources, and engineered living tissues, using ever more compact and automated systems that deposit multiple types of materials during the course of building a single object. As explicit design and manufacturing complexity and quality approaches that of the nature, it will be possible to fabricate objects previously considered too difficult or even impossible. Human-made objects will not stand apart but increasingly emulate and seamlessly integrate with the natural world.

This research and technology is sparking a transformation away from the limits of traditional manufacturing and centralized production [5] toward “Universal Desktop Fabrication” (UDF) —compact DF systems which can produce essentially any complete, finished, and functional object; not merely mechanical parts, but everything from birthday cakes, to complete cell phones (with batteries), to a human heart. Imagine an Internet of physical things, a 3D fax machine or the “replicator” from the science fiction TV series, Star Trek (Fig. 1). If such technology can be made accessible to individuals, it has the potential to revolutionize the limited ways humans construct objects, manipulate matter, and interact with the world. Individuals will not have to buy a generic, mass produced product shipped around the world to their local superstore. Instead they may choose to download an object, customize the design to fit their needs and 'print'. UDF lowers the financial cost and expertise required for invention, essentially placing an entire R&D laboratory on an individual’s desktop [6]. This will empower countless individuals to become creators of technology rather than passive consumers.

Unfortunately, significant barriers exist to the realization of UDF. The majority of intellectual property in the DF field is held by a few corporations, restricting competition and the identification of new applications, slowing innovation, and ensuring systems remain costly and complex. Commercially available systems are proprietary, and each system is optimized for one or two typically proprietary materials. Systems are not capable of varying the material composition freely through the part. Additionally, traditional human approaches to representing objects combined with intangible digital processes, having no physical limitations, have resulted in the development of popular

Fig. 1. Television's imaginary Star Trek Replicator

design software that is incapable of accurately representing real objects and thus an unsuitable platform for UDF, often proving problematic even for traditional manufacturing. DF and UDF hardware systems under development in research laboratories will soon be capable of producing functional objects with such extraordinary complexity of shape and material composition that existing digital design and engineering tools will no longer be able to represent them.

In order to surmount these barriers to the realization and dissemination of UDF, we are proposing an inexpensive and open research platform for its development, based on combining and extending several existing digital design and fabrication technologies and research projects. Inexpensive, desktop DF has demonstrably broad appeal [7]; therefore we expect that a UDF platform will readily attract intellectual capital from the flourishing online software and hardware development communities, vastly accelerating the rate of advancement and public adoption of the technology.

2 Characterization of UDF

In order to facilitate meaningful discussion, it is necessary to clearly define UDF. An important part of defining UDF is to understand the relationship between traditional manufacturing and UDF. The types and complexity of objects that these approaches can produce are quite different. In addition, it is necessary to identify and define the features and objectives required by UDF.

2.1 Simple Taxonomy of Representational Complexity

Understanding UDF means first understanding the ways in which humans have represented (and manufactured) objects historically and how with powerful computation these representations can change. To clarify for further discussion, a taxonomy of representational complexity should be defined. Using composition, construction and topology, three very general representational categories are suggested for describing real objects: *simple*, *complex* and *heterogeneous* (Fig. 2). As any real object viewed close enough can be considered extremely complex in construction, this taxonomy can also be mapped as various levels of granularity, from gross to highly detailed.

Simple representations have an explicit separation between different materials, and geometry tends to be smooth and continuous. Examples of objects that can be easily described this way might include such things as an egg or a swimming pool.

Complex representations are more natural in composition; various materials are distributed or intersect one another non-uniformly. Surfaces can be rough and detailed, often having deep valleys and peaks, as can be observed in the leaves of a tree or a geological body.

Heterogeneous representations exhibit a gradual change in material composition and associated properties throughout. Surfaces boundaries may be defused. All natural

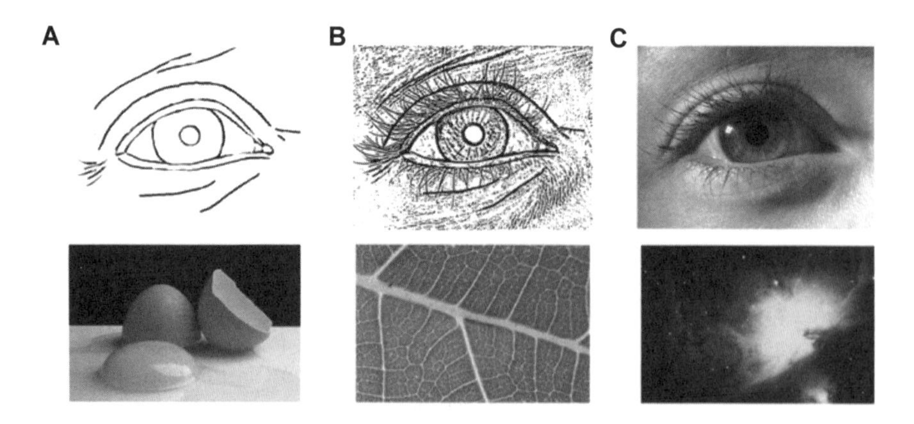

Fig. 2. The top set of images graphically illustrate the differences between the various representations: a) *simple* b) *complex* c) *heterogeneous*. The set of images below are examples of real objects that can be represented by these categories (however all real objects are truly *heterogeneous*).

and/or real objects regardless how simple they may appear are truly *heterogeneous.* It is not possible to represent many objects, such as a real human heart or diffuse nebulae, any other way.

Traditionally humans have defined objects using primarily *simple* representations and most, if not all, current industrial software modeling packages still represent objects in this way. Likewise, expensive RP and DF systems that exist primarily use a single material and are not comparable to the fantasy of a Star Trek replicator. These technologies are far from achieving a level of complexity close to that of the natural. Current DF software and hardware systems act like digitally controlled replacements for traditional manufacturing and while that alone has advantages, to fully capitalize on inherent potential in DF and the realization of UDF, robust, accurate and realistic descriptions of objects are necessary.

2.2 UDF Features

Academic and corporate research efforts are underway to develop, primarily single material, desktop DF systems [8, 9], invariably laying the foundation for UDF systems. UDF should not be considered as simply any '3D printer', but an inexpensive, personal system that, using a variety of materials, can fabricate a broad range of extremely complex and functional objects (previously thought unfeasible). It is important to identify what minimal properties such a UDF system would possess in order to be considered a viable public platform. These features should be understood as not just purely technical in nature but also reflecting the social and market aspects of such a platform. The 'Universal' nomenclature in the term signifies that it is readily available, easy to use, open to modification, and most importantly can fabricate a broad range of objects. The 'Desktop' aspects include low cost, small size and extremely low or zero toxicity and waste. A more detailed short list of features is presented here. This list is not necessarily meant to be exclusive or complete and mostly ignores the feasibility of the listed features. Instead, it serves both as a list of desirable objectives and as a reference point for further discussion.

Easy. Systems must be relatively simple to operate and use.

Free/Open. Some of the most successful, long term consumer desktop technologies today are built on free and open standards and collaboration. It has also shown to increase the rate of technological development [5, 6].

Detailed. The model and fabrication process must be fine enough that objects can obtain qualities and attributes of natural and real objects.

Heterogeneous. The system must be able to represent *heterogeneous* objects with a broad range of materials and fabricate new materials and composites.

Self-Assembly. Digital and/or self-assembly methods, including physical error-correction, will enable the fabrication of objects with tolerances superior to those of the fabrication machine itself, and the production of multiple copies of a given object with near-perfect fidelity.

Inexpensive. The complete system price, power consumption and cost of materials must be roughly similar to other desktop computing technology.

Fast. The fabrication process must be cost competitive with traditional manufacturing approaches when the design freedom provided by UDF is accounted for.

Compact. Systems must be small and lightweight (perhaps at some point it may not require a separate machine for fabrication).

Safe. Hazardous and/or toxic processes are unacceptable for low cost desktop systems as it drives up cost and more importantly can harm users and environment.

Disassemble. Systems should be able to recycle locally by disassembling objects back into raw materials.

3 Related Works

Most available DF systems, if not all, are oriented toward tightly integrating with existing and limited commercial Computer Aided Design, Engineering or Manufacturing (CAD/E/M) frameworks and representations. These commercial systems are not designed to model *heterogeneous* objects. This practical bias places focus on the fabrication of homogeneous and *simple* objects, ignoring *complex* and *heterogeneous* ones. Existing systems usually do not attempt to rethink DF as a whole, instead they rely on traditional CAD systems and independently solve hardware or software issues. The creation of a complete UDF system requires approaching the problems of DF anew and thus it becomes necessary to develop both the hardware and software components in concert. Most existing systems do not take such a holistic approach and are not interested in the same objectives as UDF. The following sections discuss various systems and research projects oriented towards inexpensive and/or *heterogeneous* fabrication.

3.1 Hardware

A common method to currently manufacture blended multi-material objects is by using complex injection molding processes whereby one material is injected into a mold, followed by another material. Specifically calculated and computer controlled temperatures and amounts yield objects with an expected smooth transition of material [10]. Although this system produces *heterogeneous* objects in some sense, there is a lack of precise control over the internal composition and complexity of the objects. In addition, injection molding requires non-reconfigurable tooling, typically very expensive and time consuming. This favors fixed manufacturing and mass production, making it unsuitable as a DF technology.

Most DF hardware systems use a Solid Freeform Fabrication (SFF) process [11], by slicing a shape into cross sectional layers and adding a layer of material at a time to build up an object. There are a wide range of SFF methods and techniques. One method is building each layer with a target material and building support structures for overhanging features in the same material or another, explicitly sacrificial material. After fabrication, the support material is removed by another process. These "fabricated support" SFF methods include Fused Deposition Modeling (FDM), stereolithography, and Laser Engineered Net Shaping (LENS). Another method involves applying a layer of typically powder or laminar target material to the entire working surface, and selectively binding or fusing the material within the cross-section of the desired part to prior layers. Rather than building a separate support structure, this method utilizes the unbound/unfused material for supporting overhanging and unattached features. Such methods include 3D printing, laminated object manufacturing, and selective laser- or electron beam-sintering. Other methods include hybrid processes such as shape deposition manufacturing [12] that uses several staged processes, including more traditional CAM processes like milling, to produce high tolerance parts. Most of these methods, due to the extensive tuning of the fabrication process for a specific material and the restrictions of existing CAD systems are limited to the fabrication of homogeneous objects. There are a few notable exceptions.

Although the Z Corp Spectrum Z510 is not actually a multi-material fabricator, it has the capacity to print any color at any point in the object. This capability is primarily used to print the surface of the object with color in the form of a 2D image texture map. However this is very useful as a way to physically visualize various properties of an object, including material, by mapping various properties to colors. The Z510 does not directly produce functional objects, as the bound powder parts are quite fragile. Infiltration with epoxy or cyanoacrylate resins can render them robust enough for light mechanical use.

Most of the systems that produce *heterogeneous* functional objects are somewhat experimental, extremely specialized and expensive, such as the Optomec's LENS 850-R. This fabricator is capable of producing metal objects from a variety of alloys, as well as fabricating composites and functional gradient materials. It has been designed as an aerospace and military solution for the limited production of new parts and rapid repair of specialized parts.

Apart from such expensive and exotic systems several commercial companies are in the process of producing SFF systems targeted at small businesses and individual users. Desktop Factory has developed a '3D printer' which is currently the lowest priced commercially available system [8]. However, it prints only single material objects with fairly low resolution from a composite plastic powder.

Other more inexpensive and dynamic systems are under development. Recent research has resulted in a few desktop SFF systems using a "do it yourself" (DIY) approach. These systems are extremely inexpensive (supplanting money by time invested) and flexible. One such system of notoriety is the RepRap Project [9]. This project's stated goal is the creation of a self-replicating machine. However, the project has produced an inexpensive FDM fabricator capable of printing usable plastic parts. Due to the DIY/free-source nature of the project, the hardware is also easily extensible and all the plans, specifications and modifications are placed on the Internet and freely downloadable. It is possible for the hardware to be adapted to use various fabrication processes and materials. This project and fabricator have specifications compatible with that of UDF.

3.2 Software

As of yet, no complete commercial 3D CAD/E/M package for *heterogeneous* objects exist. Instead designers and engineers are limited to creating homogeneous parts and multi-material assemblies. However, over the last decade, volumetric and *heterogeneous* representations of objects have received much attention in shape modeling and CAD/E/M research [13]. The rest of this section will discuss various systems for the representation of object's properties (or materials). A diverse group of solutions has been developed. In general these can be divided into two categories: *discrete* representations and *continuous* representations. Additionally, several advanced representations exist. *Composite* representations can be identified as a collection of sub-objects where each separate object can be *discrete* or *continuous*. *Hybrid* representations can use both *discrete* and *continuous* in simultaneous conjunction. *Discrete* models can produce detailed and complex property distributions, but at the cost of accuracy, practical resolution and usability. Examples of such systems include voxels and volume meshes. Such models include voxels and volume meshes. *Continuous* models are based on rigorous functions describing exact geometry and are much more accurate and compact and include control features, control points, and real functions. Several of these methods are worth discussion in more detail.

Voxel based representations are well established (especially in medical visualization) and for more than a decade have been proposed for modeling and fabrication [14]. These representations are good for *complex* objects and useful for representing volumetrically scanned data from magnetic resonance imaging or other such technology. It is also easy to implement hardware optimization and parallelization for voxels. However it is not easy to edit the large voxel sets which are required to reduce aliasing and make smooth or high quality objects.

Unlike voxels, control point based *heterogeneous* modeling is continuous, utilizing Bezier, B-spline volumes and tri-variate NURBS [15, 16, 17]. These representations are fairly compact, exact and can represent *complex* and *heterogeneous* material distributions. However the representation is only applied to property distribution. The geometry model usually relies on the standard CAD/E/M representation, Boundary Representation (B-Rep) [18, 19], and thus requires two completely separate processes when modeling geometry and composition. In addition, parameterization of the object as a whole becomes problematic, limiting the complexity and abstraction of designed objects and reducing usability.

Real function based properties can also be used to represent the distribution of materials inside B-Rep geometry [20, 21]. When applied in this way, real functions, have the same advantages and drawbacks as control point based methods, except more detailed and constructive modeling of materials and properties is possible. However, real function based properties can also be applied to real function based geometry [22, 23, 24]. In this case modeling and property assignment can happen simultaneously in a singular uniform environment. This advantage has a drawback. Compatibility with standard CAD/E/M becomes a problem, as it can be very difficult or impossible to import certain kinds of B-Rep based data into a real function model.

4 Implementation Problems

Practical research aimed at developing complete usable UDF systems is in its infancy. As such, many known (and unknown) problems face UDF technologies. Some general problems are presented in the subsequent section. Many of these problems are not discussed in detail, but are simply presented as a basis to understand the complex technical challenges involved.

Accessibility. Current fabrication systems are physically large and heavy and often require special facilities. Cost and operation of complete systems still remains prohibitive for individuals or even small research teams. Specially manufactured and expensive materials, often only available from the vendor, are required for operation. The machine maintenance and operation requires an expert. Despite the high cost of operation, many systems are slow, some take days to complete a single object.

Accuracy. Most of the current DF systems have poor resolution and aliasing can be physically observed and felt by touch. Practically, these systems operate at a scale somewhere not far below the millimeter. In order to produce truly *heterogeneous* objects with smooth details and advanced functionality, system resolutions must achieve micrometer scale. New problems present themselves at this scale, such as accurate system control, speed of fabrication, maximum size of objects, and repeatability of fabrication and data representation. Many of these problems have already been solved by the desktop printing industry. However, overall this still remains a complex suite of machine design and control, and materials science problems.

Complexity. It is important to note that while some progress has been made in developing inexpensive desktop fabricators (see section 3), *complex* or *heterogeneous* fabrication still remains elusive on low cost machines. Even using very expensive frameworks it is problematic and an active area of intense research. Most systems use a single material and operate using mesh or control point data, often limiting the ability of hardware in the complexity and/or accuracy of the objects they can build. This is because machines are painstakingly optimized for each material, different materials often have conflicting processing requirements, and because traditional 3D CAD software is difficult to use, expensive and fundamentally incapable of modeling real objects (i.e. *heterogeneous* objects).

Health/Environment. Many of the first RP processes developed used hazardous processes and/or materials, including carcinogenic resins, and high-powered lasers. There has been some progress recently, however hazards still remain an issue. In addition many of these systems, if used by millions of people, would have a profound environmental impact. Mass production facilities are often compelled and sometimes financially motivated to collect and recycle by-products of manufacturing, and economy of scale can make this relatively cost effective. It is unclear whether waste management can be cost-effectively rescaled for personal fabrication. In the Age of Information and the 'paperless society', humans are not using less paper, they are using even more as it becomes a temporary medium to exchange information [25]. Unlike the paper printing process, the objective of UDF is not to exchange information (which at some point arguably could be replaced 100% by digital processes) but to fabricate material objects effortlessly. Also, unlike the paper printing processes, using the current processes available for *heterogeneous* fabrication, there is still no clear answer on how to recycle the resulting objects or better yet disassemble them back into raw materials. Improvements to *heterogeneous* fabrication allow ever more intimately combined materials, exacerbating the disassembly/recycling problem.

Standardization. In addition to the strictly technical issues identified with the current research, there are also issues surrounding standardization and development. Little standardization or global collaboration exists at present and what does exist is poor and outdated (for example the STL file format). Even though the idea of assembling objects digitally is a very popular topic, there does not seem to be enough open development or collaboration. As it seems to be currently true for many fields in IT, much of the work done over the last decade on DF has been by corporations and now even academic institutions that are closely guarding and protecting their inventions as secrets, [26] stifling technology wide innovation.

A main focus of the current research is solving those issues related to accessibility and complexity. In the next sections more details are provided on these two topics.

4.1 Problem of Accessibility

Commercial DF technology is still focused on producing passive mechanical parts in a single material, and the emphasis of commercial R&D has been on improving the quality, resolution, and surface finish of parts, and on broadening the range of usable materials. Growth in the market for and capability of commercial DF technology has been disappointingly slow – commercial systems have been available for more than two decades, yet worldwide annual sales of systems are still measured only in thousands. At present SFF systems remain very expensive and complex, focused on production of single material mechanical parts, and used primarily by corporate engineers, designers, and architects for prototyping and visualization and a limited range of end use parts. These factors are linked in a vicious cycle which slows the development of the technology: Niche applications imply a small demand for machines, restricting commercial R&D and adoption of the new capabilities demonstrated in the laboratory, while small demand for machines keeps the machines costly and complex, limiting them to niche applications.

4.2 Problem of Complexity

Currently most, if not all, major industrial grade CAD/E/M software represents objects as a boundary or division of space, or B-Rep. In practice this means a hierarchical tree of divisions of 2D surfaces in 3D space, that for 'solid' objects (which are necessary for fabrication), should define a closed object. B-Rep models fail to define the internal composition of objects. In other words, traditional 3D models are represented as empty spaces inside a zero thickness 'shell'. Fundamentally this means it is not possible to exactly represent natural, real objects in the world. However, as long as such B-Rep models are truly closed, have correct normals and with additional modeling data (often added in a separate process), they can be used to fabricate some types of multi-material objects. The process is computationally and memory intensive; creates large data sets and can be problematic (depending on the objective). This becomes more evident when fabricating *complex* or *heterogeneous* objects. The utilization of B-Rep geometry for multi-material fabrication is most applicable for modeling *simple* objects with clear divisions between materials where each material in an object is modeled as a separate part.

If a given system or technology is provided poor input, then often the capabilities and output is also poor or at least problematic (on rare occasions input to a system can be 'improved'). Thus, if a computer can not represent real objects in a universal and functional manner, then it will be problematic to develop a general method for digitally fabricating arbitrary objects and an average technology user will not be able to practically utilize DF systems. Even with dynamic manufacturing methods like local composition control and SFF, currently available software does not operate in truly *heterogeneous* manner. These software systems, in part due to human thinking, simplify reality as sets with clear boundaries. In contrast to such simplification, natural and real objects have no such boundaries; they are *complex* and *heterogeneous* in construction. For example, humans generally describe a watermelon as a green skin with a red inside (Fig. 3a), when in fact a watermelon's 'skin' is thick, irregular, mostly white and it is unclear where it ends fading from green to white to pink, and where finally the fibrous red fruit begins (Fig. 3b).

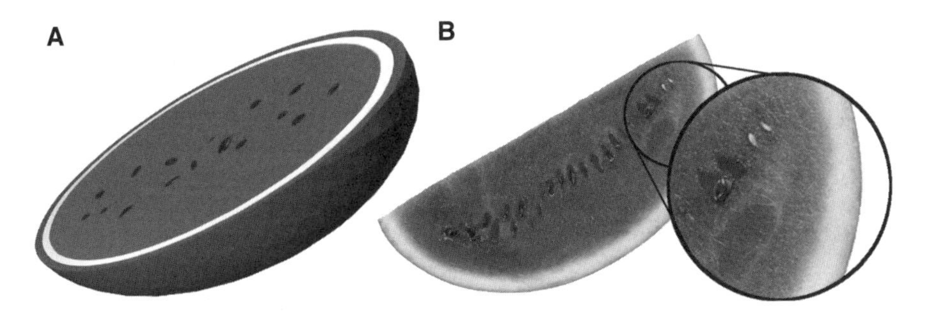

Fig. 3. Watermelon informatics: (a) traditional *simple* CAD model, (b) real object with *heterogeneous* internal material distribution

For a more practical explanation, imagine an architect using traditional CAD software and modeling a wooden house in 3D. It will finally be represented as 2D polygons set next to each other in 3D space, not as real objects and materials with connections. The small individual cuts of wood, the grain of the wood and the existence of nails or glue go unrepresented. In fact, far from modeling the material properties of a building, in practice most architects create models in 3D as a separate process (from the creation of building plans in 2D) for visualization of design only. The simplification of objects is not the fault of software but an accepted and necessary process of human thinking and design. Part of the traditional design and manufacturing processes has been to fill in the missing details or errors (often not well documented) in design as objects are built for the first time. There has always been a gap between what is designed and what is manufactured. However, due to modern information and computation technology, this needs not be the case and results of this approach are already visible in design. Moreover, as the computational ability to explicitly define exact objects increases, whole new categories of human-made objects and design previously unconsidered or improbable become possible. Explicitly designing objects that can self-repair or flora capable of generating electrical power, enter the realm of possibility. A major barrier preventing the micro-fabrication of such objects is the lack of computationally uniform and robust representations and frameworks that represent both property (including material) distribution and geometry simultaneously. A designer should be able to define the geometric boundary of an object as unclear or defused and indicate at any given point in an object, a variety or properties including but not limited to material composition. Simple and sharp interfaces are replaced by complex and smooth variations. In order for this to be practical it must also be done in a compact and accessible method. This is one of the most difficult challenges facing the fabrication of *complex* or *heterogeneous* objects.

Recently to compensate for the limitations of B-Rep, researchers have been looking at novel ways to combine B-Rep geometric data with additional data to describe material distributions (see section 3.2). Thus far extending or patching formats that fundamentally are incapable of encapsulating the nature of real objects has serious limitations. For example, even accounting for the fact that most fabricators are designed to print in one material, software prevents them from printing extremely geometrically *complex* and large objects at high resolutions, like an internally accurate skull, including the porous features inside the bones and teeth. Moreover, how would a user be able to reasonably create or modify such a data set using traditional software? To fully take advantage of digital fabrication technologies, future representations should be able to operate on both the surface geometry and internal composition in a uniform, compact, and consistent manner.

5 Approach

As stated previously, given the enormity of the task to develop a fully functional UDF system, it is not possible at this time to seek solutions to all the problems identified. For example assembling objects at a nanometer scale, which is required to make many desirable objects such as microelectronics, has not been broached by this research. Indeed many of the issues surrounding these problems remain unclear and additional problems are expected to be defined. Instead the primary objective of the research presented here is to work on the most accessible problems, develop solutions to these and most importantly to develop an inexpensive and functional open platform for collaboration and experimentation. It should be a generalized fabrication system using inexpensive, available, open technologies that exist today, resulting in a low cost, complete, usable, *heterogeneous* SFF system.

The open platform should not be limited to just companies, institutes and universities, but instead to any person or organization that has a small budget and access to the Internet. As the Internet has repeatedly demonstrated, having many diverse groups and people developing a technology is an extremely successful development model. To further rapid and diverse collaboration, Free and Open Source Systems (FOSS) methodology and licenses have been adopted.

Continuing in this line of thinking, the hardware system should be easily customizable and use an unlimited variety of inexpensive and easily attainable raw materials for printing. The design of the system should be simplified so that the construction of the hardware platform uses various inexpensive parts, is available online, and can be assembled together in a few days. In addition, users should be able to use a wide variety of easily attainable consumer materials.

Many methods now exist to fabricate objects digitally such as FDM, ink-jet deposition and photo static methods. The experimental UDF system should not be limited to a single DF method, instead it should allow a variety of tools and methods to be developed and used, perhaps even several different methods used to make a single object. The system should be able to dynamically mix or assemble several materials and/or processes together for a given resolution at any arbitrary location in the object.

To understand and finely control a *heterogeneous* object's design, users should be able to edit both the geometry and composition of *heterogeneous* objects, by a similar or identical method at the same moment. It should not require separate modeling stages for geometry and then composition, making it impossible for a user to visualize the composition of an object while modeling the geometry or forcing the user when making a modification to step though a complex processes every time. Likewise for fabrication, the system should have a framework able to identify geometric features and material composition in a uniform method. In addition the resolution and complexity of modeling and fabricating with the system should only be limited by the current computational power available.

Thus a simple, compact and uniform system that simultaneously represents both internal composition and object geometry as a so-called "implicit" model with real continuous functions is required. Function-based modeling is a necessary core technology for UDF (and perhaps the increasingly digital future), that is leading towards interactive modeling of *complex* and *heterogeneous* objects without requiring an explicit specification of the internal configuration. This will provide the means to develop and operate nanometer scale engineering, simulation, design and fabrication systems. The proposed system will use direct fabrication from an object's compact function representation and not from intermediate and degrading file formats like STL. These formats not only degrade the topology but more importantly have no way to represent real *heterogeneous* objects. Although it is theoretically possible for several STL models to be combined to represent a *complex* multi-material object, the data size would make storage and computation prohibitive. A functional UDF system must adopt a procedural, function based approach to modeling and fabrication. However, it should also be able to adequately accept discrete legacy data in a uniform way.

Several existing research efforts have already laid the foundation for UDF. Two projects, Fab@Home (FaH) and HyperFun (HF), as in "hyper-dimensional functions", are both advanced research efforts in their respective areas. It is also interesting to note, but perhaps not surprising that they are both FOSS, utilizing the concepts of peer production to simultaneously speed up production cycles and democratize innovation. The HF project is a good choice as an underling representational foundation for UDF development, able to digitally describe, create and modify any object or environment. The HF project lacks a DF hardware component, however. FaH is a good choice for a DF hardware platform, as it is simple and inexpensive, yet capable of multiple-material deposition and easily extensible. FaH includes CAM software, but lacks integrated *heterogeneous* digital design and fabrication tools. Better than either of these projects alone would be a single system integrating design, engineering, and manufacturing.

6 Previous Work

The FaH project and team has developed a usable low cost DF robot, and a uniform volumetric modeling system has been developed by the HF project. Both of these projects have years of development behind them and provide a developmental foundation for further work.

6.1 Fab@Home

The FaH Project has been inspired by the FOSS approach employed by the RepRap Project. The aim of FaH is to put DF technology into the hands of the maximum number of curious, inventive, and entrepreneurial individuals, and to help them to drive the expansion and advancement of the technology. To achieve this, we have developed an open source, low-cost, personal DF system, which we call the "FaH Model 1" (Fig. 4a), and a user-editable "wiki" website to publish the system designs and software, and to foster a collaborative user community. The parts for the Model 1 kit has a rough cost of $2300 (USD). It includes a free, open-source CAM application which controls the hardware, and processes STL files into manufacturing plans. Almost any room-temperature liquid or paste can be used as the deposition material. Only basic hobbyist tools and skills are required to assembly and use the Model 1 and its software. We have endeavored to make obtaining, assembling, using, and experimenting with the Model 1 as simple and intuitive as possible; the website provides step-by-step ordering, assembly (Fig. 4b) and operational instructions, and an interactive three-dimensional, WYSIWYG, CAM application (Fig. 5).

This custom CAM application which imports individual or assemblies of tesselated geometry (polyhedra) in the STL file format, generates hardware executable manufacturing plans, and controls their execution on the fabrication hardware. The system operator uses a Graphical User Interface (GUI) to specify with which material and tool combination each polyhedron should be fabricated. The tool path planning consists of slicing each polyhedron according to the road thickness associated with its particular material/tool combination, offsetting resulting boundary polygons by a half of the material deposit width for the material/tool, and filling enclosed areas with raster fill (hatch) paths. Slices (containing paths) are then sorted by their height and executed, with the software prompting the operator to change the material and/or tool as required. The hardware currently allows only one tool/material combination to be mounted at a time, and changes are manually executed, so although the use of multiple materials is possible, time and labor become a significant factor for detailed objects, such as batteries. To reduce this cost, we have developed a technology, dubbed Backfill Deposition. In practice, as geometry data describing component parts of a device such as a battery are imported into the fabrication system software, the operator may use the GUI to assign a sequential fabrication priority to each of the parts.

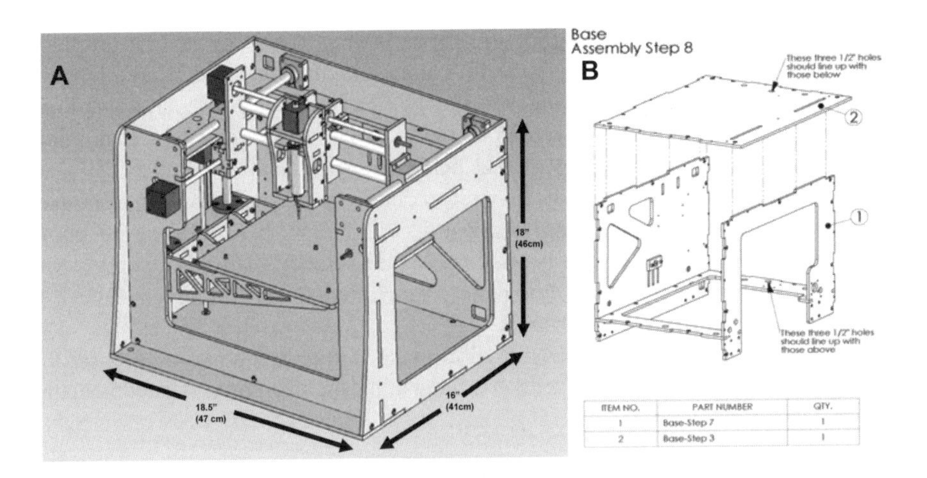

Fig. 4. The FaH Model 1 design. (a) 3D CAD model of an assembled Model 1; (b) An example of assembly instructions available via the project website (http:// www.fabathome.org).

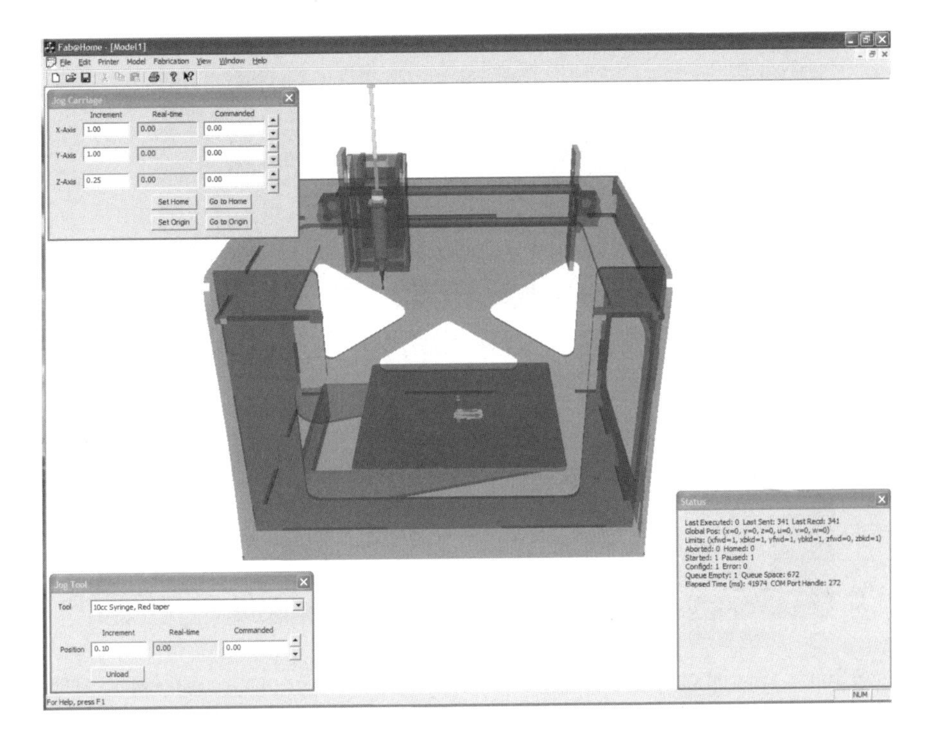

Fig. 5. The FaH CAM application displaying a model ready for fabrication, dialog boxes for positioning and real-time status information

The SFF system will fabricate higher priority parts of an assembly to their full height prior to fabricating lower priority parts, in contrast to strict layered fabrication. This can reduce the number of tool changes in some cases from one per layer, down to one per STL file (or part). The priority will be obeyed by the system except where doing so would violate the relationship of one part supporting another. Additionally, this option facilitates fabrication of objects which contain or are made from liquid materials. It allows the fabrication system to construct a container before filling it. For example, the case of a battery can be given a higher priority than the materials to be deposited into it, and it will be completely fabricated to its full height before the deposition of the other materials begins.

The Model 1 machines have been used to make *simple* functional objects (Fig. 6). A user-editable "wiki" website facilitates publishing the designs and documentation.

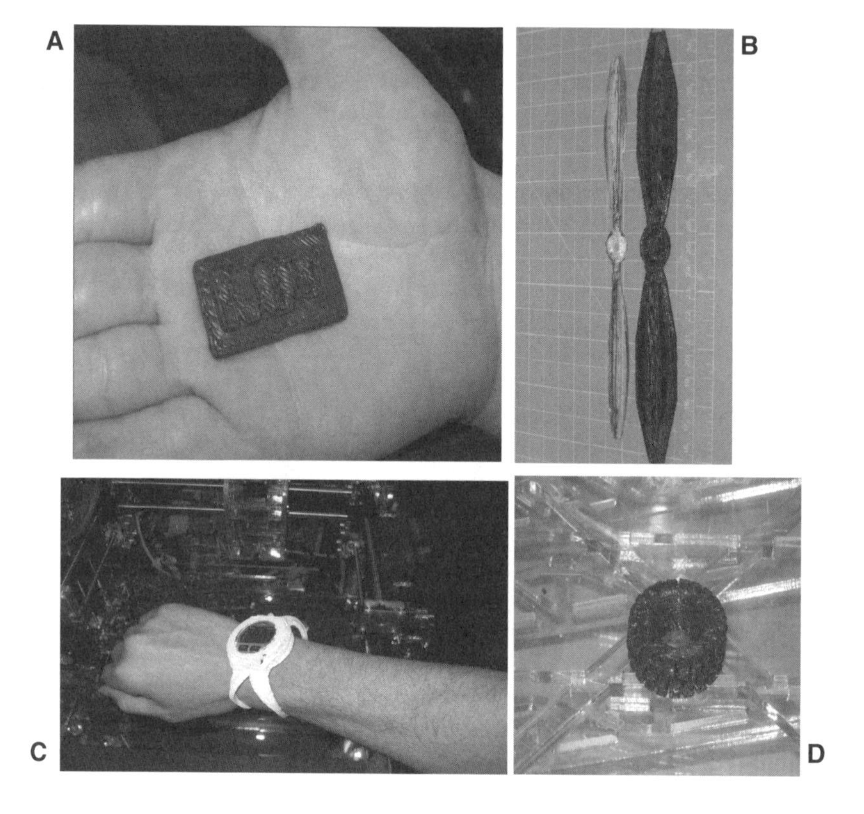

Fig. 6. Single material objects built with a FaH Model 1: (a) A personalized chocolate bar built with a modified Model 1 by Noy Schaal; (b) A mold for a model airplane propeller fabricated using 1-part RTV silicone rubber, and a propeller cast with epoxy from the mold; (c) a watch made by fabricating a silicone watchband and inserting a conventional watch body during the process; (d) a replica of a model car tire fabricated of black silicone rubber

Discussion forums are available using the free Google Groups service, and the source code for the project is shared via SourceForge, a free service which facilitates FOSS development. Through these media, participants in FaH have begun to exchange their ideas for applications and their improvements to the hardware and software with us and each other.

As evidence of the broad appeal of DF, and the potential impact of making DF more publicly accessible, in the first five months after October 2006 (when the website was first made publicly accessible), the project website had more than 3.5 million requests for pages from more than 150,000 distinct hosts in more than 150 countries. Users have begun to make contributions to the FaH wiki, the Google Group, and the SourceForge project in the form of new deposition process ideas, bug reports, questions, feature requests, alternative vendors, group purchasing arrangements, and more.

6.2 HyperFun

At the beginning of the personal computer revolution in the early 1980s there was a need to have a standard, generalized language for digital and desktop printing; the same need exists for DF today. The PostScript language was invented to answer the needs of desktop printing. PostScript is so noteworthy because it goes beyond typical printer control formats and is a complete self-contained programming language, allowing it to implement on-the-fly rasterization using interpreters (PostScript Raster Image Processors), making it extremely compact and device-independent. Like PostScript, HF is a completely self-contained, compact, and device-independent programming language for representing and constructing real objects. This feature as well as others makes HF well suited to become a "3D PostScript" for DF technologies.

In addition to being a programming language HF is a robust software framework, used to create, visualize, and fabricate volumetric 3D models. The platform includes several on-line, Web based rapid interfaces for accessible, collaborative and flexible modeling (Fig. 7). Unlike other modeling packages, it can easily model *heterogeneous* objects in infinite detail. HF is able to represent imaginary objects or capture real existing objects with all the properties and details found in reality and nature. Making this possible, HF is built using a new approach to computing with geometry called the Function Representation (FRep) (see other papers of this volume for more details). In contrast to other existing geometric models, FRep provides a uniform method to model both surface geometry and internal composition simultaneously. It is also a compact and precise framework that can represent objects with unlimited complexity and properties.

Formally a HyperFun object is defined by a vector-function, where each component is a real continuous function of point coordinates. The first component defines object geometry by the inequality $F(x1, x2, x3, ..., xn) \geq 0$. Other components of the vector-function define object attributes representing object's properties at the given point. The HyperFun language allows the user to define a geometric object and its attributes with the help of assignment statements (using auxiliary local variables and arrays, if necessary) as well as conditional selection and iteration statements in a

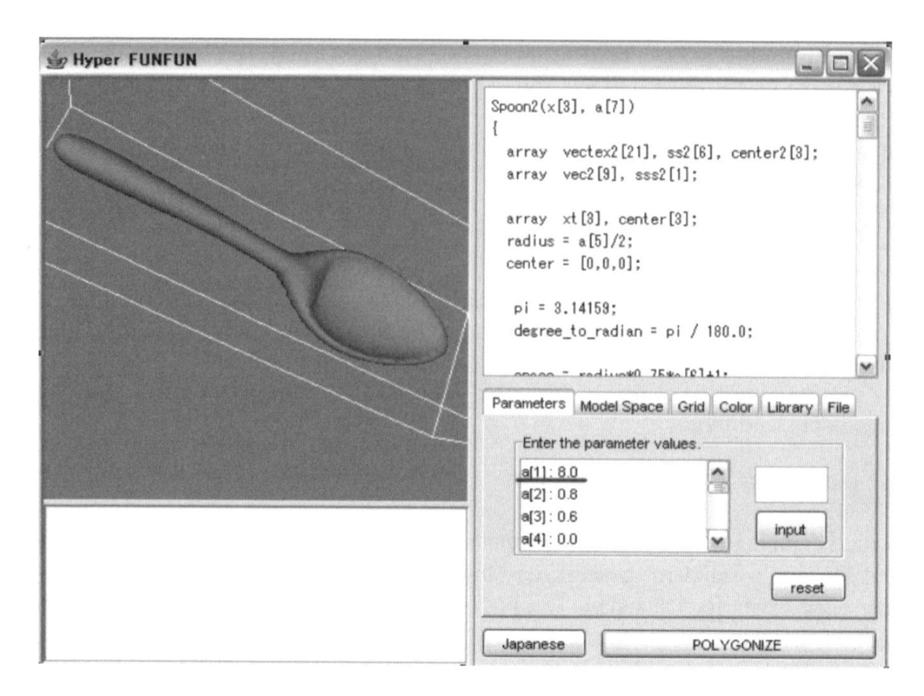

Fig. 7. A development environment for modeling on the Web

single function evaluation procedure. The functional expressions are built with using conventional arithmetic and relational operators, standard mathematical functions, built-in special geometric transformations and FRep library functions for primitives, operations, and attributes.

To our knowledge FRep/HyperFun is currently the only generalized framework and language for easily extensible, *heterogeneous,* volumetric modeling. In recent years it has gained popularity as the need for *heterogeneous* modeling grows. HyperFun.org develops tools for FRep modeling using the HyperFun language. It is an international, non-profit, FOSS organization. Members of the HyperFun team make a freely associated group of researchers and students from different countries all over the world (UK, USA, Russia, France, Japan, Norway, and others). The group has published more than 100 papers in academic journals and conferences, and develops and distributes software under a special FOSS license addressing human and environmental issues surrounding the dissemination of DF technology. Software tools supporting the HyperFun language are freely available at the HyperFun Project Web site (www.hyperfun.org) and source code can be found at SourceForge.net. To date the HF language and framework has been used to model a large number of single material, *simple* objects that have been fabricated using different techniques, from stereolithography to objects milled in wood (Fig. 8). In addition a variety of *complex* and *heterogeneous* objects have been modeled for visualization using HF.

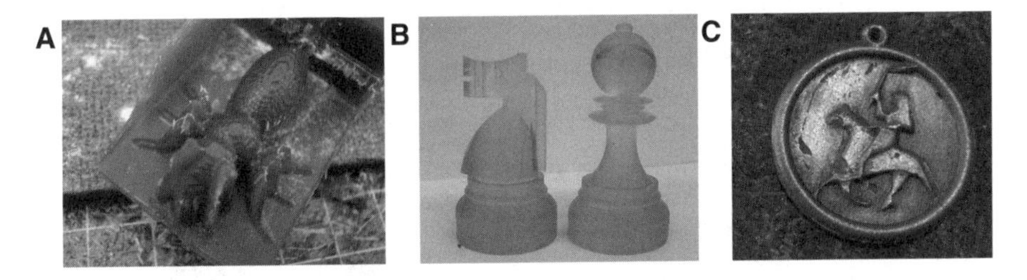

Fig. 8. Various objects fabricated using HyperFun: a) an ant milled in wax (about 3 mm long); b) a chess set fabricated using stereolithography; c) a Norwegian horse and rider crest cast in pure silver

As vector graphics and ideas behind PostScript made 2D desktop publishing and graphical interfaces possible and widespread, the ability of Frep and HyperFun to completely and compactly describe any 3D object has the potential to simplify complex desktop fabrication and physical interfaces, making them viable public technologies.

7 Experimental Work

Much of the goal of the research presented in this section is to bridge HF and FaH with additional development and in doing so, rapidly solve some of the issues outlined herein. The current objective is to directly drive and control the FaH equipment from HF software. Several FaH fabricators have been constructed by the HF and FaH researchers to enable this objective and additional development, both in Japan and Norway. Providing an easy means of directly driving the FaH from HF will mean that individuals can go from fabricating single material or *simple* (multi-material) objects to being able to fabricate *complex* and *heterogeneous* objects using the FaH.

The current FaH CAM software uses the STL file format to import objects and it is internally designed to operate on mesh based boundary data. As discussed above, this is an inadequate representation when fabricating *heterogeneous* objects, however the STL file format can be used to print *simple* objects.

7.1 Extending the Hardware for Multi-material Fabrication

The default FaH system is designed so that, besides the three axes required for Cartesian control, a fourth axis controls a plunger and a syringe with a single material, depositing exact amounts of material at a given location. It is possible to change out syringes during the fabrication process to create an object with more than one material; however this can be slow, very time consuming and is only practical for simple divisions of material. Recently the ability to add a fifth axis and a second syringe to the FaH has been developed (Fig. 9) along with an update to the FaH CAM software platform.

Work is underway to incorporate inkjet material deposition capability along with a single or dual syringe system (Fig. 10). An Inkjet Printing (IJP) head deposits material by ejecting small droplets of a solution at a given spatial frequency onto a substrate, allowing precise placement of relatively small volumes of these materials.

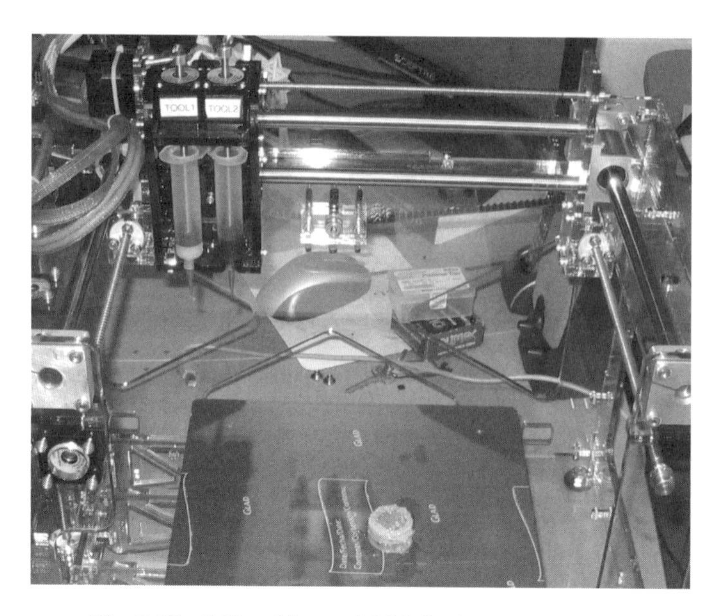

Fig. 9. The FaH multi-material fabrication tool and space

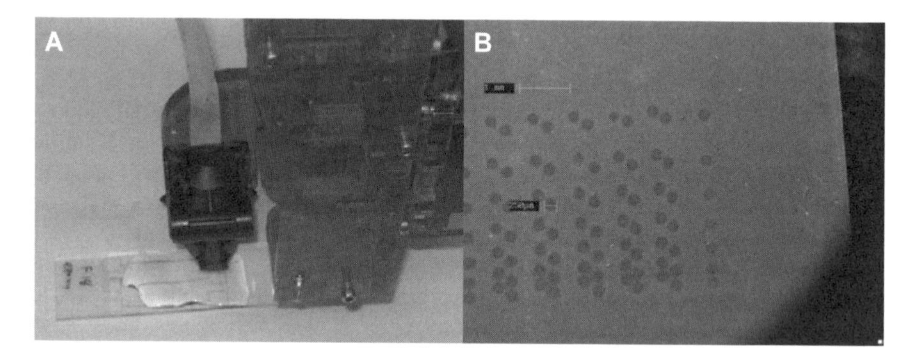

Fig. 10. (a) Inkjet head mounted on FaH along with syringe tool; (b) SEM image of droplet patterns deposited by inkjet with FaH

IJP has two main advantages over syringe deposition. First, such a tiny volume of material can be deposited (picoliters to nanoliters) with such high repeatability that material dries or solidifies very quickly, and lateral positional accuracy is determined almost entirely by the positioning system, rather than by material relaxation or flow. Second, achieving precise control of material flow from a syringe requires the syringe needle remain very close to, but not touching, the substrate, so that the deposited flow does not break irregularly into droplets and the needle does not collide with previously deposited material. This is exceedingly difficult to achieve without sophisticated sensing and feedback control. An IJP head, however, can remain several millimeters above the substrate, and hence is much less susceptible to destructive interactions with minor flaws in the object being fabricated.

Inkjet printing should be able to produce smaller and better-defined patterns of a material thus achieving greater object complexity or even heterogeneity than is possible with a syringe tool. The inexpensive inkjet system currently explored for FaH has lateral resolution (solidified droplet diameter) of 200-250 micrometers, but depending on the solids concentration of the ink used, we have observed vertical resolution (solidified droplet thickness) of 30-100 nanometers. More sophisticated systems can achieve lateral resolution of 25-50 micrometers. However, inkjets are restricted in the range of materials that can be deposited – materials must have a low and well-controlled viscosity, must be filtered to the micron-level, and materials must not solidify or precipitate solid phases within the head, or it will be destroyed. For this reason, it is clearly understood that the inkjet capability complements, but does not supplant, the syringe tool deposition method for the fabrication of *complex* or *heterogeneous* objects.

Due to these additions, the FaH is currently able to fabricate *simple* objects using the STL file format and the FaH CAM software. It is also now possible to fabricate arbitrary multi-material *complex* and *heterogeneous* objects given appropriate representations and control. Development is underway to control all five axes of the machine directly from code generated using the HyperFun framework.

7.2 HF Models Fabricated Using the FaH

Using the HF and FaH frameworks, several test objects have been fabricated including a horse modeled after a traditional Norwegian carving and a model of Darth Vader's head from Star Wars. Both of these objects were fabricated using a single material (Fig. 11).

Fig. 11. Single material objects fabricated using HF and FaH

The fabrication path was generated by FaH CAM after importing STL files generated by the HyperFun framework. The CAM software had to be modified to handle the more complicated topology of Darth Vader's bust and generate the correct tooling paths. However, the short term development goal is to drive the FaH from HF for direct *hetrogenous* fabrication. The material used to fabricate these objects is a Norwegian construction adhesive, Ebofix which performed nicely in the FaH and resulted in nice semi-translucent objects. Both of these models where fabricated in the period of a day, Darth Vader's bust being, to date, the longest build for the FaH at almost nine hours.

7.3 Functional Multi-material Objects

A variety of functional models have been fabricated using the FaH including: batteries that are producing power even before the fabrication process is over, LED flashlight with a working switch (Fig. 12), toys that light up when pushed and electro active polymer actuator able to respond to electrical current by physical motion. Extensive use is made of the priority feature of the path planning software to reduce the number of tool changes in complicated multi-material object. For example, when fabricating a standard cylindrical battery, the battery case and the node conductor are set to the same priority, but higher than that of the other materials. The case and the anode conductor are deposited in a normal (non-backfill) layer-wise fashion, which allows the conductor to extend through an opening in the wall of the case for ease of connection. The FaH is extremely versatile multi-material fabricator capable of creating a wide variety of objects and utilizing a broad range of materials including epoxy, Ag-filled silicone, polyvinyl alcohol, alginate hydrogels and even chocolate.

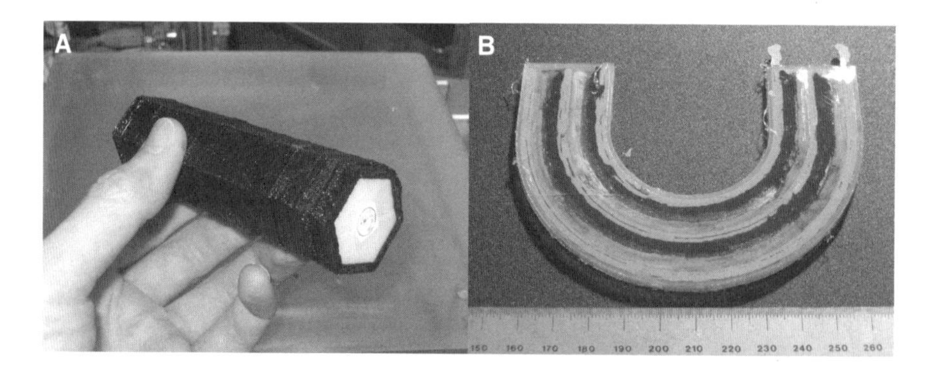

Fig. 12. (a) A fabricated LED flashlight with a working switch (b) and a fabricated zinc air battery

7.4 Complex and Heterogeneous Objects

Several models have been made on a Z Corp Spectrum Z510, a 3D color fabricator, to clearly demonstrate HyperFun's ability to represent and fabricate *heterogeneous* objects (Fig. 13). Although the object in Fig. 13 may look similar to objects modeled using traditional modeling software, it is important to understand that no texture maps were used to define or fabricate this object. It is purely a function-based model (see its detailed description elsewhere in this volume). Cross sections of the model are made at the resolution of the Spectrum Z510 and surface color (or even internal color) is derived from the composition of material modeled using HF.

Fig. 13. HF Multi-material geology model fabricated on Z Corp Spectrum Z510

As the modifications to the FaH currently allow only to print in two materials, several new test models using two materials have been created to print with specific materials as good examples to demonstrate the difference between *simple*, *complex* and *heterogeneous* objects. Even using such a uniform and precise framework as HF, solving the issue of how to fabricate smoothly blended, *heterogeneous* objects on any given system can still require technological choices and solutions to be made. There are two approaches for using the FaH to fabricate such objects: blend the materials during fabrication or use some method of dithering between materials. Generalized code is being developed allowing for any method or type of dithering to be used. As HF allows for any number of properties to be assigned to any location it is also interesting to consider not only controlling the material distribution but the method of fabrication where several very different tools and/or processes can be used, each perhaps with several materials.

8 Discussion

Many of the implementation problems facing UDF remain unsolved and there is much work yet to do and new problems to discover. However, it is interesting to see just how far we can already come by simply adding a bit of glue between existing technologies and projects. It is also interesting to note the extremely low cost of the system proposed. With active work and a few additions, the current system can start to verge on a functional UDF.

The FaH, as of yet, does not blend materials on-the-fly and the deposition size of a given material is relatively large, so in some sense it can only practically fabricate *complex* and not *heterogeneous* objects. However, a variety of higher resolution and more sophisticated fabrication methods can be developed including blending materials before depositing them. Similarly the FaH tool options do not yet include a digital self-assembly tool. While the current hardware and software platform is not designed to digitally assemble objects, it will be possible in the near future to add such capabilities.

At present there are no accessible GUI tools for the modeling of *heterogeneous* objects with complex parts and relationships. HF lacks such an interface and users can not import or use existing models from traditional CAD systems with the standard HyperFun tools, so everything must be modeled anew. However, research and development exists that solves this problem. It will be possible in the near future to functionalize STL or other mesh data and use with the standard HyperFun tools as any other geometric primitive in the system. However, to truly make HF a usable part of a UDF system, additional GUI based design tools will be required. This is an active area of development.

The very real and serious issues of human and environmental hazards, although not covered here, will require earnest discussions and more active research. Input materials should avoid delivery to a UDF system in the form of powders or gases. Low toxic and bio-plastics should be considered for use in fabrication. It maybe possible to leverage the design freedom and complexity provided by UDF to redesign objects using biologically neutral and/or biodegradable materials and abandon the use of rarefied and toxic materials. Finally, the most important long term question to pose is can these complicated multi-material objects be disassembled back into parts or be recycled in some way. Environmentally, without the appropriate research and development behind better, smarter materials and fabricators, UDF could prove to be unsustainable.

9 Conclusion and Future Work

Inexpensive digital computation is allowing us to change the way we see and interact with the world and each other—to understand the world as heterogeneous and operate in and modify the world as such. We can now use computation to control matter, to design and fabricate "natural" solutions and objects. This has the potential to create products which are universally superior physiologically, environmentally, and functionally. Increasingly this is so because digital computation also makes it possible to instantly collaborate globally and share complex information, resulting in peer-based and localized designs. It puts the power of innovation into the hands of the few and the many at the same moment. UDF and similar technologies will change the way humans produce and consume goods, allowing individuals access, not to a factory or a superstore, but their own inexpensive and limitless digital workshop.

This digital epoch has already begun to take place, as a growing number of people invest in this technology and put it into action. We are continuing the development of the software and hardware of the Fab@Home Model 1 to provide performance and usability enhancements in anticipation of an onslaught of questions and complaints as the first wave of Model 1 users finish assembling and start using their machines. This will be a critical test of the survival of FaH, and we must ensure that we do not discourage these brave early adopters. Efforts will be directed to continued development of direct *heterogeneous* fabrication on the FaH, utilizing the HF framework. Modification to the FaH, sample object and code will be available on-line to let developers and users explore the new possibility.

Future development will be focused on improving the capabilities and integration of both hardware and software systems bringing together a complete UDF system. Development to build the next generation UDF fabricator and design tools are already underway. Ongoing UDF hardware research is developing digital self-assembly processes for the fabrication of objects by using materials with known properties and geometries. Software development will continue focusing on the creation of a complete UDF GUI modeling and fabrication software suite based on FRep and HyperFun technologies.

Acknowledgments

The authors would like to thank the Digital Materialization Group and Alexander Pasko Jr. for the funding that made this research possible; Cherie Stamm for her construction of several Fab@Home printers and tireless editorial assistance; Benjamin Schmitt for the use of his geology model, Yichiro Goto and Mio Hilaga for the implementation of the HyperFun Java applet shown in this paper; Fab@Home co-creator Daniel Periard for his major contributions to the Fab@Home wiki, software and hardware testing, and evangelizing the project at tech events; Jennifer Yao for her assistance in the earliest stages of hardware and documentation testing; Haakon Karlsen for his close collaboration with the Norwegian Fab@Home and HyperFun silver jewellery projects, and most importantly all the early adopters and users of Fab@Home and HyperFun for providing critical feedback to our research.

References

[1] Mach, E.: Space and Geometry in the Light of Physiological, Psychological and Physical Inquiry. In: McCormack, T.J. (ed.) Trans., The Open Court Publishing Company, Chicago (1906)
[2] Farah, M.: Visual Agnosia. MIT Press/Bradford Books, Cambridge, MA (1990)
[3] Bak, D.: Rapid prototyping or rapid production? 3D printing processes move industry towards the latter. Assembly Automation 23(4), 340 (2003)
[4] Burns, M.: Automated fabrication. Prentice Hall, Englewood Cliffs, NJ (1992)
[5] Benkler, Y.: Coase's penguin, or, Linux and The Nature of the Firm. Yale Law Journal 112 (2002)
[6] Von Hippel, E.: Democratizing innovation: The evolving phenomenon of user innovation. Journal für Betriebswirtschaft 55(1), 63–78 (2005)
[7] Malone, E., Lipson, H.: Fab@Home: The Personal Desktop Fabricator Kit. Rapid Prototyping Journal 13, 245–255 (2007)
[8] The Desktop Factory 3D Printer, http://www.desktopfactory.com/
[9] RepRap: The Replicating Rapid-Prototyper, http://reprap.org

[10] Goodship, V., Love, J.: Multi-material Injection Moulding. Rapra Review Reports 13(1) (2002)
[11] Beaman, J., Marcus, H., Bourell, D., Barlow, J.: Solid Freeform Fabrication: A New Direction in Manufacturing. Kluwer Academic Publishers, Norwell, MA (1997)
[12] Weiss, L., Merz, R., Prinz, F., Neplothink, G., Padmanabhan, P., Schultz, L., Ramaswami, K.: Shape Deposition Manufacturing of Heterogeneous Structures. Journal of Manufacturing Systems 16(4), 239–248 (1997)
[13] Kou, X., Tan, S.: Heterogeneous object modeling: A review. Computer-Aided Design 39, 284–330 (2007)
[14] Chandru, V., Manohar, S., Prakash, C.: Voxel-based modeling for layered manufacturing. IEEE Computer Graphics and Applications 15, 42 (1995)
[15] Siu, Y.: Modeling and prototyping of heterogeneous solid CAD models. PhD thesis, University of Hong Kong (2003)
[16] Kou, X., Tan, S.: A hierarchical representation for heterogeneous object modeling. Computer-Aided Design 37, 307 (2005)
[17] Samanta, K., Koc, B.: Feature-based design and material blending for free-form heterogeneous object modeling. Computer-Aided Design 37, 287 (2005)
[18] Baumgart, B.: Winged edge polyhedron representation. Stanford University, Stanford, CA (1972)
[19] Braid, I.: The synthesis of solids bounded by many faces. Communications of the ACM 18(4), 209–216 (1975)
[20] Zhu, F.: Visualized CAD modeling and layered manufacturing modeling for components made of a multiphase perfect material. PhD thesis, University of Hong Kong (2004)
[21] Shin, K., Dutta, D.: Constructive representation of heterogeneous objects. Journal of Computing and Information Science in Engineering 1, 205 (2001)
[22] Pasko, A., Adzhiev, V., Schmitt, B., Schlick, C.: Constructive hypervolume modeling. Graphical Models 63(6), 413–442 (2001)
[23] Pasko, A., Adzhiev, V., Sourin, A., Savchenko, V.: Function representation in geometric modeling: concepts, implementation and applications. The Visual Computer 11(8), 429–446 (1995)
[24] Biswas, A., Shapiro, V., Tsukanov, I.: Heterogeneous material modeling with distance fields. Computer Aided Geometric Design 21(3), 215–242 (2004)
[25] Sellen, A., Harper, R.: The Myth of the Paperless Office. MIT Press, Cambridge, MA (2003)
[26] Warshofsky, F.: The Patent Wars: The Battle to Own the World's Technology. Wiley, Chichester (1994)

Author Index

Printing: Mercedes-Druck, Berlin
Binding: Stein + Lehmann, Berlin

Lecture Notes in Computer Science

Sublibrary 3: Information Systems and Application, incl. Internet/Web and HCI

For information about Vols. 1–4601
please contact your bookseller or Springer

Vol. 5021: S. Bechhofer, M. Hauswirth, J. Hoffmann, M. Koubarakis (Eds.), The Semantic Web: Research and Applications. XIX, 897 pages. 2008.

Vol. 5017: T. Nanya, F. Maruyama, A. Pataricza, M. Malek (Eds.), Service Availability. XII, 225 pages. 2008.

Vol. 5013: J. Indulska, D.J. Patterson, T. Rodden, M. Ott (Eds.), Pervasive Computing. XIV, 315 pages. 2008.

Vol. 5006: R. Kowalczyk, M. Huhns, M. Klusch, Z. Maamar, Q.B. Vo (Eds.), Service-Oriented Computing: Agents, Semantics, and Engineering. X, 154 pages. 2008.

Vol. 4997: B. Monien, U.-P. Schroeder (Eds.), Algorithmic Game Theory. XI, 363 pages. 2008.

Vol. 4993: H. Li, T. Liu, W.-Y. Ma, T. Sakai, K.-F. Wong, G. Zhou (Eds.), Information Retrieval Technology. XIII, 685 pages. 2008.

Vol. 4976: Y. Zhang, G. Yu, E. Bertino, G. Xu (Eds.), Progress in WWW Research and Development. XVIII, 699 pages. 2008.

Vol. 4956: C. Macdonald, I. Ounis, V. Plachouras, I. Ruthven, R.W. White (Eds.), Advances in Information Retrieval. XXI, 719 pages. 2008.

Vol. 4952: C. Floerkemeier, M. Langheinrich, E. Fleisch, F. Mattern, S.E. Sarma (Eds.), The Internet of Things. XIII, 378 pages. 2008.

Vol. 4947: J.R. Haritsa, R. Kotagiri, V. Pudi (Eds.), Database Systems for Advanced Applications. XXII, 713 pages. 2008.

Vol. 4936: W. Aiello, A. Broder, J. Janssen, E.. Milios (Eds.), Algorithms and Models for the Web-Graph. X, 167 pages. 2008.

Vol. 4932: S. Hartmann, G. Kern-Isberner (Eds.), Foundations of Information and Knowledge Systems. XII, 397 pages. 2008.

Vol. 4928: A.H.M. ter Hofstede, B. Benatallah, H.-Y. Paik (Eds.), Business Process Management Workshops. XIII, 518 pages. 2008.

Vol. 4903: S. Satoh, F. Nack, M. Etoh (Eds.), Advances in Multimedia Modeling. XIX, 510 pages. 2008.

Vol. 4900: S. Spaccapietra (Ed.), Journal on Data Semantics X. XIII, 265 pages. 2008.

Vol. 4892: A. Popescu-Belis, S. Renals, H. Bourlard (Eds.), Machine Learning for Multimodal Interaction. XI, 308 pages. 2008.

Vol. 4889: A. Pasko, V. Adzhiev, P. Comninos (Eds.), Heterogeneous Objects Modelling and Applications. VII, 285 pages. 2008.

Vol. 4882: T. Janowski, H. Mohanty (Eds.), Distributed Computing and Internet Technology. XIII, 346 pages. 2007.

Vol. 4881: H. Yin, P. Tino, E. Corchado, W. Byrne, X. Yao (Eds.), Intelligent Data Engineering and Automated Learning - IDEAL 2007. XX, 1174 pages. 2007.

Vol. 4877: C. Thanos, F. Borri, L. Candela (Eds.), Digital Libraries: Research and Development. XII, 350 pages. 2007.

Vol. 4872: D. Mery, L. Rueda (Eds.), Advances in Image and Video Technology. XXI, 961 pages. 2007.

Vol. 4871: M. Cavazza, S. Donikian (Eds.), Virtual Storytelling. XIII, 219 pages. 2007.

Vol. 4858: X. Deng, F.C. Graham (Eds.), Internet and Network Economics. XVI, 598 pages. 2007.

Vol. 4857: J.M. Ware, G.E. Taylor (Eds.), Web and Wireless Geographical Information Systems. XI, 293 pages. 2007.

Vol. 4853: F. Fonseca, M.A. Rodríguez, S. Levashkin (Eds.), GeoSpatial Semantics. X, 289 pages. 2007.

Vol. 4836: H. Ichikawa, W.-D. Cho, I. Satoh, H.Y. Youn (Eds.), Ubiquitous Computing Systems. XIII, 307 pages. 2007.

Vol. 4832: M. Weske, M.-S. Hacid, C. Godart (Eds.), Web Information Systems Engineering – WISE 2007 Workshops. XV, 518 pages. 2007.

Vol. 4831: B. Benatallah, F. Casati, D. Georgakopoulos, C. Bartolini, W. Sadiq, C. Godart (Eds.), Web Information Systems Engineering – WISE 2007. XVI, 675 pages. 2007.

Vol. 4825: K. Aberer, K.-S. Choi, N. Noy, D. Allemang, K.-I. Lee, L. Nixon, J. Golbeck, P. Mika, D. Maynard, R. Mizoguchi, G. Schreiber, P. Cudré-Mauroux (Eds.), The Semantic Web. XXVII, 973 pages. 2007.

Vol. 4823: H. Leung, F. Li, R. Lau, Q. Li (Eds.), Advances in Web Based Learning – ICWL 2007. XIV, 654 pages. 2008.

Vol. 4822: D.H.-L. Goh, T.H. Cao, I.T. Sølvberg, E. Rasmussen (Eds.), Asian Digital Libraries. XVII, 519 pages. 2007.

Vol. 4820: T.G. Wyeld, S. Kenderdine, M. Docherty (Eds.), Virtual Systems and Multimedia. XII, 215 pages. 2008.

Vol. 4816: B. Falcidieno, M. Spagnuolo, Y. Avrithis, I. Kompatsiaris, P. Buitelaar (Eds.), Semantic Multimedia. XII, 306 pages. 2007.

Vol. 4813: I. Oakley, S.A. Brewster (Eds.), Haptic and Audio Interaction Design. XIV, 145 pages. 2007.

Vol. 4810: H.H.-S. Ip, O.C. Au, H. Leung, M.-T. Sun, W.-Y. Ma, S.-M. Hu (Eds.), Advances in Multimedia Information Processing – PCM 2007. XXI, 834 pages. 2007.

Vol. 4809: M.K. Denko, C.-s. Shih, K.-C. Li, S.-L. Tsao, Q.-A. Zeng, S.H. Park, Y.-B. Ko, S.-H. Hung, J.-H. Park (Eds.), Emerging Directions in Embedded and Ubiquitous Computing. XXXV, 823 pages. 2007.

Vol. 4808: T.-W. Kuo, E. Sha, M. Guo, L.T. Yang, Z. Shao (Eds.), Embedded and Ubiquitous Computing. XXI, 769 pages. 2007.

Vol. 4806: R. Meersman, Z. Tari, P. Herrero (Eds.), On the Move to Meaningful Internet Systems 2007: OTM 2007 Workshops, Part II. XXXIV, 611 pages. 2007.

Vol. 4805: R. Meersman, Z. Tari, P. Herrero (Eds.), On the Move to Meaningful Internet Systems 2007: OTM 2007 Workshops, Part I. XXXIV, 757 pages. 2007.

Vol. 4804: R. Meersman, Z. Tari (Eds.), On the Move to Meaningful Internet Systems 2007: CoopIS, DOA, ODBASE, GADA, and IS, Part II. XXIX, 683 pages. 2007.

Vol. 4803: R. Meersman, Z. Tari (Eds.), On the Move to Meaningful Internet Systems 2007: CoopIS, DOA, ODBASE, GADA, and IS, Part I. XXIX, 1173 pages. 2007.

Vol. 4802: J.-L. Hainaut, E.A. Rundensteiner, M. Kirchberg, M. Bertolotto, M. Brochhausen, Y.-P.P. Chen, S.S.-S. Cherfi, M. Doerr, H. Han, S. Hartmann, J. Parsons, G. Poels, C. Rolland, J. Trujillo, E. Yu, E. Zimányie (Eds.), Advances in Conceptual Modeling – Foundations and Applications. XIX, 420 pages. 2007.

Vol. 4801: C. Parent, K.-D. Schewe, V.C. Storey, B. Thalheim (Eds.), Conceptual Modeling - ER 2007. XVI, 616 pages. 2007.

Vol. 4797: M. Arenas, M.I. Schwartzbach (Eds.), Database Programming Languages. VIII, 261 pages. 2007.

Vol. 4796: M. Lew, N. Sebe, T.S. Huang, E.M. Bakker (Eds.), Human–Computer Interaction. X, 157 pages. 2007.

Vol. 4794: B. Schiele, A.K. Dey, H. Gellersen, B. de Ruyter, M. Tscheligi, R. Wichert, E. Aarts, A. Buchmann (Eds.), Ambient Intelligence. XV, 375 pages. 2007.

Vol. 4777: S. Bhalla (Ed.), Databases in Networked Information Systems. X, 329 pages. 2007.

Vol. 4761: R. Obermaisser, Y. Nah, P. Puschner, F.J. Rammig (Eds.), Software Technologies for Embedded and Ubiquitous Systems. XIV, 563 pages. 2007.

Vol. 4747: S. Džeroski, J. Struyf (Eds.), Knowledge Discovery in Inductive Databases. X, 301 pages. 2007.

Vol. 4744: Y. de Kort, W. IJsselsteijn, C. Midden, B. Eggen, B.J. Fogg (Eds.), Persuasive Technology. XIV, 316 pages. 2007.

Vol. 4740: L. Ma, M. Rauterberg, R. Nakatsu (Eds.), Entertainment Computing – ICEC 2007. XXX, 480 pages. 2007.

Vol. 4730: C. Peters, P. Clough, F.C. Gey, J. Karlgren, B. Magnini, D.W. Oard, M. de Rijke, M. Stempfhuber (Eds.), Evaluation of Multilingual and Multi-modal Information Retrieval. XXIV, 998 pages. 2007.

Vol. 4723: M. R. Berthold, J. Shawe-Taylor, N. Lavrač (Eds.), Advances in Intelligent Data Analysis VII. XIV, 380 pages. 2007.

Vol. 4721: W. Jonker, M. Petković (Eds.), Secure Data Management. X, 213 pages. 2007.

Vol. 4718: J. Hightower, B. Schiele, T. Strang (Eds.), Location- and Context-Awareness. X, 297 pages. 2007.

Vol. 4717: J. Krumm, G.D. Abowd, A. Seneviratne, T. Strang (Eds.), UbiComp 2007: Ubiquitous Computing. XIX, 520 pages. 2007.

Vol. 4715: J.M. Haake, S.F. Ochoa, A. Cechich (Eds.), Groupware: Design, Implementation, and Use. XIII, 355 pages. 2007.

Vol. 4714: G. Alonso, P. Dadam, M. Rosemann (Eds.), Business Process Management. XIII, 418 pages. 2007.

Vol. 4704: D. Barbosa, A. Bonifati, Z. Bellahsène, E. Hunt, R. Unland (Eds.), Database and XML Technologies. X, 141 pages. 2007.

Vol. 4690: Y. Ioannidis, B. Novikov, B. Rachev (Eds.), Advances in Databases and Information Systems. XIII, 377 pages. 2007.

Vol. 4675: L. Kovács, N. Fuhr, C. Meghini (Eds.), Research and Advanced Technology for Digital Libraries. XVII, 585 pages. 2007.

Vol. 4674: Y. Luo (Ed.), Cooperative Design, Visualization, and Engineering. XIII, 431 pages. 2007.

Vol. 4663: C. Baranauskas, P. Palanque, J. Abascal, S.D.J. Barbosa (Eds.), Human-Computer Interaction – INTERACT 2007, Part II. XXXIII, 735 pages. 2007.

Vol. 4662: C. Baranauskas, P. Palanque, J. Abascal, S.D.J. Barbosa (Eds.), Human-Computer Interaction – INTERACT 2007, Part I. XXXIII, 637 pages. 2007.

Vol. 4658: T. Enokido, L. Barolli, M. Takizawa (Eds.), Network-Based Information Systems. XIII, 544 pages. 2007.

Vol. 4656: M.A. Wimmer, J. Scholl, Å. Grönlund (Eds.), Electronic Government. XIV, 450 pages. 2007.

Vol. 4655: G. Psaila, R. Wagner (Eds.), E-Commerce and Web Technologies. VII, 229 pages. 2007.

Vol. 4654: I.-Y. Song, J. Eder, T.M. Nguyen (Eds.), Data Warehousing and Knowledge Discovery. XVI, 482 pages. 2007.

Vol. 4653: R. Wagner, N. Revell, G. Pernul (Eds.), Database and Expert Systems Applications. XXII, 907 pages. 2007.

Vol. 4636: G. Antoniou, U. Aßmann, C. Baroglio, S. Decker, N. Henze, P.-L. Patranjan, R. Tolksdorf (Eds.), Reasoning Web. IX, 345 pages. 2007.

Vol. 4611: J. Indulska, J. Ma, L.T. Yang, T. Ungerer, J. Cao (Eds.), Ubiquitous Intelligence and Computing. XXIII, 1257 pages. 2007.

Vol. 4607: L. Baresi, P. Fraternali, G.-J. Houben (Eds.), Web Engineering. XVI, 576 pages. 2007.

Vol. 4606: A. Pras, M. van Sinderen (Eds.), Dependable and Adaptable Networks and Services. XIV, 149 pages. 2007.

Vol. 4605: D. Papadias, D. Zhang, G. Kollios (Eds.), Advances in Spatial and Temporal Databases. X, 479 pages. 2007.

Vol. 4602: S. Barker, G.-J. Ahn (Eds.), Data and Applications Security XXI. X, 291 pages. 2007.